Flora of the Venezuelan Guayana

GENERAL EDITORS

Julian A. Steyermark, Paul E. Berry, and Bruce K. Holst

Flora of the Venezuelan Guayana

VOLUME 1: INTRODUCTION

VOLUME EDITORS

Paul E. Berry, Bruce K. Holst, and Kay Yatskievych

MISSOURI BOTANICAL GARDEN
St. Louis

TIMBER PRESS
Portland, Oregon

ISBN 0-88192-313-3

Printed in Hong Kong

Published in 1995 by
Timber Press, Inc.
The Haseltine Building
133 S.W. Second Avenue, Suite 450
Portland, Oregon 97204, U.S.A.

Library of Congress Cataloging-in-Publication Data

Flora of the Venezuelan Guayana / [general editors, Julian A. Steyermark, Paul E. Berry, and Bruce K. Holst].
p. cm.
Includes bibliographical references and indexes.
Contents: v. 1. Introduction / editors, Paul E. Berry, Bruce K. Holst, and Kay Yatskievych.
ISBN 0-88192-313-3 (v. 1)
1. Botany—Venezuela—Guayana Region—Identification. 2. Botany—Venezuela—Guayana Region—Pictorial works. I. Steyermark, Julian A. (Julian Alfred), 1909–1988. II. Berry, Paul E. III. Holst, Bruce K.
QK273.F56 1995
581.987′6—dc20 95-3323
CIP

This flora is dedicated with respect and admiration to Julian Alfred Steyermark (1909–1988), who initiated the project and devoted many years to the exploration and study of the Venezuelan Guayana. He is shown here at the beginning of an expedition to Sierra de la Neblina in October 1970, when he made plant collections with Charles Brewer-Carías and G. C. K. Dunsterville. Julian particularly liked this photograph, which shows a moriche oriole (*Icterus chrysocephalus*) about to peck him on the head.

Institutional Affiliations and Addresses of Authors and Editors

Paul E. Berry
Missouri Botanical Garden
P.O. Box 299
St. Louis, MO 63166, U.S.A.

Bruce K. Holst[1]
Missouri Botanical Garden
P.O. Box 299
St. Louis, MO 63166, U.S.A.

Otto Huber
Fundación Instituto Botánico de Venezuela
Apartado 2156
Caracas 1010-A, Venezuela

Julian A. Steyermark (deceased)
Missouri Botanical Garden
P.O. Box 299
St. Louis, MO 63166, U.S.A.

Kay Yatskievych
Missouri Botanical Garden
P.O. Box 299
St. Louis, MO 63166, U.S.A.

[1]Current address: Marie Selby Botanical Gardens
811 South Palm Avenue
Sarasota, FL 34236, U.S.A.

Contents

Foreword

Peter H. Raven

Julian Steyermark was a remarkable man who loved plants, studied them, rejoiced in their beauty and diversity, and communicated about them all his life. The most prolific plant collector ever, Julian gathered some 138,000 samples for scientific study—a permanent record of the world's botanical riches in the middle decades of the 20th century. Blessed with a keen sense of humor and a deep interest in people, Julian was an outstanding companion and a sincere and devoted friend.

His life work was carried out in both temperate and tropical regions. Expert on the plants of his native state, he made the *Flora of Missouri* a model of regional treatments in the United States, its pages packed with remarks that reflected the insights gained during years of study of the plants that he loved. As his professional life developed, Julian moved on to study and collect the plants of Guatemala, ranging through the much more extensive forests that occurred in that Central American country half a century ago. During World War II, he was sent to South America to search for new sources of quinine. Visiting the isolated tabletop mountains of southern Venezuela, he became deeply fascinated by the region. He resided in Venezuela for a quarter of a century, becoming a citizen, and devoting much of his attention to the plants of the southern part of the country, Venezuelan Guayana.

I first met Julian in the 1960s, although he had long since become a botanical legend. In the late 1970s, I learned of his plans to publish a flora of the Venezuelan Guayana. Since he was nearing retirement from the Instituto Botánico in Caracas, it seemed to me that it might be advantageous for him to return to St. Louis, where we could provide the necessary support to help him to complete his work as rapidly and efficiently as possible. In 1984, he accepted my offer, moving to the Missouri Botanical Garden, but returning frequently to Venezuela to gather additional data in the herbarium and explore selected regions from which the available specimens were insufficient for full treatment in his flora. He collaborated extensively on those trips, and at times in St. Louis, with Bruno Manara, whose outstanding illustrations are in many cases the first of these species to be published, and which are such an adornment to these volumes. The close collaboration between these two men assured the botanical accuracy and appropriateness of these illustrations.

The *Flora of the Venezuelan Guayana* has been a major enterprise at the Missouri Botanical Garden since the 1980s, involving 180 botanists worldwide. An introductory volume was conceived after Julian's death in 1988, once Paul Berry, Bruce Holst, and Kay Yatskievych began working together to bring the project to completion. This volume owes much of its content to Otto Huber, a talented plant ecologist and close colleague of Julian's, who shared with him and still maintains a profound interest in and knowledge of the Guayana region and its remarkable flora. With the companion Volume 2, Volume 1 initiates a work that will eventually describe and, in many cases, illustrate the approximately 10,000 species of plants that occur in the core area of the ancient Guayana Shield, whose organisms are characterized by extensive endemism, a unique evolutionary history, and surprising biogeographical connections to other regions.

Policy makers, conservationists, botanists, and other interested parties will be able to use the lovingly prepared and detailed accounts in this work to manage, study further, appreciate, and conserve the plants of this extraordinary region. Hitherto, they have been as poorly known as any set of plants on Earth; now, thanks to the devotion of Julian Steyermark, Otto Huber, and those who are carrying this work forward to completion, we have for the first time a substantial base of knowledge that will make possible further advances in a wide variety of fields. We at the Missouri Botanical Garden are proud that Julian Steyermark, a native of St. Louis, graduate of Soldan High School and Washington University, who first came to love nature in the fields and forests of this area, was able to find the means to complete his life work and to realize his dreams. We delight in the memory of his friendship, and rejoice in the shared enterprise that we enjoyed together.

Peter H. Raven
Director, Missouri Botanical Garden

Preface

The Venezuelan Guayana has fascinated biologists since the explorations of Humboldt and Bonpland and the Schomburgk brothers in the early 19th century. This area in northern South America lies in the center of the geologically ancient Guayana Shield and is home to more than 50 tabletop mountains, each known as a *tepui*. From these imposing mountains fall the highest waterfalls in the world, such as Angel Falls on Auyán-tepui. It is a region of great beauty, still largely pristine, which includes other diverse habitats such as the swamps of the Orinoco Delta, the upland Gran Sabana, large extensions of lowland tropical forest, and the peculiar white-sand savannas and shrublands of the upper Río Negro.

Each tepui holds the promise of unique plants, and this, along with the area's richness and beauty, lured Julian Steyermark there for the first time in 1944. He eventually made the Venezuelan Guayana the focus of his botanical research over several decades. In an area where no comprehensive inventory of the plant life had ever been attempted, Steyermark began such an enterprise in 1983. His idea was to publish an illustrated flora that would describe and facilitate the identification of all the vascular plants in southern Venezuela.

Subsequent botanical explorations in the Guayana region have resulted in the recognition of hundreds of new species, dozens of new genera, and several new families of plants, many of them only known to occur there. Although there is still much descriptive work to be done, a comprehensive flora will encourage more analytical and synthetic studies, such as the biogeographical and phylogenetic history of endemic plant groups, the rarity and conservation status of different plant species, and the dynamics of different vegetation types.

Since the size of this flora approaches 10,000 species, and illustrations for more than half of these will be provided, it will require a number of volumes and several years to publish all 230 plant families that occur spontaneously in the flora area. After the death of Julian Steyermark in 1988, the editors decided to add this introductory volume to the series to provide background information on the Venezuelan Guayana and an overview of the flora, vegetation, and conservation importance of the region. It should provide a useful aid to all who are interested in the biodiversity and natural history of the Guayana Shield.

Acknowledgments

The *Flora of the Venezuelan Guayana* project has been strongly supported since its inception by the Missouri Botanical Garden and especially by its director, Peter H. Raven. The Herbario Nacional de Venezuela, now part of the Fundación Instituto Botánico de Venezuela, has steadily supported this project in many ways, and it was there that Julian Steyermark conceived the idea of the flora and did most of his background research.

This volume is based upon work supported by the National Science Foundation under Grants Nos. BSR-8717303, BSR-9045532, and BSR-9201044. Any opinions, findings, and conclusions or recommendations are those of the authors and do not necessarily reflect the views of the National Science Foundation.

The Julian A. Steyermark Fund, established by the late Dr. Steyermark at the Missouri Botanical Garden, provided major funding for the flora. Funding for the maps accompanying this volume was kindly provided by Electrificación del Caroní, C.A. (CVG-EDELCA). The Armand G. Erpf Fund also provided financial support for the map production and helped fund some of the nearly 5000 species illustrations in the remaining volumes of the flora. Another series of illustrations was funded by Bruce K. Holst.

The most important collaborators in seeing this project through to completion have been Otto Huber, who wrote most of Volume 1 and reviewed many of the floristic treatments, and Bruno Manara, who completed drawings for more than half the species in the flora area and assisted in many other ways. The project has also benefited from the efforts of two postdoctoral researchers, Denis Kearns and John MacDougal.

Numerous institutions and individuals in Venezuela have contributed significantly to different aspects of the flora. These include the Corporación Venezolana de Guayana and its affiliate Electrificación del Caroní, C.A., especially through the assistance of Alfredo Lezama, Hermán Róo, Luis Castro, and Antonio Ahogado; the Ministerio del Ambiente y de los Recursos Naturales Renovables; the Instituto Nacional de Parques and the Dirección General de Parques Nacionales; the Servicio Autónomo para el Desarrollo Ambiental del Estado Amazonas; and the Consejo Nacional de Investigaciones Científicas y Tecnológicas. The Fuerzas Aéreas de Venezuela, Fundación Terramar, Charles Brewer-Carías, and the late Parker Redmond also gave important logistical support on various trips to southern Venezuela.

Several Venezuelan herbaria and associated botanists provided considerable assistance over an extended period of time, including the Herbario Nacional de Venezuela, especially through Francisco Delascio, Francisco Guánchez, and Gilberto Morillo; the Universidad Central de Venezuela, through the Herbario Ovalles at the Facultad de Farmacia and its director Stephen Tillett, and the herbarium of the Facultad de Agronomía in Maracay, particularly with the help of Carmen Emilia Benítez de Rojas; the Universidad Nacional Experimental de los Llanos Ezequiel Zamora in Guanare and its active team of botanists including Gerardo Aymard, Nidia Cuello, and Basil Stergios; the herbarium at Puerto Ayacucho, now associated with the Centro Amazónico de Investigaciones Ambientales Alejandro de Humboldt of the Servicio Autónomo para el Desarrollo Ambiental del Estado Amazonas; and the Universidad de Los Andes in Mérida, mainly through the herbaria of the Facultad de Ciencias Forestales and the Facultad de Farmacia.

In the United States, the National Geographic Society provided grants for several explorations associated with this project. The Missouri Botanical Garden, the New York Botanical Garden, and the Smithsonian Institution, in particular, lent their facilities, specimens, and the expertise of their staff to assist in the preparation of numerous floristic treatments.

Lewis Johnson, Scott Mori, and Gustavo Romero reviewed a draft version of Volume 1 and provided valuable comments to improve it. Individual chapters in this volume were also reviewed by Gerardo Aymard, Henry Briceño, Nidia Cuello, Rafael García, Timothy Killeen, Roberto Lizarralde, Gabriel Picón, John Pruski, Richard Schargel, Franco Urbani, Franz Weibezahn, Anna Weitzman, John Wurdack, and Alfred Zinck. Other people who worked part time or volunteered to assist on the flora include Daniel Berry, William Betz, Lois Brako, Germán Carnevali, Luther Raechal, Ronald Liesner, Ivón Ramírez, and George Yatskievych. Tomás Rodríguez prepared many of the technical drawings and the base maps for the topographical and vegetation maps.

Permission to use photographs was kindly granted by the Biblioteca Nacional de Venezuela, *The Chicago Sun-Times,* the Hunt Botanical Library, Celia Maguire, and Karl Weidmann. Charles Brewer-Carías offered a large selection of color slides for use in this volume, and his contribution is gratefully acknowledged. We also credit the following persons for providing photographs appearing in Volume 1: Antonio Ahogado, Joseph Beitel, Paul Berry, Antoine Cleef, Nelda Dezzeo, Andreas Gröger, Bruce Holst, Otto Huber, Ronald Liesner, Bruno Manara, Roy McDiarmid, Gabriel Picón, John Pruski, Gustavo Romero, and Julian Steyermark.

Finally, we wish to thank the staff of Timber Press, Inc., for their assistance and understanding throughout the production of this volume.

Introduction

Paul E. Berry, Bruce K. Holst, and Kay Yatskievych

The Venezuelan Guayana includes three states located south of the Río Orinoco: Amazonas, Bolívar, and Delta Amacuro. Together they comprise nearly half the land area of Venezuela, or about 454,000 km^2 (see Figure 1-1). Except for several towns and a few cities along the Orinoco, southern Venezuela is still a sparsely populated wilderness. The area is endowed with spectacular mountains and rivers and is characterized by diverse and often unique biological communities renowned for their high levels of endemism. With the abundance of herbarium collections and ecological data that has accumulated, especially since the 1940s, there is now sufficient information available to attempt a comprehensive survey of the area's plant life. The *Flora of the Venezuelan Guayana* takes a major step toward this objective by providing an abridged and illustrated vascular plant flora of Amazonas, Bolívar, and Delta Amacuro states. At a broader level, this is significant because the Venezuelan Guayana is the core area of the Guayana region, a major physiographic and biogeographic region of South America that coincides largely with the geologically defined Guayana Shield. Besides southern Venezuela, the Guayana region covers Guyana, Suriname, French Guiana, southeastern Colombia, and parts of northern Brazil.

Floristic treatments for the *Flora of the Venezuelan Guayana* will be published in a series of volumes that include all vascular plants known to occur outside of cultivation in the Guayana region of Venezuela. The treatments provide full taxonomic descriptions for each family and genus. For each of the nearly 9400 species and their subordinate taxa, there is a synoptical entry that provides information on the plant's habit, habitat, elevational range in the flora area, detailed locality data, and common names and uses. Identifications are facilitated by dichotomous keys to the genera, species, and infraspecies; close to 5000 species are illustrated by line drawings, including at least one species from each genus in the flora. The families are arranged alphabetically in two main categories, the ferns and fern allies, and the seed plants.

History of the Flora Project

The *Flora of the Venezuelan Guayana* was conceived by the late Julian Steyermark. He proposed the project to Otto Huber in 1979 shortly after they completed the *Flora del Avila,* a floristic study of a coastal mountain range in northern Venezuela. By that time, both researchers had accumulated extensive first-hand knowledge of the Venezuelan Guayana. Steyermark first visited the area from August to December 1944, when he ascended Cerro Duida, Roraima-tepui, Ptari-tepui, and Sororopán-tepui as part of a war-effort search for natural sources of quinine. He returned to the area in 1953 for an extended collecting trip to the Chimantá massif, which he visited again with John Wurdack in 1955. After establishing his residence in Venezuela in 1959, Steyermark made frequent trips throughout the Venezuelan Guayana during the next 25 years, assembling a total of 27,939 numbers of plant collections from the flora area.

Otto Huber has specialized in the ecology and floristics of the nonforested biomes of the Venezuelan Guayana since 1976 and has visited nearly every major mountain in the area, as well as many shrublands and Amazon lowland savannas; he has gathered approximately 13,000 numbers of plant collections in the flora area during this period.

Work on the *Flora of the Venezuelan Guayana* was initiated in 1983, when Steyermark began accumulating specimen data in Venezuela and commissioned Bruno Manara to start the line drawings. Upon his formal retirement from the staff of the Instituto Botánico in Caracas in 1984, Steyermark accepted an offer from Peter Raven at the Missouri Botanical Garden to continue work on the flora in St. Louis. Later in the same year, the Garden enlisted Bruce Holst as project assistant. During the next few years, exhaustive surveys for specimens from the Guayana region were made at the major Venezuelan herbaria, and selected families were studied at the Missouri Botanical Garden, New York Botanical Garden, and Smithsonian Institution. Steyermark and the Missouri Botanical Garden also sponsored a series of field trips aimed at collecting specimens from poorly known sites in southern Venezuela. After Steyermark succumbed to cancer in late 1988, Paul Berry became principal investigator of the project, and Kay Yatskievych joined the project team as editorial assistant in February 1990. Otto Huber began active collaboration on the flora in 1989.

A key element that contributed valuable information for the compilation of this flora was the thirteen-part series published in the *Memoirs of the New York Botanical Garden* under the title of *The Botany of the Guayana Highland* (summarized and indexed in Buck 1990). This was the result of a series of pioneering expeditions to remote mountains of southern Venezuela and other parts of the Guayana Shield led by Bassett Maguire and collaborators between 1944 and 1981.

Steyermark originally planned to complete all the floristic treatments himself, or at most with a small group of colleagues. As the magnitude of the project became evident, he gradually invited other botanists to participate. Many more specialists were enlisted after Steyermark's death, until eventually 180 contributors in many countries were involved.

Summary Data from the Flora Area

Although a comprehensive floristic analysis of the Venezuelan Guayana will be possible once all the family treatments are published, about three-fourths of the floristic treatments have been completed, and we have assembled reliable checklists for the remaining groups. This has allowed us to make detailed counts of the taxa and to quantify levels of endemism. A listing of the families with their numbers of genera and species in the flora area is presented in Appendix A.

Using Cronquist's (1981) delimitation of flowering plant families and Alan R. Smith's delimitation of ferns and fern allies (see Volume 2), which is similar to that of Kramer and Green (1990), we recognize 230 vascular plant families in the flora area. The total number of vascular plant species known outside of cultivation in the Venezuelan Guayana is 9411, in 1786 genera (see Table I for a breakdown of these figures by three major groupings).

Table I. Number of taxa in the Venezuelan Guayana by major plant groups.

	Families	Genera	Species
Pteridophytes	29	92	671
Gymnosperms	3	3	18
Angiosperms	198	1691	8722
	230	1786	9411

Only 123 species, or 1.3 percent of the flora, are considered to be non-native (escaped or naturalized) in the Venezuelan Guayana. This very low percentage attests to the still largely undisturbed nature of most of the area. Some of these species have only locally escaped from cultivation, and 25 of the introduced species are grasses.

The number of endemic species in the flora is 2136, or 22.7 percent of the total (see Table 4-4 for a summary of the levels of endemism). When compared to the larger and biogeographically more significant area of the Guayana Shield (see Figure 1-2 and Chapter 1), the level of endemism increases, with 40 percent (3763) of the species in the Venezuelan Guayana restricted to the confines of the shield. At the generic level, 34 genera are found only in the flora area, whereas 118 genera are endemic to the Guayana Shield. Although no family is entirely endemic to the Venezuelan Guayana, four families occur exclusively within the limits of the Guayana Shield, namely, Euphroniaceae (three species), Hymenophyllopsidaceae (eight species), Saccifoliaceae (one species), and Tepuianthaceae (six species).

In the Venezuelan portion of Pantepui, which we consider the same as the Guayana highlands and define as any area in the Guayana Shield generally above 1500 m elevation (see Chapters 1 and 4), 2322 species and 630 genera are known. A high percentage of these taxa are endemic either to the flora area or to the Guayana Shield (see Table 4-5). There are a small number of additional Pantepui genera and species outside of Venezuela, mostly in Guyana and northernmost Brazil; these taxa are discussed in Chapter 4.

Whence the Name Guayana?

Guayana and Guiana are both spelling variants of a word derived from an Amerindian linguistic source. *The New Encyclopaedia Britannica* (Goetz 1985) ascribes this word's derivation to a root word for water, with Guayana meaning land of waters. This interpretation was followed by botanists such as Mori and Prance (1987) and Ek (1990). Robert Schomburgk (1840a), the famous 19th century geographer, heard reports that Guayana received its name from a small tributary of the Río Orinoco. A much better-documented explanation is that the name came from an Amerindian (presumably Arawak) tribe called the Guayanos, who lived between the lower Río Caroní and the Serranía de Imataca (Baralt 1841; Codazzi 1841; Tavera-Acosta 1905). According to Codazzi (1841), the Guayanos maintained contact for nearly a century and a half with the Spanish Capuchin missionaries, who protected them against attacks by Carib Amerindians. Tavera-Acosta (1905) indicated that Guayana is derived from the indigenous word *uayana,* meaning white, pale, or yellow. Supposedly the use of this name by European colonists dates back to at least 1532, when Don Diego de Ordaz sent a group of Spanish settlers to the mouth of the Río Caroní. There they were greeted by Indians who called out "uayana, uayana," referring either to the skin color of the Spaniards or to the pieces of gold and silver they showed to inquire where more could be sought. One of the earliest Spanish settlements in the area was subsequently called Santo Tomé de los Guayanos, or alternatively Santo Tomás de la Guayana (Tavera-Acosta 1905).

Whatever the name's derivation, Guayana and Guiana first came to be used in a broader geographical sense in the early 16th century, following the Dutch occupation of what is now coastal Guyana and Suriname. Both names have since been used to distinguish a large region of northern South America south and mainly east of the Río Orinoco, including most of southern Venezuela, Guyana, Suriname, French Guiana, and parts of northern Brazil and southeastern Colombia.

Guiana is the spelling that was initially adopted by the British and other northern Europeans (see Plate 1), whereas Guayana has been used primarily by Spanish speakers. As a result of the geopolitical influences of the different colonial powers, the two names have sometimes received differing connotations. Guayana is most commonly applied to the Venezuelan part of the region, as the title of this flora implies. Perhaps for this reason, Maguire adopted the term *Guayana Highland,* since the high-elevation habitats of the region are concentrated in southern Venezuela. Guiana, on the other hand, has sometimes been used to cover the mainly lowland area that lies largely to the east of Venezuela and extends into Brazilian Amapá state (e.g., Maguire 1966). In most cases, however, the names should be considered synonymous, as in Guayana Shield and Guiana Shield.

Because of the potential for confusion among similar names, the following glossary explains geographical terms that are derived from the word *Guayana*:

Guayana, Guiana: a physiographic region of northern South America centered on one of the continent's two Precambrian crystalline shields and including much of southern Venezuela, Guyana, Suriname, French Guiana, and a portion of southeastern Colombia and northern Brazil.

Guayana Shield (also Guiana Shield, Guyana Shield, Guayanan Shield): the ancient crystalline shield area of northern South America.

Venezuelan Guayana: the Venezuelan part of the Guayana region; politically (and in this flora) the Venezuelan states of Amazonas, Bolívar, and Delta Amacuro.

Guayana Highland(s) (also Guiana Highlands): the high-elevation portions of the Guayana Shield, usually above 1500 m (also known as Pantepui), although previously used in a broader sense to include lower mountains and even lowland areas.

Guianas: the countries of Guyana and Suriname, and the Department of French Guiana.

Guyana: an independent country, formerly the colony of British Guiana.

Dutch Guiana: now the independent country of Suriname (alternately spelled Surinam).

Guyane Française: French spelling of French Guiana, an overseas Department of France.

Summary of Volume 1

This first volume of the flora provides a conceptual framework in which to analyze the floristic information contained in the family treatments. Chapter 1 gives a detailed overview of the physical geography of the region and its human occupation. The history of botanical exploration and collecting in the Venezuelan Guayana is covered in Chapter 2. Chapter 3 provides a detailed classification of vegetation types, emphasizing the importance of altitudinal zonation and dividing the vegetation into forest, shrubland, herbaceous, and pioneer formations. One of the accompanying maps is a 1:2,000,000-scale color vegetation map; the other is a topographical and toponymical map at the same scale. Chapter 4 provides a more complete floristic analysis of the flora area than that given here; it also examines the phytogeographical situation of the Venezuelan Guayana and develops a definition and delimitation of the Guayana region with four component provinces. The significance of the Venezuelan Guayana in national and international conservation efforts is covered in Chapter 5. A section of color photos in this volume illustrates plants, vegetation types, and landscapes in the flora area. Finally, Appendix A lists the vascular plant families of the Venezuelan Guayana, and Appendix B provides a dichotomous key to the families of seed plants.

List of Acronyms for Organizations Involved with the Venezuelan Guayana

AsoVAC. Asociación Venezolana para el Avance de la Ciencia (Caracas)
BAUXIVEN. Bauxitas de Venezuela, Filial de la CVG (Los Pijiguaos)
CAIAH. Centro Amazónico de Investigaciones Ambientales Alejandro de Humboldt, a branch of SADA-Amazonas (Puerto Ayacucho and La Esmeralda)
CBR. Consejo de Bienestar Rural (Caracas)
CODESUR. Comisión para el Desarrollo del Sur de Venezuela, part of MOP (Caracas)
CONICIT. Consejo Nacional de Investigaciones Científicas y Tecnológicas (Caracas)
COPLANARH. Comisión de Planificación Nacional de Recursos Hidráulicos, part of MOP (Caracas)
CVG. Corporación Venezolana de Guayana (Ciudad Guayana)
DGSIIA. Dirección General Sectorial de Información e Investigación del Ambiente, part of MARNR (Caracas)
EDELCA. Electrificación del Caroní, Compañía Anónima, Filial de la CVG (Ciudad Guayana)
FAO-UNSF. Food and Agriculture Organization, United Nations Special Funds (Rome)
FERROMINERA. Filial de la CVG (Ciudad Guayana)
FUDECI. Fundación para el Desarrollo de las Ciencias Físicas, Matemáticas y Naturales (Caracas)
INPARQUES. Instituto Nacional de Parques, part of MARNR (Caracas)
IVIC. Instituto Venezolano de Investigaciones Científicas (Caracas)
MAB. Man and the Biosphere Program of UNESCO (Paris)
MAC. Ministerio de Agricultura y Cría (Caracas)
MARNR. Ministerio del Ambiente y de los Recursos Naturales Renovables (Caracas)
MOP. Ministerio de Obras Públicas [former ministry] (Caracas)
OCEI. Oficina Central de Estadística e Informática (Caracas)
ORSTOM. Office de la Recherche Scientifique et Technique Outre-Mer (Paris)

PIRNRG. Proyecto Inventario de los Recursos Naturales de la Región Guayana, CVG-TECMIN (Ciudad Bolívar)
SADA-Amazonas. Servicio Autónomo para el Desarrollo Ambiental del Estado Amazonas, part of MARNR (Caracas and Puerto Ayacucho)
SEFORVEN. Servicio Autónomo Forestal Venezolano, part of MARNR (Caracas)
SVCN. Sociedad Venezolana de Ciencias Naturales (Caracas)
TECMIN. Técnica Minera, Compañia Anónima, Filial de la CVG (Ciudad Guayana)
UCV. Universidad Central de Venezuela (Caracas)
ULA. Universidad de los Andes (Mérida)
UNELLEZ. Universidad Nacional Experimental de los Llanos Occidentales Ezequiel Zamora (Barinas and Guanare)
UNESCO. United Nations Educational, Scientific and Cultural Organization (Paris)

CHAPTER 1

Geographical and Physical Features

Otto Huber

The Venezuelan Guayana covers the three southern- and easternmost states of Venezuela, namely, Amazonas, Bolívar, and Delta Amacuro (Figure 1-1). Until the early 1990s, Amazonas and Delta Amacuro were both federal territories, but Delta Amacuro was declared a state in August 1991 and Amazonas in July 1992. Government statistics list the total surface area of the Venezuelan Guayana at 453,950 km^2 (OCEI 1993a), very close to half the area of the entire country (912,050 km^2). According to official census figures, the total number of inhabitants in the Venezuelan Guayana in 1990 was 1,141,205 (including an indigenous population of 100,614), or 2.51 inhabitants per square kilometer, compared to 19.9 inhabitants per square kilometer for the entire country and 37.7 inhabitants per square kilometer for the extra-Guayanan portion of the country (OCEI 1993a, 1993b).

Of the three states, Delta Amacuro has a surface area of 40,200 km^2 (4.4 percent of Venezuela) and a population of 105,689 (2.63 inhabitants per square kilometer), including an indigenous population of 21,125. Bolívar covers 238,000 km^2 (26.1 percent of the country), with a population of 935,287 (3.93 inhabitants per square kilometer), including an indigenous population of 34,977. Finally, Amazonas covers 175,750 km^2 (19.3 percent of Venezuela), with 100,229 inhabitants (0.57 inhabitants per square kilometer), including an indigenous population of 44,512. These data are from official government statistics (OCEI 1993a, 1993b).

Geographically, the Venezuelan Guayana is delimited to the north by the Río Orinoco and its delta, with the northernmost point at 10°04' N at Punta Bernal in Delta Amacuro state. To the south, the region is ringed by a series of mountain chains including the Sierra de la Neblina (Plate 49), Sierra Parima, and Sierra Pakaraima; the southernmost point is at 0°39' N in the headwaters of the Río Arari in Amazonas state. To the east, the border of the Venezuelan Guayana follows part of the courses of the Venamo and Cuyuní rivers, the headwaters of the Río Barima, and the eastern part of the Gran Sabana; the easternmost point of the region is 59°48' W at La Línea, in Delta

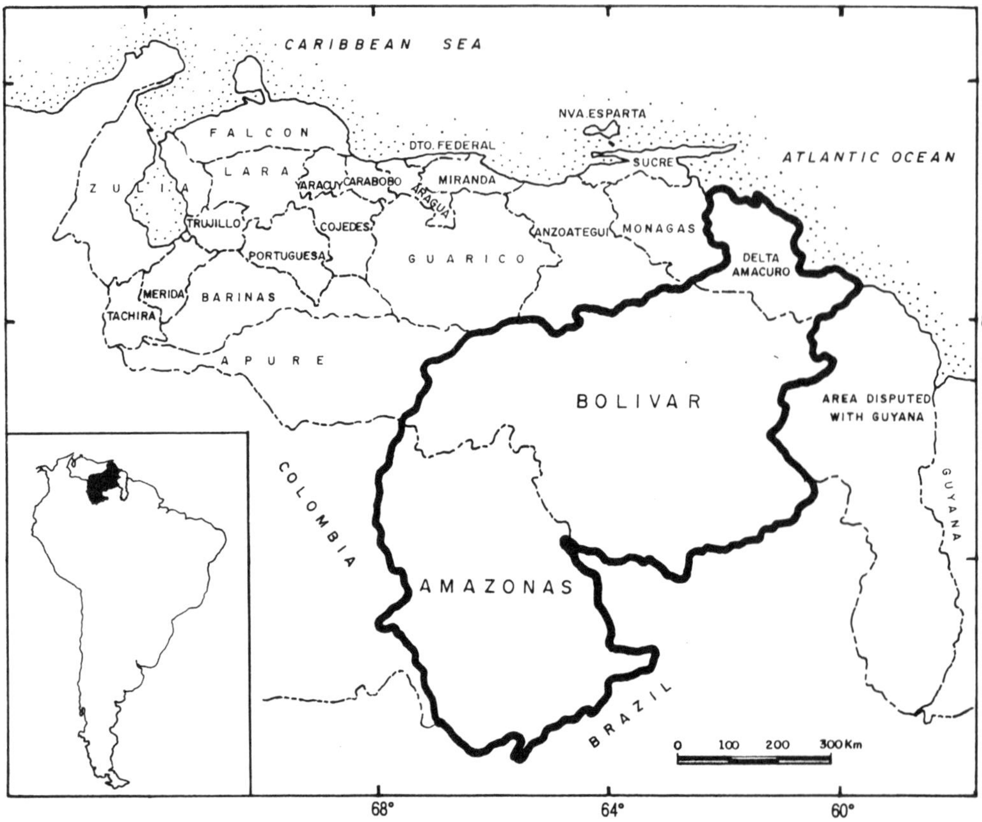

Figure 1-1. Main political divisions of Venezuela with the three states comprising the Venezuelan Guayana inside the heavy line.

Amacuro state. Finally, the western extreme of the Venezuelan Guayana occurs along the Orinoco, Atabapo, Guainía, and Negro rivers, with the westernmost point at 67°50' W near Victorino, in Amazonas state. The Venezuelan Guayana covers approximately 45 percent of the surface area of the Guayana Shield, which extends from French Guiana west to southeastern Colombia and from central Venezuela to northern Brazil (see Figure 1-2).

Numerous botanical collections were made between 1950 and 1990, when different administrative units were in effect in the Venezuelan Guayana. The former federal territories differed from states in their sparse population and in the absence of a legislative assembly; they were divided into departments (Departamentos). As states, they are now divided into counties (Municipios). Over the past several years, there have been major changes in the subordinate administrative units within Bolívar state. The former districts (Distritos) and counties were initially reorganized into units called Municipios Autónomos and Municipios Foráneos, but as of 1992 the state was divided only into ten Municipios. To interpret reports and herbarium labels made during this period, Table 1-1 provides summary information on the prior administrative units in the Venezuelan Guayana (based on MARNR 1979b; OCEI 1985).

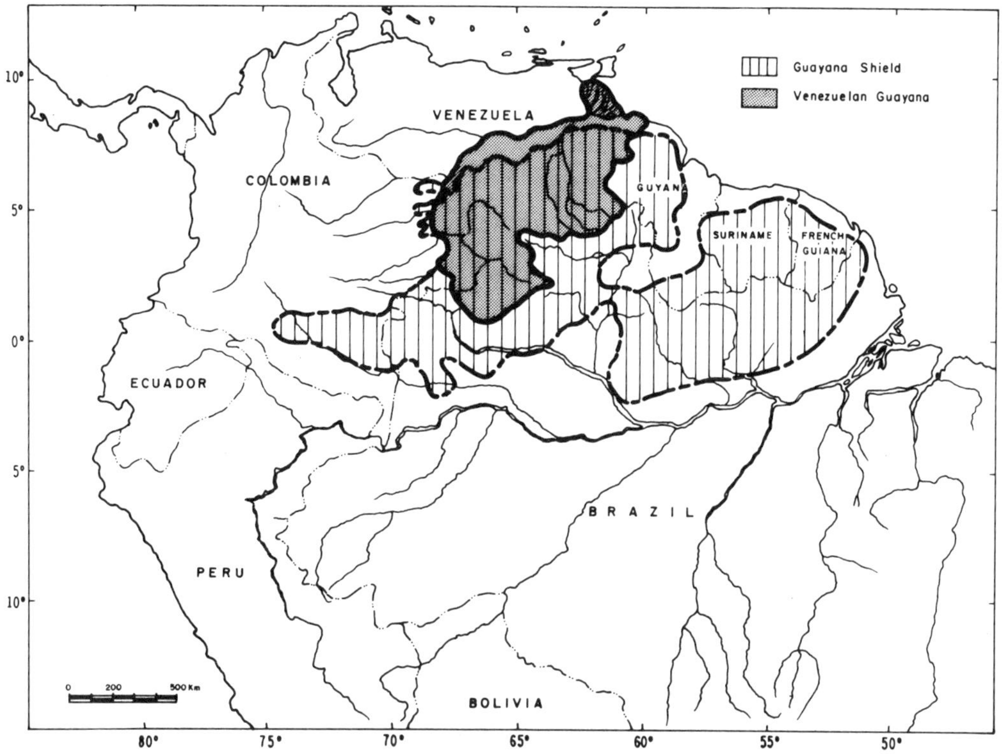

Figure 1-2. Position of the Venezuelan Guayana and the Guayana Shield in northern South America. The outline of the Guayana Shield follows Gibbs and Barron (1993).

Geology

Because of its ancient origins, the Guayana Shield has produced a complex mosaic of lithological units. Unlike the northern half of Venezuela, which is well explored geologically because of intensive oil explorations there during much of the 20th century, the tectonic and geological history of the southern part of the country is only superficially understood.

In the late 1900s, remote sensing techniques helped to produce a much better characterization of the physical features of the Venezuelan Guayana. The first broadly applied technique was *side-looking airborne radar* (SLAR), which was used from 1970 to 1972 to generate images of much of the flora area. SLAR pictures are important for the analysis of landforms and associated geomorphological phenomena. Starting in 1972, satellite imagery using different wave bands has provided information that is particularly useful for geologic, hydrologic, edaphic, and vegetation studies (see Plate 2). Aerial photographic coverage of the region began in the 1950s, but it is still fragmentary and is restricted mainly to areas of particular economic or strategic interest.

In a simplified view, the Venezuelan Guayana consists of three main lithological groups: an igneous-metamorphic basement, a sedimentary cover,

Table 1-1. Administrative units of the Venezuelan Guayana in effect when most modern plant collections were made.

Subdivision	Area (km^2)	Capital	Description
Estado Bolívar (capital, Ciudad Bolívar)			
Distrito Caroní	1,901	Ciudad Guayana	Includes Ciudad Guayana, which comprises the older city of San Félix, the newly created Puerto Ordaz, and the industrial area of Matanzas.
Distrito Cedeño	62,170	Caicara del Orinoco	All the state west of the Río Caura.
Distrito Heres	60,681	Ciudad Bolívar	The area between the Caroní and Icabarú rivers to the east and the Río Caura watershed to the west.
Distrito Piar	40,422	Upata	The western part of the Gran Sabana and the eastern part of the lower Río Caroní basin.
Distrito Roscio	8,897	Guasipati	The mining areas around Guasipati, El Callao, and El Miamo, in the northern headwaters of the Cuyuní and Yuruari rivers.
Distrito Sifontes	36,867	Tumeremo	Most of the Serranía de Imataca, area around Tumeremo and El Dorado, and the eastern sector of the Gran Sabana. This district was created in 1982 and previously belonged to the Distrito Roscio.
Distrito Sucre	27,062	Maripa	The entire eastern side of the Río Caura watershed.
Territorio Federal Amazonas (changed to Estado Amazonas in July 1992 by decree, effective January 1993; capital, Puerto Ayacucho)			
Departamento Atabapo	66,828	San Fernando de Atabapo	The area south of the Río Ventuari, north of the upper Río Orinoco, and east of the Río Atabapo.
Departamento Atures	40,564	Puerto Ayacucho	Area north of the Orinoco and Ventuari rivers.
Departamento Casiquiare	18,252	Maroa	The area south of the Río Atacavi, east of the Río Guainía, and north and west of the Río Casiquiare.
Departamento Río Negro	50,106	San Carlos de Río Negro	The area south of the upper Río Orinoco and Río Casiquiare.
Territorio Federal Delta Amacuro (changed to Estado Delta Amacuro in August 1991; capital, Tucupita)			
Departamento Antonio Díaz	26,808	Curiapo	The area mostly south of Caño Araguao.
Departamento Pedernales	2,308	Pedernales	The area mostly north of Caño Capure.
Departamento Tucupita	11,084	Tucupita	The area mostly between Caño Capure to the north and Caño Araguao to the south.

and younger intrusive rocks (Figure 1-3). Although each group represents distinct geologic events and chronologies, the area's current geology is the result of strong interactions between them.

Igneous-Metamorphic Basement

The entire Guayana Shield consists of a rock basement with a variety of igneous and metamorphic rock types, especially granites and gneisses. This basement was formed during different orogenetic phases characterized by large and long-lasting tectonic-thermal events that occurred repeatedly during Archean and Proterozoic times. Generally, four such events are recognized (Mendoza 1977; Schubert and Huber 1990):

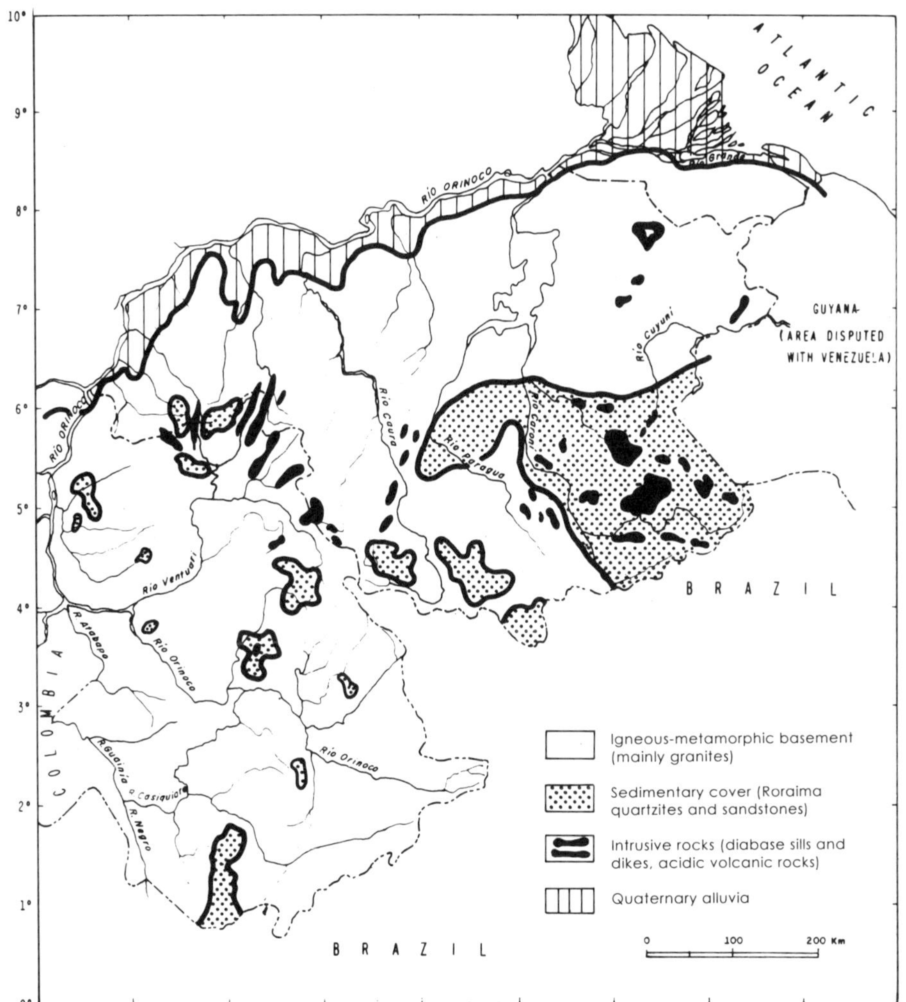

Figure 1-3. Simplified geologic map of the Venezuelan Guayana. Adapted from MARNR-ORSTOM (1987), Sidder (1990), and Gibbs and Barron (1993).

1. *Gurian Orogenesis* (3.6–2.7 billion years ago), when the oldest granites of the Imataca Group were formed
2. *Pre-Transamazonian Orogenesis* (2.6–2.1 billion years ago), when granites of the Supamo Group originated
3. *Transamazonian Orogenesis* (2.0–1.7 billion years ago), when granites and gneisses of the Cuchivero Group were formed
4. *Orinocan Orogenesis* (1.2–0.8 billion years ago), when granites and gneisses of the Suapure and Parguaza regions originated

Most of these granitic nuclei were subsequently covered by thick layers of sedimentary rock. Only where this cover was completely removed by erosion can the basement rock be seen on the surface, as in the Serranía de Imataca (northeastern Bolívar), the Sierra Parima (eastern Amazonas), or the many granitic hills (also known as inselbergs or *lajas;* see Plate 5) and low ranges of Parguaza granite in northwestern and north-central Amazonas. However, one of the largest mountain massifs in the Guayana, the Sierra de Maigualida (Plate 47), is granitic, although it is not yet clear to which group it belongs.

This mosaic of large igneous and metamorphic rock nuclei is one of the earliest shields of the earth's crust. It formed the western section of the ancient supercontinent of Gondwana, which also included an eastern (Africa) and a southern section (the Brazilian Shield). Gondwana began to break up into several units during the Late Jurassic, approximately 150 million years ago, with the Atlantic Ocean forming between the South American and African land blocks. The igneous-metamorphic basement did not approach its present configuration until near the end of the Tertiary (25 million years ago).

Sedimentary Cover

Between 1.6 and 1 billion years ago, from the end of the Transamazonian Orogenesis to the beginning of the Orinocan Orogenesis, most of the granitic basement of the Guayana Shield was overlain with layer upon layer of sand. This sand probably came from the adjacent, easterly uplands of the ancient Gondwana continent during intermittent but intensive sedimentation phases. The mainly horizontally layered strata of sand were heavily compressed and cemented together by silica during many successive thermal events until they reached thicknesses of several thousands of meters. The resulting quartzite and sandstone rocks are known today as the Roraima Group. The frequent occurrence of ripple marks and cross-bedded sections on freshly exposed rock surfaces of tepui summits indicates that the sandstone strata of the Roraima Group were probably deposited in shallow seas or in large inland lakes (Briceño et al. 1990). Geochronological measurements (Teggin et al. 1985) and the complete lack of fossil spores or macrofossils in the rocks support the Precambrian age.

Leonard Dalton (1912) first named and described the Precambrian Roraima quartzites and sandstones east of Santa Elena de Uairén. These are very hard rocks that are usually whitish to reddish when freshly exposed. Vast extensions of this once continuous, although not uniform, rock cover were heavily weathered and fragmented by more than a billion years of erosion cycles that left behind just a few strikingly isolated mountains. These characteristic table mountains, with sheer vertical walls and mostly flat summits, are the outstanding physiographic feature of the Venezuelan Guayana. In southeastern Venezuela, the Pemón Amerindians call this kind of mountain a *tepui,* a term used as a suffix to the proper name of each particular mountain, such as Auyán-tepui or Ilú-tepui (Armellada and Gutiérrez 1981). This term has been widely adopted in both technical and popular literature (Huber 1987).

Since the phases of deposition and sedimentation that produced the rock strata of the Roraima Group were intermittent, various types of quartzites and sandstones were produced. Based on the stratigraphy of the southeastern tepuis of Venezuela, especially Roraima-tepui, Reid (1974) described four formations of the Roraima Group, listed here from the youngest, uppermost formations to the lowest, oldest ones: (1) Matauí Formation, 600–900 m thick; (2) Uaimapué Formation, to 650 m thick; (3) Kukenán Formation, 50–400 m thick; and (4) Uairén Formation, > 850 m thick. The rocks of each formation differ in characters such as color, hardness, and acidity, which reflect different compositions of silica, cherts, jaspers, and other components.

It is still not clear how closely the stratigraphic sequence of the eastern tepuis outlined above correlates with other tepuis in central and southern Venezuelan Guayana. Geologists once assumed that the entire basement of the Guayana Shield was buried under a continuous block of sandstone layers, but Gansser (1974) and Gosh (1985) postulated that sedimentation basins of several different sizes deposited quartzites and sandstones of the Roraima Group during different time periods. They further suggested that the different basins filled in first in the east, and later to the west and south. This implies that the eastern rock strata are the oldest, and the western and southern strata are the youngest in the Roraima Group. If this hypothesis is correct, then the roughly north–south axis of the large, high massif of older granitic rocks in the Maigualida and Parima mountains would have acted as a divide between these different sedimentation basins.

Intrusive Rocks

The last major rock group in the Guayana Shield comprises the intrusive rocks, which have repeatedly penetrated the metamorphic basement and sedimentary cover during Paleozoic and Mesozoic times. These rocks are present in almost all the larger tepui massifs, where they form stocks and sills of varying sizes (Briceño et al. 1990).

The predominant intrusive rocks are different kinds of diabases and, to a lesser degree, granites. Soils originating from these rocks are lower in silica content than the older granites of the basement formations and the quartzites of the Roraima Group. As a result, they are usually more fertile and are mostly covered by dense forests.

Geomorphology

The highly structured landscapes of the Guayana Shield are vivid evidence of the different erosion forces and cycles that have acted in the past and continue today. Since there are no convincing fossil records from any rock type in the Guayana Shield, it is difficult to assign definite dates to the various rock strata. Using other methods, however, geologists have constructed a general time sequence for the major geological phases and the main erosion cycles associated with them.

Modern geomorphological research in the Guayana Shield has shown that there is a large core area that has apparently never been covered by marine waters after completion of the sedimentary phases. In this area of long and uninterrupted continentality, the main forces of erosion have been climate and tectonics, which produced a clearly recognizable sequence of erosion surfaces at different altitudinal levels (Schubert and Briceño 1987). Because of the horizontally layered quartzite and sandstone strata that predominate in the highland areas, the principal erosion surfaces there are extensive plateaus or peneplains. At lower elevations, where granitic rocks prevail, the peneplains are characterized by a hilly or undulating landscape.

For the Venezuelan Guayana, which includes fluvio-deltaic fringe areas of the Guayana Shield, the following series of broad geomorphological units can be recognized. Listed from higher to lower altitudinal units, they are plateaus and peaks, escarpments, lower slopes, valleys, hill-lands, peneplains, and plains. (Figure 1-4).

Plateaus and peaks (*altiplanicies* and *picos*). Plateaus and peaks form the uppermost portion of the characteristically flat-topped tepui summits, mainly at elevations between 2000 and 2600 m. Most tepuis reach their highest plateau level between 2000 and 2400 m (Auyán-tepui, Los Testigos, Ptari-tepui, parts of the Chimantá massif, Sierra de la Neblina, and Cerros Jaua, Yaví, Coro Coro, Parú, and Duida), but several have considerably higher summit levels, such as Roraima-tepui (Plates 26–28), Kukenán-tepui, Yuruaní-tepui, Ilú-tepui, and Eruoda-tepui (2400–2700 m), and Cerro Marahuaka (2800 m). On Sierra de la Neblina, the flat summit area is surmounted by higher conical peaks reaching just over 3000 m.

Escarpments (*acantilados*). Escarpments are the upper vertical walls of tepuis, usually between 300 and 700 m high, but sometimes reaching a vertical drop of 1000 m or more.

Lower slopes (*laderas, vertientes*). Lower slopes form the base of the tepuis (debris talus) and are usually moderately to steeply inclined and covered by soil and rock that have fallen from the upper walls.

Valleys. Most valleys are deeply incised river courses, often following fault lines, rock crevices, or contact zones between different rock types. On the plateaus or level uplands, the valleys are barely incised and are often subject to lateral displacements. For the large rivers in the lowlands, the valley structure is often difficult to discern, as river beds there are broad, with meandering courses.

Hill-lands (*colinas* and *lomas*). The northern to western border of the Guayana Shield in Venezuela is characterized by numerous hills that are mostly rounded and of low elevation (200–600 m). Most consist of the underlying igneous basement rocks, such as granites or granitoid gneisses. The typical exfoliation of the exposed surfaces of these rocks by chemical and thermic weathering also occurs in other tropical regions with similar orography, such as the Tumuc Humac mountains of French Guiana and the mountains around Rio de Janeiro, Brazil. Since these rounded hills are often isolated by intervening plains, they have been called *inselbergs* and are known locally as *lajas*.

Peneplains (*peniplanicies, penillanuras*). Vast extensions of the lowlands in Bolívar and in Amazonas states consist of a gently undulating landscape with interspersed low hills. These areas are called peneplains because of their transitional character between hill topography and a level plain, at elevations mainly between 100 and 300 m.

Plains (*planicies, llanuras*). These land surfaces consist of nearly level plains and are typical of the Orinoco Delta in Delta Amacuro state. They also occur in the Río Atabapo drainage basin, in western Amazonas. In the Venezuelan Guayana, their elevation ranges between 0 and 100 m.

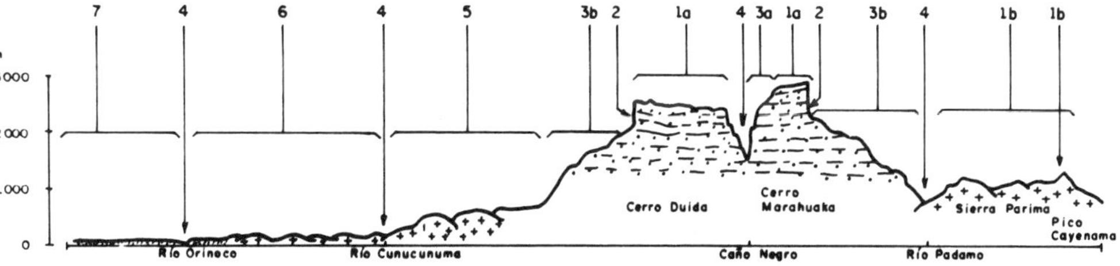

Figure 1-4. Schematic representation of the main geomorphologic units found in the Venezuelan Guayana. Elevations above sea level (altitudinal scale exaggerated). 1 = summit plateaus and peaks (highlands, 1a; uplands, 1b), 2 = escarpments, 3 = slopes (upper, 3a; lower, 3b), 4 = valleys, 5 = hill-lands, 6 = peneplains, 7 = plains.

Soils

The long weathering processes of different rock types under variable climatic conditions and changing vegetation cover in the Venezuelan Guayana have produced an impressive series of different soil types. The first large-scale soil inventory of the region (at a scale of 1:250,000) was made in the early 1970s in Amazonas state, using side-looking radar techniques (Aero-Service Corporation 1972). This preliminary study was further refined between 1976 and 1983 by a French and Venezuelan team that combined accurate radar and satellite interpretation techniques with extensive field surveys. This team later published a soil atlas of Amazonas state (MARNR-ORSTOM 1987). A major soil inventory of the entire Venezuelan Guayana began in 1986; it has begun to produce an accurate set of soil maps at a working scale of 1:250,000 and published at a scale of 1:500,000 (Zinck 1986; CVG-TECMIN 1987, 1989, 1991a, 1991b, 1991c, 1991d, 1991e).

The soils of the Venezuelan Guayana comprise most of the orders known to occur in tropical latitudes, with a clear predominance of entisols, oxisols, ultisols, and histosols; smaller amounts of inceptisols, spodosols, and alfisols are also present.

Histosols consist mainly of decomposing organic matter, such as peats and bogs. They occur either in some permanently flooded lowland areas of the Orinoco Delta, or else as peats to 2 m thick directly overlying the rocky substrate of many tepui summits. The remaining soil orders in the Venezuelan Guayana all have humus-rich layers (horizons) overlying mineral materials, with different degrees of sand, silt, and clay components. Spodosols are usually deep sandy soils with a dark-colored organic and/or ferric horizon within 2 m of the surface; they occur only in the Atabapo and Negro river drainages of southwestern Amazonas state and underlie sclerophyllous Rio Negro caatinga forests. Among the entisols, the quartzipsamments are widespread in central Amazonas state, where characteristic white-sand savannas occur on deep, white, fine- to coarse-grained, sandy soils that are saturated with water during most of the year. Alfisols, oxisols, and ultisols are mainly forest soils with a variable degree of clay content in the mineral horizon. Oxisols are frequently found under savanna vegetation in northern or southeastern Bolívar state, where the upper red soil layer often contains dense ferruginous concretions called ironstone (local terms are *coraza* or *arrecife*) or small iron nodules (locally called *ripio*). Inceptisols occur extensively in alluvial plains and on mountain slopes, where they usually form shallow soils derived from the underlying rock strata.

Because of the generally low mineral content of the parent rocks of the Guayana Shield and the high weathering rates that occur under tropical conditions, soils are generally poor to very poor in nutrients (especially phosphorus, calcium, and nitrogen). They also have low to very low cation exchange capacities (CEC) and are generally highly acidic, with pH values between 4 and 5. High accumulations of toxic aluminum compounds have often been measured in the subsoils, which severely affect the nutrient balance of the vegetation growing on them (Fölster and Huber 1984; Fölster 1986). Only where the soils originate from basic to intermediate igneous parent rock, such as intrusive diabases, gabbros, or volcanic rock types, do more favorable nutrient conditions with lower soil acidity occur. In these cases, a more vigorous forest cover prevails, and it is apparently more resistant to the impact of shifting cultivation and fire than forests that overlie the predominantly poorer soils of the region.

Climate

The local and regional climate types of the Venezuelan Guayana are strongly influenced by two major global phenomena: the northeastern trade winds (*vientos alisios*) and the Intertropical Convergence Zone (ITCZ). The trade winds blow almost constantly from the Atlantic Ocean toward the continent, carrying moisture-rich air masses from the cooler ocean surface above the warmer inland surface. The ITCZ consists of regular, annual oscillations of large continental air currents to the north and south of the equator, causing heavy disturbances in local weather dynamics.

The climate of the Venezuelan Guayana varies greatly according to distance from the sea, topography, and altitude. Since there is still only a sparse network of meteorological stations in this remote region, it is premature to attempt a detailed overview of the climatic pattern of the area. However, a simplified climatic classification scheme was developed by a multidisciplinary team working on a detailed natural resource inventory of the region and is adopted in this treatment (García 1987; see Figure 1-5 and Table 1-2 for definitions of the climate types). This system describes the basic climatic parameters associated with the main vegetation types and ecosystems of the region, using three measures: average annual air temperature, average annual rainfall, and seasonality (measured as the number of dry months with less than 50 mm rainfall per month). Since most meteorological stations in the Venezuelan Guayana record only rainfall, temperatures must be estimated from altitudinal temperature gradient tables, which calculate a standard temperature decrease of 0.6°C per 100 m altitude.

Special care should be taken in this section not to confuse latitudinal terms with altitudinal life zones. Since tropical sites at any altitude have nearly constant day lengths throughout the year, they typically have *day climates,* with pronounced daily temperature oscillations and very slight monthly oscillations (tropical isothermy). In the Venezuelan Guayana, which lies between 0 and 10° north latitude, the longest and the shortest days of the year differ by less than an hour. In contrast, climates in temperate or subtropical zones are typically *year climates,* where the daily temperature oscillations are small compared with the yearly seasonal oscillations (high-latitude heterothermy). Thus, it is incorrect to call cool, upper montane climates in the tropics *subtropical,* only because of a lower average temperature.

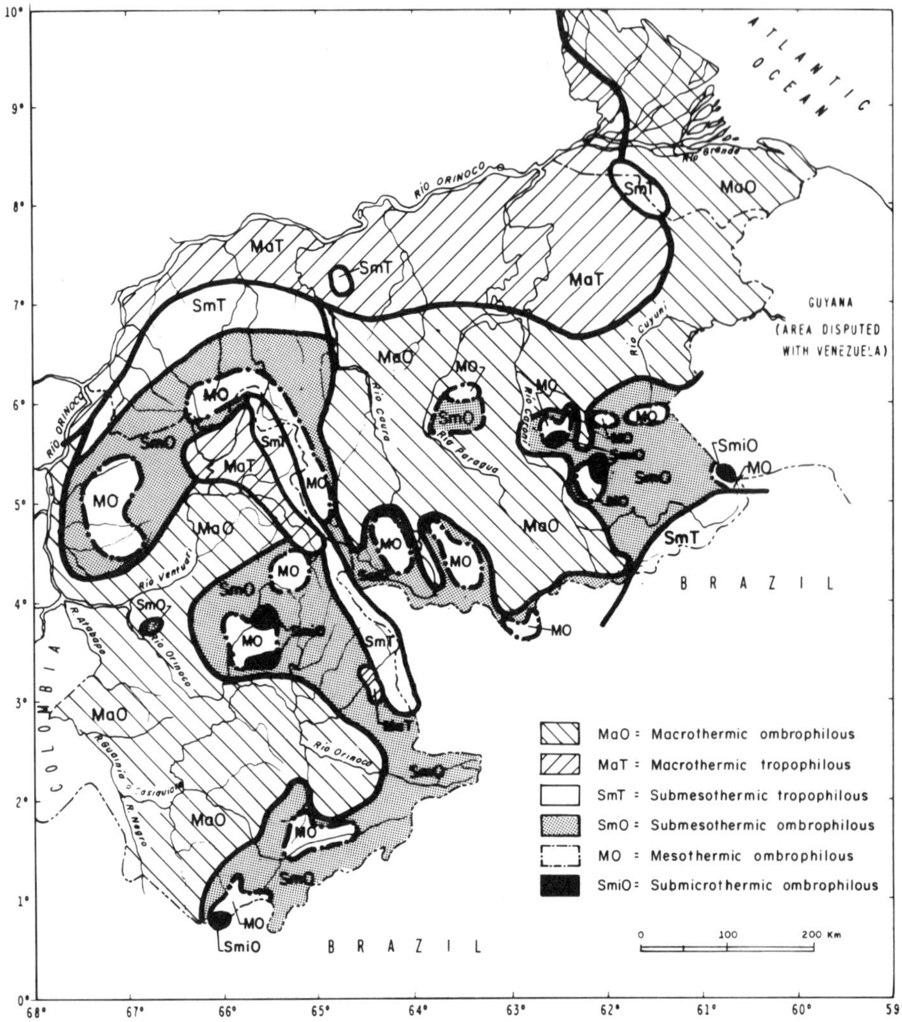

Figure 1-5. Distribution of main climatic types in the Venezuelan Guayana.

Table 1-2. Climatic subdivisions adopted for the Venezuelan Guayana (following García 1987). Dry months are those with less than 50 mm mean precipitation.

Altitude (m above sea level)	Mean annual temperature (°C)	Mean annual precipitation (mm)	Number of dry months
0–500	macrothermic (> 24°)	ombrophilous (> 2000)	< 2
0–500	macrothermic (> 24°)	tropophilous (1000–2000)	2–5
500–1400	submesothermic (18–24°)	ombrophilous (> 2000)	< 2
500–1400	submesothermic (18–24°)	tropophilous (1000–2000)	2–4
1400–2400	mesothermic (12–18°)	ombrophilous (> 2000)	< 1
2400–3000	submicrothermic (8–12°)	ombrophilous (> 2000)	< 1

Macrothermic Ombrophilous Climate

Several large lowland areas of the Venezuelan Guayana have macrothermic ombrophilous climates, that is, they are hot (average annual air temperature always more than 24°C) areas with more than 2000 mm of annual rainfall and either a short or no dry season at all. The three areas are as follows:

1. *The outer section of the Orinoco Delta and its southern extension east of Serranía de Imataca, including part of the Río Cuyuní basin.* A short dry season is noticeable from December to February (Figure 1-6). This area is strongly influenced by the moisture-rich trade winds that constantly blow off the nearby Atlantic Ocean.
2. *The middle and upper Caroní, Caura, and Paragua river basins.* The highest annual rainfalls of the entire Guayana region occur here (> 4200 mm), causing a perhumid climate with an almost imperceptible dry season in December and January (Figure 1-7). Evaporation is high and produces cloud cover during most of the year; winds are strong during the rainy season, sometimes causing windfalls in the lowland forests.
3. *The upper Orinoco and Casiquiare peneplains.* Average rainfall varies between 2500 mm in the north, at the confluence of the Orinoco and Ventuari rivers, to almost 4000 mm in the southwest (San Carlos de Río Negro). Another rainfall gradient that increases from east to west extends from the western base of the Sierra Parima to the Río Negro (Huber et al. 1984). Evaporation is also high, especially during the barely pronounced dry season from December to February (Figure 1-8).

Macrothermic Tropophilous Climate

These hot (average annual air temperature always more than 24°C) climate types with a moderate dry season occur mainly along the northern and northwestern border of the Guayana Shield and extend in a continuous belt from the inner Orinoco Delta westward through northern Bolívar state into Amazonas as far as Puerto Ayacucho. Locally, they also occur in the northern Río Ventuari basin in northeastern Amazonas. The five areas are as follows:

1. *Inner Orinoco Delta.* This area has 1500–1800 mm of annual rainfall and a marked dry season of up to 4 months from December to March (Figure 1-9). There is a strong influence of trade winds especially during the dry season.
2. *Lower Orinoco-Caroní drainage.* The area between the Río Orinoco in the north, Tumeremo in the east, La Paragua in the south, and Caicara to the west, has a strongly biseasonal climate with a pronounced

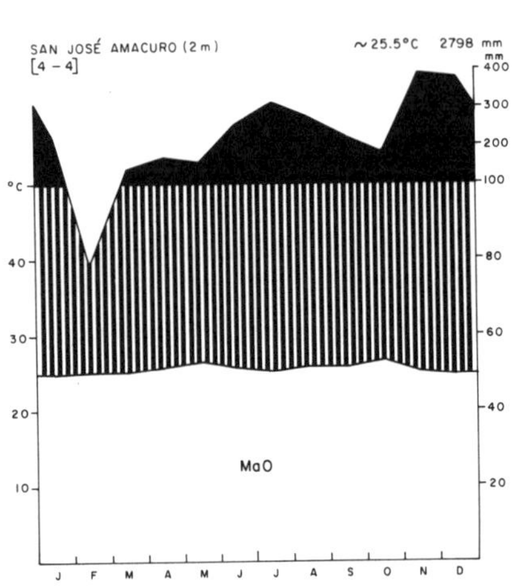

Figure 1-6. Climatic diagram of San José de Amacuro (Delta Amacuro state).

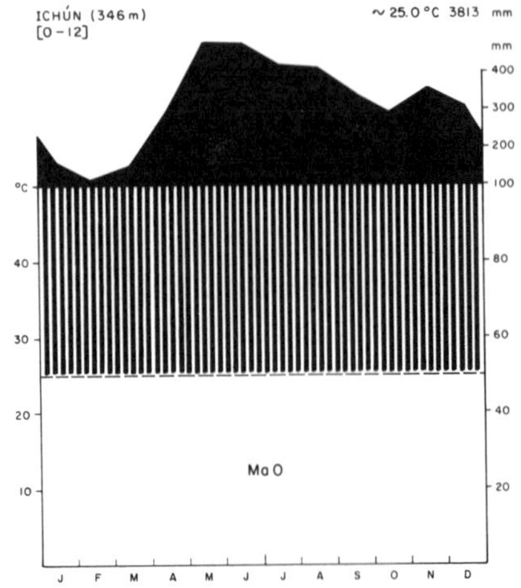

Figure 1-7. Climatic diagram of Ichún (upper Río Paragua, Bolívar state).

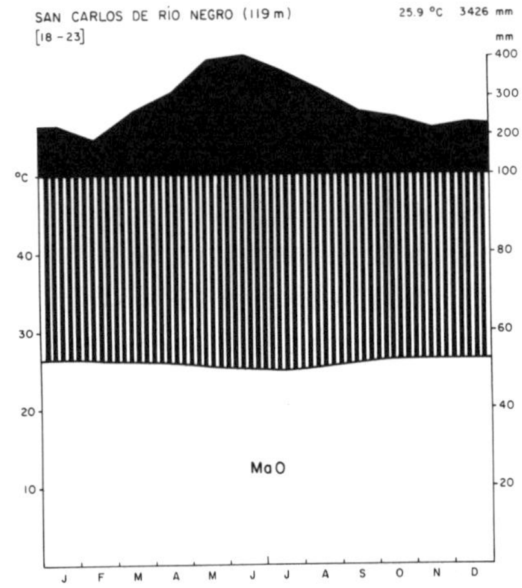

Figure 1-8. Climatic diagram of San Carlos de Río Negro (Amazonas state).

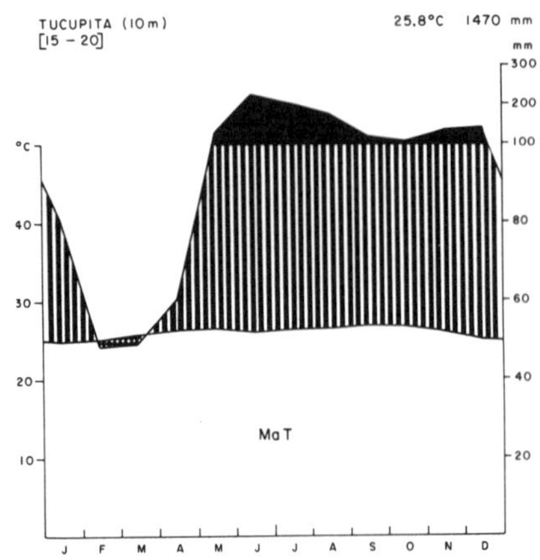

Figure 1-9. Climatic diagram of Tucupita (Delta Amacuro state).

Figures 1-6 through 1-15. Climatic diagrams from the Venezuelan Guayana, following the conventions of Walter (1979). Top line, left, is the locality of the climatic station, followed in parentheses by the height above sea level; to the right is the mean annual temperature and the mean annual precipitation. On the second line, in brackets, the first number is the number of full years of observation for temperature, the second for precipitation. The lower line on the diagram is the mean monthly temperature, whereas the upper and more variable line is the mean monthly precipitation. Dotted line is the estimated mean temperature when records are lacking. Black = periods of high precipitation (> 100 mm per month); striped = relatively humid season; checkered = period of relative drought. MaO = macrothermic ombrophilous climate; MaT = macrothermic tropophilous climate; SmO = submesothermic ombrophilous climate; SmT = submesothermic tropophilous climate; MO = mesothermic ombrophilous climate (see Table 1-4, García 1987).

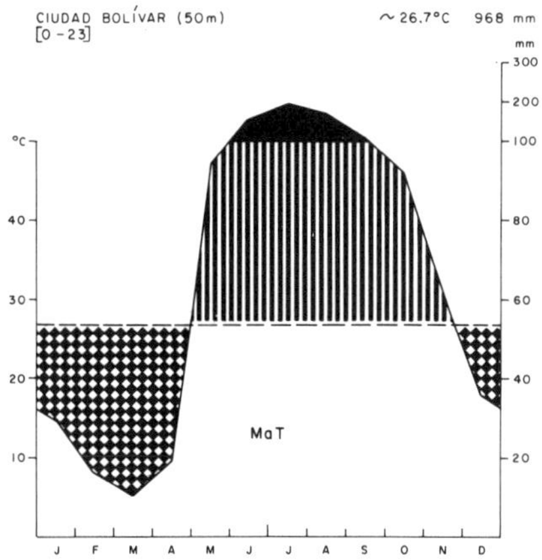

Figure 1-10. Climatic diagram of Ciudad Bolívar (lower Río Orinoco, Bolívar state).

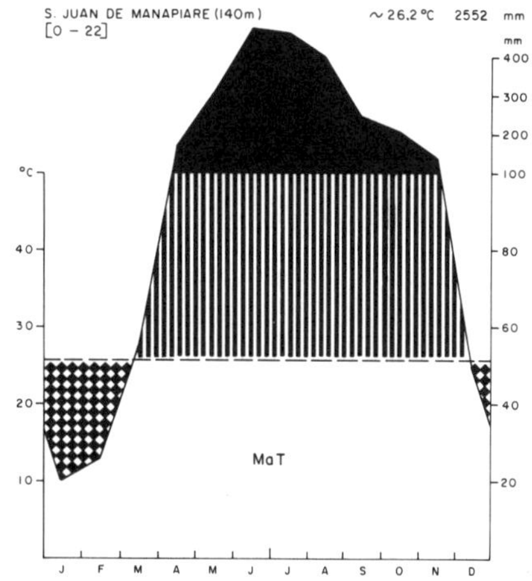

Figure 1-11. Climatic diagram of San Juan de Manapiare (Amazonas state).

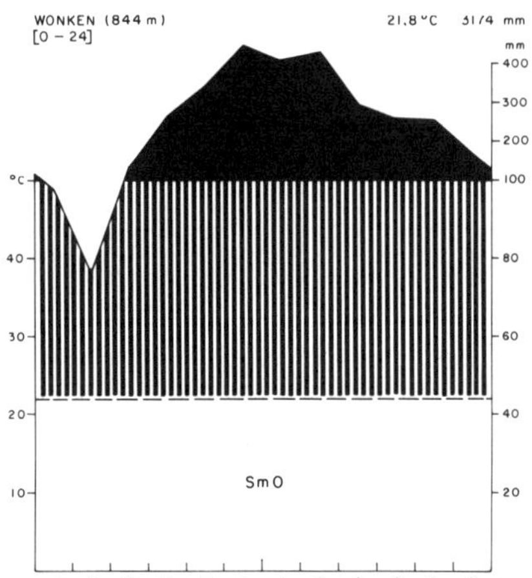

Figure 1-12. Climatic diagram of Wonkén (Gran Sabana, Bolívar state).

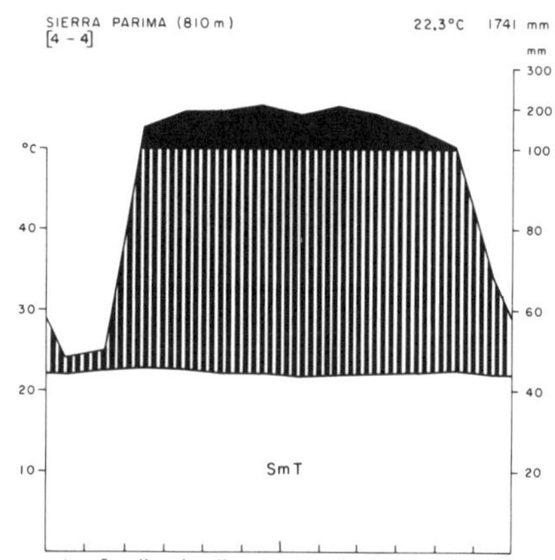

Figure 1-13. Climatic diagram of Sierra Parima (Amazonas state).

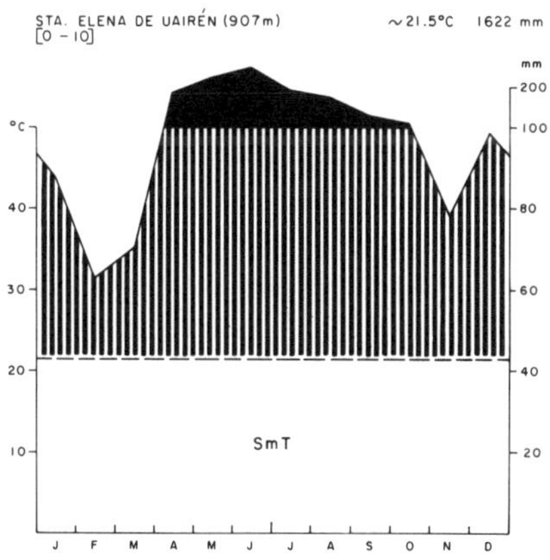

Figure 1-14. Climatic diagram of Santa Elena de Uairén (Gran Sabana, Bolívar state).

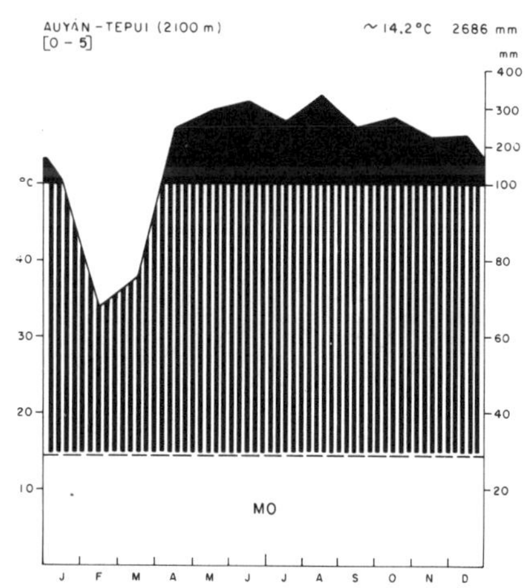

Figure 1-15. Climatic diagram of Auyán-tepui (Bolívar state).

dry season of 4 to 5 months between November and March or April. There are only 950–1400 mm of annual rainfall during the rest of the rainy season (Figure 1-10). The area around Ciudad Bolívar on the lower Río Orinoco is the driest of the Venezuelan Guayana.

3. *Northwestern Guayana Shield between Caicara and Puerto Ayacucho.* Along this belt following the Río Orinoco, there is a steady increase in average annual rainfall from 1400 mm in Caicara to 2200 mm in Puerto Ayacucho. There is still a marked seasonality, however, with a dry season of 3 to 4 months lasting from December to March or April.
4. *Northern Río Ventuari basin.* The region of the upper Ventuari and Manapiare rivers has a tropophilous climate, with a dry season from December to March and a rainy season during the remaining months of the year. The average annual rainfall at Cacurí and San Juan de Manapiare is 2000–2500 mm (Figure 1-11). This relatively dry area is probably caused by a rain shadow of the high mountains of the Sierra de Maigualida in the east and the Yaví-Yutajé massif in the north.
5. *A small, isolated area at the western base of the southern Parima range on the lower Río Ocamo.* Although there are no weather data from this location, the high proportion of deciduous trees there indicates the existence of a dry climate with alternating seasons.

Submesothermic Ombrophilous Climate

Most Guayanan uplands between 500 and 1200 m elevation have humid, submontane climate types, with average annual air temperatures that range between 18 and 24°C and a dry season (if present at all) that does not exceed two months. Five main areas with such climates are identifiable.

1. *The Gran Sabana* in southeastern Bolívar state. Average temperatures are around 20°C and average rainfalls are between 2000 and 3000 mm. A weak dry season occurs from December to March (Figure 1-12).
2. *The lower (south-central) section of the Guaiquinima massif* between 800 and 1300 m elevation. Average temperatures are between 20 and 22°C and rainfall is estimated above 2000 mm.
3. *The lower mountains and piedmont hills* between 500 and 1500 m elevation, from the Pakaraima range in southern Bolívar to the northwestern Cuao-Sipapo massif in Amazonas. There are no climate data for this region, but annual rainfall is high to very high, and no dry season has been observed.
4. *The extensive, mostly 600- to 1200-m high hill-lands and piedmonts of the upper Río Orinoco and the Río Matapire drainages* in central and southern Amazonas. No climate data exist from this region, but rain-

fall appears to be high, with a very short dry season in January and February.

5. *The upper slopes and summit area of Cerro Yapacana,* a low tepui of approximately 1300 m elevation southeast of the confluence of the Orinoco and Ventuari rivers in central Amazonas state.

Submesothermic Tropophilous Climate

These lower montane climate types, with a pronounced dry season 2–4 months long from December to March or April, occur in scattered portions of the uplands of the Venezuelan Guayana. Sometimes they are transitional between macrothermic tropophilous, lowland climates and submesothermic or mesothermic ombrophilous, montane climates, as in northwestern Bolívar state or in the uplands of the Sierra Parima (Figure 1-13). Submesothermic tropophilous climates also appear to be caused by local rain shadows, as in the extreme south of the Gran Sabana (Figure 1-14) or on the western slopes of the Sierra de Maigualida. Other isolated areas with this climate type are the Serranía de Imataca in northeastern Bolívar and the hills east of the lower Río Caura in north-central Bolívar.

Mesothermic Ombrophilous Climate

Most upper slopes and plateaus of the Venezuelan Guayana at elevations between 1500 and 2400 m have cool, montane climates with average annual air temperatures between 12 and 18°C and high to very high (2500–3500 mm) average rainfall. Additional moisture is supplied by the frequent occurrence of dense mists. Rainfalls are less common from December to March, when short dry periods with sunny days and high solar radiation may occur, but a true dry season is virtually absent (Figure 1-15). Winds and thunderstorms are frequent during the normal rainy season in Venezuela (March to November), when the Intertropical Convergence Zone strongly influences the weather conditions of the entire region.

Submicrothermic Ombrophilous Climate

In the Venezuelan Guayana, this upper montane climate has an average annual air temperature around 10°C or less. It is also characterized by heavy rainfalls and dense cloud and mist formation almost all year, with frequent strong winds and high solar radiation. It occurs only on the highest summits (> 2350 m) of some tepuis, such as in the Roraima-Ilú chain, the southern tip of Auyán-tepui, the northeastern portion of Chimantá-tepui, Cerros Duida and Marahuaka, and the southern peaks of Sierra de La Neblina. Since minimum air temperatures of 1 to 2°C have been measured on Roraima-tepui and on the Chimantá massif, it is likely that occasional freezing temperatures occur there during the year. However, frost has never been reported on any of these mountains; the constantly high air moisture may prevent the air from reaching the freezing point. It is thus unclear if any high-elevation tepui plants can be considered frost resistant.

Paleoecology

Since the Guayana Shield has been geologically stable since at least the late Tertiary, the region has often been regarded as being climatically stable during this period as well. Consequently, the mountains of the Venezuelan Guayana were considered by several authors as biotic refugia from otherwise severe climatic fluctuations during the Pleistocene (Haffer 1974; Prance 1982; Steyermark 1982). Similar arguments have also been used to explain the presence of an ancestral flora in both the forested lowlands and the tepui summits (Halliburton 1952; Maguire 1956). Not until the 1980s were scientific data available to test these hypotheses, when studies were made on alluvial and lacustrine deposits of the Gran Sabana and in peat deposits of several high-tepui summits (Schubert and Fritz 1985; Schubert et al. 1989; Rull 1991).

Radiocarbon measurements of the age of the lowermost peat layers sampled on the summits of the Chimantá massif (at 1900–2300 m elevation), Auyán-tepui (at 1980 m), and of Cerro Guaiquinima (at 900–1400 m), gave values ranging between 6000 and 3000 years ago (Rull 1991). Surprisingly, no older peat deposits were found on any of these tepuis. At least two explanations have been advanced. One possibility is that peat deposition actually started at the maximum age recorded (approximately 6000 years ago); in this case it could be assumed that prior to this age, drier climatic conditions, unfavorable for peat formation, prevailed and that the present peat layers began to be deposited as a consequence of increasingly humid climatic conditions. According to a second hypothesis, peat deposits are continuously decomposed at their (lower) contact zone with the underlying rock surface and are steadily washed out by internal water flows. This would not allow an indefinite growth of the peat cover beyond a certain thickness. In this scenario, the maximum age found at the bottom of the peat column would only indicate the maximum age of the deposition cycle, not the actual beginning date of the present peat layer.

Since it is not yet clear which of these possibilities may be valid, paleoecological interpretations are best limited to the Holocene period from 6000 years to the present. A careful analysis of the palynological record from various depths in the peat samples has shown the existence of a series of more or less intensive climatic oscillations (Rull 1991). These changes coincided with and were similar to the Holocene changes reported by van der Hammen (1982) in the northern Andes. They are documented by the alternating dominance of pollen from herbs and pollen from shrubby or forest vegetation, especially the displacement of *Stegolepis* meadows by *Bonnetia* forests (presumably during wetter periods) and vice versa. Such vegetation changes occurred repeatedly on the Chimantá massif and on Auyán-tepui.

In the Gran Sabana, the palynological record for the last 3000 years shows repeated, major shifts in the grass, palm, and forest pollen levels, which clearly indicates strong climatic oscillations during this period (Rull 1991).

Although scientists are just beginning to study the paleoecological history of the Venezuelan Guayana, the available data indicate that the Guayana Shield has not been spared from significant climatic oscillations during the last five millennia. The different vegetation belts on the slopes and summits of these mountains have probably been subjected to vertical displacements or migrations similar to those recorded for many other tropical and extra-tropical mountains. The numerous peat islands occurring between 1000 and 1500 m in the Gran Sabana with typically high-tepui elements such as *Stegolepis* and *Orectanthe* are possibly relicts of past Holocene migration phases.

Hydrography

Rivers are the traditional transportation route to the interior regions of the Amazon and Guayana and have always been important to both inhabitants and visitors. Since Humboldt and Bonpland's visit in 1800, almost all major botanical explorations in the Guayana lowlands have been conducted along the rivers; only during the last two decades has exploration by air (mainly helicopters) become more important.

The river system draining the Venezuelan Guayana is very complex because of the numerous water courses and the strikingly different hydrologic characteristics often found within small drainage basins. The waters of the Amazon and surrounding tropical river basins vary considerably in their color, sediment load, and physical and chemical parameters, properties that are strongly influenced by the geology, vegetation cover, and climatic regimes of each particular basin (see Plate 8). According to Sioli (1965) and emended by Franz Weibezahn (personal communication), the Amazon and Orinoco basins can be divided into the following three categories:

1. *White-water:* characterized by high turbidity; grayish white color; a high content of inorganic suspended solids (fine sands and clays), resulting in very low transparency (Secchi disk measurements of 0.1 to 0.5 m); rich in nutrients; pH near neutral (7.0)
2. *Clear-water:* low amounts of suspended solids; color yellow-brown, sometimes greenish; high transparency (Secchi disk measurements of 1.0 to 3.5 m); low nutrient content; moderate acidity (pH between 5.0 and 7.2)
3. *Black-water:* very low amounts of suspended solids; color dark brown or dark reddish brown, similar to a tea infusion; high transparency (Secchi disk measurements of 1.3 to 3.0 m); low to extremely low nutrient content; high acidity (pH 3.8–5.0)

It is generally agreed that the dark brown color of the black-water rivers is caused by high contents of humic and tannic acids dissolved in the ground water during the decomposition process of certain plant material (Paolini 1978). In the Guayana Shield, the drainage of characteristic vegetation types that grow either on extremely nutrient-poor substrates (such as quartzites, sandstones, or white-sand soils), or on thick organic materials (such as peat), invariably produce effluents with typical black-water characteristics (for further discussion see Janzen 1974).

Although a detailed hydrographic map of the entire Guayana region has yet to be made, side-looking radar images at a scale of 1:250,000 have provided useful information. For this reason, an updated topographic map has been produced (see accompanying map) with particular care given to the river systems. The small scale (1:2,000,000) of this map, however, does not allow an exhaustive inventory of all the water courses in this vast mountainous area. The following section provides an overview of the larger river basins and their main hydrologic characteristics.

Major River Basins of the Venezuelan Guayana

The Venezuelan Guayana is drained by a myriad of water courses that belong to three great river basins that eventually flow into the Atlantic Ocean. Within the flora area, the Orinoco basin drains 360,670 km^2, the Amazon basin drains 53,280 km^2, and the Essequibo basin drains 40,000 km^2 (see Figure 1-16).

With an approximate coverage of 79 percent of the Venezuelan Guayana, the Río Orinoco basin is by far the most important river system in the area. The Orinoco also receives important affluents from the Colombian and Venezuelan Andes and, to a minor degree, from the Colombian and Venezuelan Llanos and from the Venezuelan Coastal Cordillera.

The river system belonging to the Amazon basin is located in southern and southwestern Amazonas state, where it drains approximately 12 percent of the Venezuelan Guayana; its most remarkable water course is the Casiquiare channel, which branches off from the upper Río Orinoco through the Orinoco-Casiquiare bifurcation near Tamatama.

Finally, an area covering approximately 9 percent of the Venezuelan Guayana in the easternmost and southeastern portions of Bolívar state belongs to the Essequibo river system. The larger part of this area corresponds to the Río Cuyuní drainage and the smaller part to the Río Mazaruni.

A remarkable geographic feature of southern Venezuela is the presence of a natural, direct water connection between the Amazon and the Orinoco basins. These are the first and the third largest river systems in the world as measured by average discharge at their mouths. Although reported since the mid-17th century, this strange river connection continued to be the subject of controversial interpretations among various schools of European geographers (with several contrasting maps produced during this period) up to the 18th

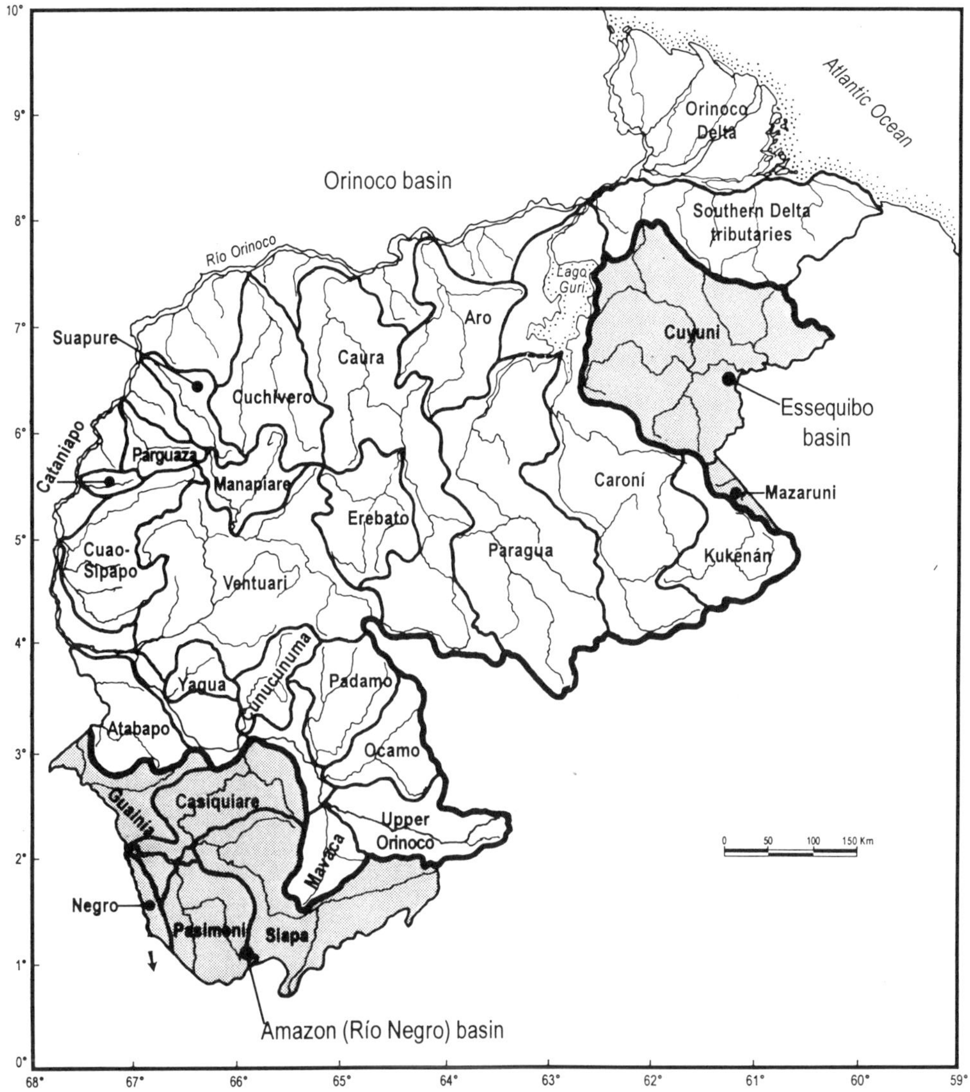

Figure 1-16. Major river basins in the Venezuelan Guayana. The most extensive is the Orinoco basin (unshaded). Some of its immediate drainage and minor tributaries are not named on this map. The other two major basins, the Essequibo and the Amazon, are shaded.

century. The existence of a stream divide of the Río Orinoco near Cerro Duida in Amazonas state was not irrefutably recognized and accepted until a Jesuit missionary, Manuel Román, made the uninterrupted river journey from Carichana on the middle Orinoco to the Portuguese settlements established on the upper Río Negro in 1744 (Useche 1987). The water course branching off from the Orinoco and flowing toward the Amazon basin was then identified as the Casiquiare channel, often called Brazo ("arm") Casiquiare in Spanish. After some 300 km, this water course joins the Río Guainía coming from the

Colombian interior to form the Río Negro, which eventually joins the Amazon River near the city of Manaus in the heart of the Brazilian Amazon.

Although the flow of a single river into two different hydrological basins had been known previously from other parts of the world, the case of the Orinoco-Casiquiare divide has long been considered a major geographic puzzle because of the enormous sizes of the two river basins involved. Furthermore, it was difficult to explain why a river of already considerable size with a well-developed stream bed should suddenly divide its course without an evident intervening geographical obstacle. Not until 1958 did ecologist Volkmar Vareschi discover that the water divide is caused by an irregular configuration of the Orinoco stream bed, which diverts approximately 25–28 percent of the Orinoco's water flow into the Casiquiare (Vareschi 1963a, 1963b). In the recent geologic past, the extremely level, partially inundated terrain of the area favored the river capture activity of the ancient Río Casiquiare, whose northern headwaters are only a few hundred meters away from the Orinoco (Stern 1970).

Río Orinoco Basin

Both in length and volume, the Río Orinoco is the main water course of the Venezuelan Guayana. Its total length is approximately 2150 km, and its average discharge is 36,000 m^3 per second (Weibezahn 1990). Of its total drainage surface of 1,080,000 km^2, which is shared by Venezuela and Colombia, approximately one third (360,000 km^2) corresponds to the Venezuelan Guayana. The course of the Orinoco resembles a giant fishhook that surrounds most of the Venezuelan Guayana and its main mountain systems. The river's source was only discovered in 1951 by a joint French-Venezuelan expedition and lies at the southeastern end of the Sierra Parima (Amazonas state) on the upper slopes of Cerro Delgado Chalbaud, at 2°19' N and 63°21' W and 1047 m elevation (Rísquez-Iribarren 1962).

In terms of its water chemistry and sediment load, Weibezahn (personal communication) considers the Orinoco to be essentially a white-water river, although subject to strong seasonal changes in its inorganic and organic suspended solids along various parts of its course. Almost all suspended inorganic sediments come from the Andean and Llanos tributaries, whereas those originating in the Guayana Shield (mostly black-water rivers) yield only about 5 percent of suspended sediments (Meade et al. 1990; see Plate 8).

Along its extensive course, the Orinoco has an average gradient of less than 0.01 percent, indicating that its waters flow rather slowly and have a long transit time (Weibezahn 1990). Still, there are numerous spots with rapids along the river (Plate 10). The three main series of rapids that are not navigable even during the rainy season are the Raudales de Atures, near Puerto Ayacucho (Plate 7); the Raudales de Maipures (Maypures), near the mouth of the Río Samariapo (approximately 50 km south of Puerto Ayacucho); and the Raudal de los Guaharibos, approximately 150 km downstream from the source of the river where the Orinoco leaves the Sierra Parima.

Historically, the river has been divided into three main sections, the lower (*bajo*) Orinoco, the middle (*medio*) Orinoco, and the upper (*alto*) Orinoco. Because these sections are widely cited and their limits are not uniformly defined, their main interpretations are described here (taken partly from Weibezahn 1990 and sources therein). According to Vila (1960), the lower Orinoco runs from the river's mouth to the mouth of the Río Apure. The middle Orinoco extends from the Río Apure to the mouth of the Atabapo and Guaviare rivers, and the upper Orinoco from there to the river's source. Weibezahn (1990), the foremost Venezuelan expert of Orinoco limnology and hydrography, considers this to be the hydrographically most precise subdivision of the river.

A somewhat different subdivision was adopted by the governmental hydrological commission COPLANARH (1969, 1972). In this system, the lower Orinoco extends as far as the mouth of the Atabapo and Guaviare rivers, and the upper Orinoco begins at the bifurcation of the river at the Casiquiare channel at Tamatama. This subdivision was adopted by official government agencies such as CODESUR in 1972 (see Aero-Service Corporation 1972) and by the environmental ministry in its current master plan of Amazonas state (MARNR 1983–1984), with a transition sector called the middle-upper Orinoco between the Atabapo and Casiquiare rivers.

Early literature reports, including that of Humboldt (1818–1829), Chaffanjon (1889), and Jahn (1909), had the middle Orinoco starting somewhere between Ciudad Bolívar and the mouth of the Río Caura, or even to the Atures rapids. The upper Orinoco comprised the entire river above the Atures rapids near present-day Puerto Ayacucho, an interpretation that has persisted among most botanical explorers until the present day.

Orinoco tributaries. The multiple mouths of the Río Orinoco form a classical tropical river delta with a wide, fan-shaped net of rivers, channels, and temporary lakes and lagoons, covering an extensively flooded area of some 20,000 km^2. The first branching of the main stem of the Orinoco occurs about 150 km inland from the external limit of the delta, near the town of Barrancas. The southernmost branch of the Orinoco in the delta is called the Río Grande, which flows almost due east, whereas all other major delta rivers flow in a predominantly north to northeasterly direction. All the water courses in this flat and swampy area are interconnected through a large web of anastomosing channels and rivulets. The whole system is strongly influenced by tidal sea level oscillations in the order of 1.5 to 2.5 m. The most important rivers of the Orinoco Delta are (from south to north) the Río Grande, Caño Sacupana, Caño Araguao, Caño Araguabisi, Caño Güiniquina, Caño Araguaito, Caño Mariusa, Caño Caiguara, Caño Macareo, Caño Cocuina, Caño Capure, and Caño Mánamo.

The lowermost course of the Orinoco is usually divided into two sections. The first, still called the Río Orinoco, extends from the mouth of the Río Caroní downstream to the town of Barrancas. The second section extends from Barrancas eastward to the Atlantic Ocean and follows the southernmost branch of the Orinoco Delta, receiving the name Río Grande. Many tributaries from the plains and hill-lands of southernmost Delta Amacuro flow into the lower Orinoco and the Río Grande, draining an area of approximately 23,600 km^2. From east to west these southern tributaries are Río Barima, Río Amacuro, the anastomosing river system formed by Caño Arature–Caño Basama and its upper tributary Río Cuyubini, Río Aguirre (Río Acure), Río Imataca, Río El Toro (also called Río Grande), Río Sacoroco, Río San José, and finally Río Upata, which flows into the Orinoco just east of Ciudad Guayana. Most of these rivers are navigable at least in their lower half.

The Río Caroní basin covers almost 95,000 km^2 and is by far the largest river basin fully contained within the Venezuelan Guayana. It is drained by two large river systems, the Caroní and the Paragua rivers. Between 1960 and 1990, a series of hydroelectric power plants was built in its lower course. The main one is the gigantic dam at Guri (officially called the Represa Raúl Leoni), and a reservoir at Guri, Lago Guri, which covers close to 4250 km^2 (Plate 78). Because of the dam's great economic importance, this river basin is the region's best known in terms of topography, climate, hydrography, and vegetation.

The Río Caroní (total length 760 km) is a typical black-water river with its source at 2700 m elevation on the high tepuis of the Roraima-Ilú chain. The upper Caroní is called the Río Kukenán (Cuquenán) and extends from the summit of Kukenán-tepui to the confluence with the Río Caruay at the western end of the Gran Sabana (200 km long, drainage area 11,920 km^2); only after these two rivers merge is the name Caroní applied to its downstream course. Sometimes, however, the name Caroní is extended upstream to the confluence with the Aponguao, or even the Río Yuruaní. The major left-hand tributaries going downstream on the Kukenán-Caroní rivers (and their alternate spellings) are Río Arabopó (Arabupu), Río Uairén, Río Icabarú (Ikabarú), all in the Gran Sabana, and Río Paragua, which joins the Caroní approximately 180 km upstream from its confluence with the Orinoco. The major right-hand tributaries of the Caroní are Río Yuruaní, Río Aponguao (Apanwao), and Río Caruay (Karuai), all in the Gran Sabana; Río Tírica (Tirika), and Río Apacará (Apakara), both from the Chimantá massif; Río Urimán, from Aprada-tepui and Uaipán-tepui; Río Cucurital from Auyán-tepui; Río Carrao (Karao, Karrao) from Sierra de Lema and Auyán-tepui; and Río Antavari (Antabare).

The Río Paragua (Paravá, Paraua) is 540 km long, with a drainage basin covering 39,320 km^2. Its source is at 1500 m elevation in the Pakaraima range and in Serranía Marutaní (Urutaní or Pia-Zoi), along the Brazilian border. Like the Caroní, it is a black-water river, but it has far fewer rapids and is navigable during the rainy season. The main tributaries of the Río Paragua are Río Paramichi (Paravá-musí, Paramushí), from Serranía Marutaní; Río Ichún (Ichúm), from Cerro Ichún; Río Curutu (Curumú, Kurutú, Kurutí) and Río Mari (Marik, Guaina), both from Cerro Guanacoco; Río Karún (Carún), from the Caroní-Paragua interfluvium; and Río Carapo, Río Tonoro, Río Asa (Aza), and Río Chiguao (Chiwao), all from Cerro Guaiquinima.

The Río Aro is a small, little-known river basin of 14,120 km^2, west of the lower Caroní basin. The main river, Río Aro, is 220 km long and is a predominantly clear-water river; it drains the low hill-lands between the lower courses of the Caroní, Paragua, and Caura rivers.

The Río Caura forms the second largest river basin in the Venezuelan Guayana (47,000 km^2). Its main river, the Río Caura, is 680 km long and originates at 2000 m elevation in the southern Sierra de Maigualida and in the Uainama and Aribana ranges along the Brazilian border. It is a black-water river, and its basin is one of the least disturbed areas in the Venezuelan Guayana. All major tributaries of the Caura flow from the west, and include Río Merewari (Merewere, Merevari), from Sierra Uasadi and Sierra Aribana; Río Erebato, from the Jaua massif and from eastern Sierra de Maigualida; Río Nichare (Icutu), from Serranía Nichare in the northern Sierra de Maigualida; and Río Mato, from Cerro Mato. The only major eastern tributaries are the Río Yuruáni (not to be confused with the Río Yuruaní of the Gran Sabana) and the Río Chanaro, both draining the central Caura-Paragua interfluvium.

The Río Cuchivero basin drains the central part of northwestern Bolívar state (the former Distrito Cedeño). Its area of 15,000 km^2 is crossed roughly from south to north by two large rivers, the Río Cuchivero (290 km) and the Río Guaniamo. The Cuchivero has its source at approximately 2000 m elevation in the northern Sierra de Maigualida and on Cerro Ualipano, whereas the Río Guaniamo arises on slopes north of Cerro Yaví, on the northern border of Amazonas state. The Río Cuchivero was originally a clear-water river, but due to the intense mining activities that started in its upper sections and in the Río Guaniamo in the 1970s, it now carries a high load of inorganic suspended materials.

The Río Suapure basin (4720 km^2) is located in the northwest corner of Bolívar state. Its main river, the Río Suapure, with a total length of 180 km, arises from the northern slopes of the Yutajé-Coro Coro massif, at the northern border of Amazonas state. The Río Suapure was originally a clear-water river; however, sediments from the large bauxite mine of Los Pijiguaos (about 30 km upstream from its junction with the Orinoco) have altered the sediment load of the lower reaches of the river.

The westernmost river basin of Bolívar state corresponds to the Río Parguaza, a clear-water river with its headwaters on the summit of Cerro Guanay, at approximately 1800 m elevation. The total basin area is 3920 km^2, and the Parguaza is nearly 140 km long.

A small, though important river basin lies to the southeast of Puerto Ayacucho in northwestern Amazonas state and covers an area of 1750 km^2. Its main river, the clear-water Río Cataniapo, arises at about 1000 m elevation on Cerro Pailón, in the hilly Cuao uplands, then flows for 108 km roughly westward toward the Orinoco. The lower part of the basin has been heavily deforested for agriculture in the 20th century, and there are plans to build a dam in the upper valley to supply hydroelectric energy to Puerto Ayacucho.

The Río Sipapo basin covers an area of 13,760 km^2 and is located in northwestern Amazonas state, draining almost all the large Cuao-Sipapo mountain massif. The largest river is the Río Cuao (200 km long), a predominantly clear-water river; the other major rivers of the basin are (from north to south) the Autana (Plate 11), Sipapo, and Guayapo, all black-water rivers.

The major part of the Río Atabapo basin (9760 km^2) lies in Venezuela in the western section of Amazonas state, with the remaining part extending into Colombia (Departamento del Guainía). All rivers of this extremely flat basin are black-water rivers. The main river is the Río Atabapo (120 km long), formed by the confluence of three tributaries, Río Atacavi (110 km long), Río Temi, and Río Guasacavi, which originates in Colombia. The only major lower tributary of the Río Atabapo is Caño Caname.

The Río Ventuari is 440 km long and originates in the southern Sierra de Maigualida at 1800 m elevation and in the northern Sierra Parima at elevations of 1000 to 1200 m, also draining the southern slopes of Cerro Yaví, the Yutajé-Coro Coro massif, and the eastern slopes of the Cuao-Sipapo massif. It is the largest river basin in Amazonas state and occupies an area of approximately 41,760 km^2 (Plate 9). The main right-hand tributaries of the Ventuari going downstream are Río Asita, Río Manapiare, Río Marieta, Caño Picure, and Caño Guapuchí; the principal left-hand tributaries are Río Yatiti, Río Hacha, Río Parú, Río Marueta, Río Yureba, and Caño Maraya. Although many of its tributaries are black-water rivers, the Ventuari itself is essentially a clear-water river with a sometimes beautiful emerald-green color. It is quite navigable during the rainy season as far as the Salto Tencua, a spectacular waterfall located about 40 km east of the mouth of the Río Manapiare. The Manapiare is the northernmost tributary of the Ventuari and drains an area of 6640 km^2; it is navigable in its lower basin during the rainy season, when large parts of the surrounding plains become deeply flooded.

The Caño Yagua, a small black-water river approximately 100 km long, drains an essentially flat area of 3320 km^2 to the east and south of Cerro Yapacana, near the junction of the Orinoco and Ventuari rivers. A small lagoon, the Laguna Yapacana (sometimes also called Laguna Yagua), is located about 30 km east of Cerro Yapacana; it is one of the few lagoons present in lowland Venezuelan Amazonas.

The Río Cunucunuma (Kunukunuma) is one of the most beautiful black-water rivers of Amazonas state. It originates between 1200 and 1400 m elevation in the Kikirisha mountains of northern Sierra Parima, but it also receives important tributaries from Cerros Duida, Huachamacari, and Marahuaka along its lower and middle course. The main river draining the Duida highlands is the Caño Negro, which joins the Río Cunucunuma about 10 km east of the Yekwana Amerindian settlement of Culebra. The lower course of the Cunucunuma is interrupted by a series of spectacular rapids, including the Raudal Picure and Raudal Guarinuma. The basin drained by the Cunucunuma covers an area of 5760 km^2.

The Río Padamo basin covers 10,280 km^2 and drains the northwestern slopes of the Sierra Parima in east-central Amazonas state. A clear-water river 160 km long, it has three important left-hand tributaries going downstream, Río Cuntinamo (Kuntinamo), Río Uotamo (Botamo), and Río Matacuni (Metacuni, Matakuni).

The clear-water Río Ocamo is the main river draining the central and southwestern slopes of Sierra Parima in east-central Amazonas. It is 200 km long, and the basin covers 8560 km^2. It has only one important southern tributary, Río Putaco, which drains the southern Sierra Parima and joins the Río Ocamo about 20 km below the great rapids of Raudal Arata.

The Río Mavaca basin is the only important southern tributary of the upper Orinoco, covering an area of 5560 km^2. Its main river, the clear-water Río Mavaca, is 120 km long and originates on the northern slopes of Sierra Unturán.

The Major Venezuelan Tributaries of the Amazon River

Because of the bifurcation of the upper Orinoco into the Casiquiare channel, a direct fluvial connection exists between the Orinoco and Amazon rivers. The portion of the Amazon basin lying in the Venezuelan Guayana covers 53,280 km^2 and is formed by the following five river basins.

The Casiquiare channel (Río, Brazo, or Canal Casiquiare) extends approximately 300 km from its origin in the Río Orinoco to its junction with the Río Guainía at the southwestern border of Amazonas state. Along its sinuous traverse through an almost completely level terrain (the Casiquiare peneplain), the Casiquiare changes steadily from clear-water near the Orinoco to black-water as it flows into the Río Negro, due to the influx of several important black-water tributaries. The total area drained by the Casiquiare is 44,160 km^2, but excluding its two most important tributaries, the Río Siapa and the Río Pasimoni (see below), the Casiquiare basin is reduced to 13,200 km^2. Other notable tributaries of the Casiquiare are Río Pamoni, from Cerro Vinilla, and Río Pasiba (Vasiva, Pasiva), with a beautiful and biologically fascinating black-water lagoon near its mouth (Laguna Pasiba or Vasiva). The Casiquiare channel is navigable during the rainy season, when a shortcut to the Río Guainía basin exists through the Caño Deshecho, which then connects with the upper Caño San Miguel. During the short dry season, however, navigation on the Casiquiare is greatly limited by several treacherous rapids, especially the Raudal Murciélago.

The Río Siapa basin covers 21,040 km^2 and is the largest subbasin of the Casiquiare drainage. The Río Siapa is 360 km long, with its headwaters between 1000 and 1800 m elevation in Sierra Curupira and Sierra Tapirapecó along the Brazilian border. The toponymy of this river is confusing; its lower course is usually called Siapa or Idapa, while its upper course, which extends upstream from the rapids of Raudal Gallineta east of Cerro Aracamuni, appears on most maps as the Río Matapire. This little-known upper river section also receives tributaries from the southern slopes of Sierra Unturán and from the northeastern slopes of the Neblina massif. Almost all of the more important tributaries of the Río Matapire come from the south, such as Río Cajigal, Río Codazzi, Río Castaño, and Río Arari (from east to west). The most important tributary of the lower Siapa is the Río Manipitare, which drains the western slopes of Cerro Vinilla and Cerro Aratitiyope.

The second largest tributary of the Casiquiare is the Río Pasimoni (Pacimoni), another typical black-water river of the Venezuelan Amazon. Its headwaters lie in the summits of the Neblina massif at 2400–3000 m elevation, and it is 170 km long, draining 9920 km^2. The Pasimoni is formed by the confluence of Río Yatúa, which drains the western slopes of the Aracamuni and Avispa massif, and Río Baría, which drains part of the Neblina massif. The Río Baría is also connected to the Brazilian side of the Río Negro basin by the Canal Maturacá, a waterway frequently used by Portuguese Amerindian-slave hunters during the 17th and 18th centuries.

The Venezuelan portion of the Río Guainía is only 160 km long, and the basin drained by this typical black-water river in Amazonas state is 7360 km^2. The headwaters of this large border river lie in the dissected lowland plains of the Departamento del Guainía in neighboring Colombia. At its junction with the Casiquiare channel at 2°N latitude, the Río Guainía changes name to the Río Negro. The Venezuelan tributaries of the Río Guainía are (from north to south) Caño Pimichín (Plate 13), Caño San Miguel (Conorochite), and Caño Tirinquín, all black-water rivers draining the Casiquiare peneplains. During the 18th century, a rudimentary road was constructed from a small port half way up the Caño Pimichín to the town of Yavita on the Río Temi. This produced a convenient terrestrial shortcut that connected the Atabapo and Guainía river basins (the historical portage used by Alexander von Humboldt). In 1970, this road was prolonged as far as Maroa, the most important town on the Río Guainía. A fluvial shortcut between the Río Guianía and the Casiquiare channel also exists during the rainy season, via the Caño San Miguel and Caño Deshecho.

The Río Negro extends along the Venezuelan-Colombian border for 90 km in the southwestern corner of Amazonas state before entering Brazil near the widely visible granite outcrop or inselberg of Piedra Cocuy (Cocui, Cucui, Cucuyh). This is a navigable, black-water river with a small drainage basin in Venezuela of 1760 km^2 formed by a few west-flowing tributaries.

Cuyuní and Mazaruni River Basins in Venezuela

In the eastern Venezuelan Guayana south of the Orinoco Delta, a third major river system is formed by the upper Río Cuyuní basin and the western headwaters of the Mazaruni River. Both are major tributaries of the Essequibo River in lowland Guyana. The Venezuelan portion of this drainage basin covers 40,000 km^2 and is further subdivided into two subbasins, the Cuyuní basin and the Kamoyrán basin.

The headwaters of the Río Cuyuní are located in the Sierra de Lema and on Cerro Venamo, between 1300 and 1500 m elevation. After flowing for approximately 90 km in a northerly direction, the Cuyuní receives two major tributaries from the west, Río Supamo (230 km long) and Río Yuruán (220 km long), near the town of El Dorado. From El Dorado, the Cuyuní changes course to the east and flows about 180 km before entering Guyana at the mouth of the Río Acarabisi. The Río Cuyuní and its tributaries are predominantly clear-water rivers, but they are scarcely navigable because of the many rapids present in the drainage area.

The Río Kamoyrán is a small black-water river and the westernmost tributary of the upper Mazaruni River. Its Venezuelan section flows through the northeastern Gran Sabana at elevations between 1000 and 1300 m and covers an area of less than 2000 km^2. The name of this river is spelled in different ways, including Kamoyrán, Kamürán, Kamurán, Kamarang, Camoirán, and Camarán.

Physiography and Landscapes

The presence of wide plains, deep valleys, and impressive rock massifs makes the Venezuelan Guayana one of the most diversified assemblages of landscape types in South America. Three large physiographic units can easily be identified in the geography of the Venezuelan Guayana: lowlands, uplands, and highlands (Figure 1-17). Since each unit is associated with characteristic biological features, a more detailed description of this physiographic system and its component units is presented below.

Lowlands (0–500 m)

Lowlands are defined as all land between sea level and approximately 500 m elevation, with a hot (macrothermic) climate and a level or low to hilly physiography. In the Venezuelan Guayana, there are four major lowland regions, each with notable differences.

The lower Orinoco-Cuyuní lowlands is a vast region encompassed roughly by the Caño Mánamo and the Río Orinoco to the north, the Río Caroní and Lago Guri to the west, and the Río Cuyuní to the south; it forms the easternmost lowland region of the Venezuelan Guayana. Most of this area consists of extensive level plains (the Orinoco Delta), gently undulating peneplains (the Cuyuní basin), and a few hills and low mountain ranges (the Serranía de Imataca). According to the level of flooding, the alluvial plains of the delta include three landscape units, the nearly permanently flooded lower or outer delta zone, the periodically flooded middle delta zone, and the nonflooded upper or inner delta. The densely forested Orinoco delta and the adjacent Amacuro plains to the south are both part of the larger physiographic unit of the Guianas coastal plains, an area stretching in a continuous belt of variable width along the Atlantic coast southeastward until reaching the mouth of the Amazon River in Brazil. The Cuyuní peneplains, which extend almost

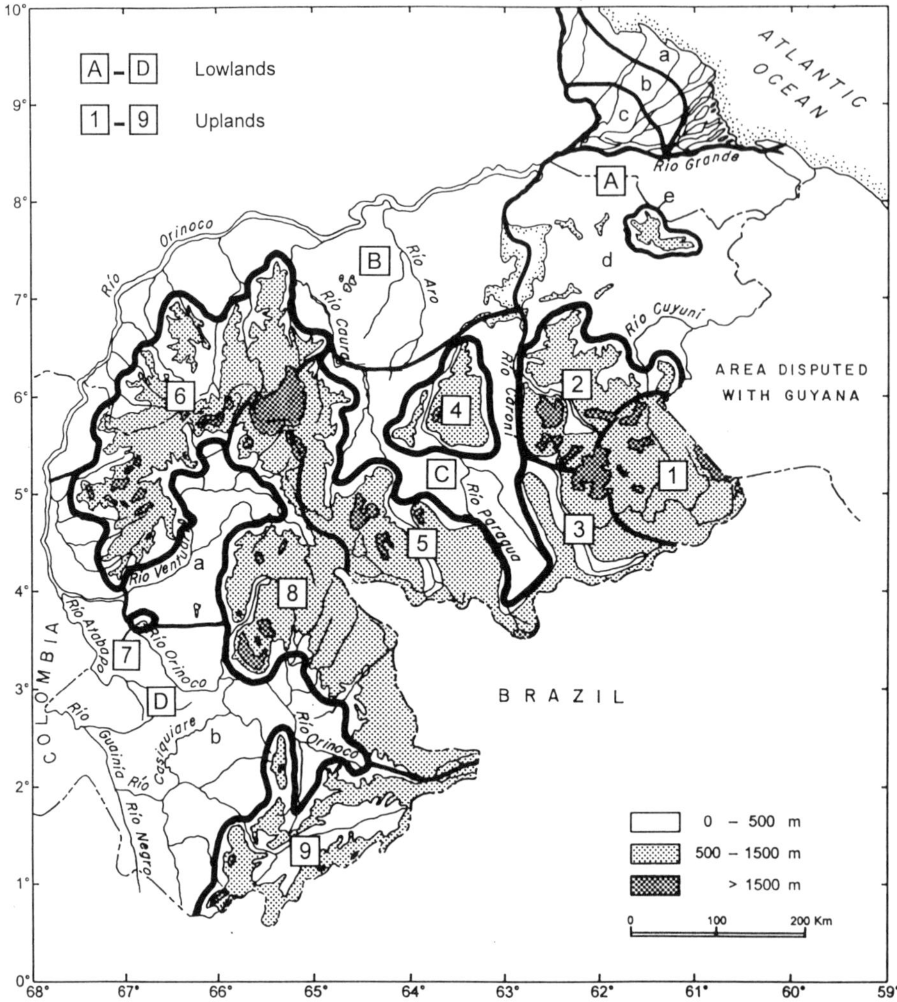

Figure 1-17. Main physiographic regions of the Venezuelan Guayana. A–D are lowlands, 1–9 are uplands, and the shaded area over 1500 m elevation corresponds closely to highlands. A = lower Orinoco-Cuyuní lowlands (a = outer Orinoco Delta, b = middle Orinoco Delta, c = inner Orinoco Delta, d = Cuyuní basin, e = Serranía de Imataca), B = middle Orinoco lowlands, C = Caura-Paragua peneplains, D = Ventuari-Orinoco-Casiquiare plains (a = Ventuari plains, b = Casiquiare peneplains). 1 = Gran Sabana, 2 = Sierra de Lema and Sierra Senkopirén, 3 = Chaco-tepui, 4 = Cerro Guaiquinima, 5 = Ichún and Erebato uplands, 6 = northwestern uplands, 7 = Cerro Yapacana, 8 = Asita-Parima uplands, 9 = southern uplands. Highland areas are treated in greater detail in Figures 1-23 through 1-35 and in Figure 4-3. Elevations above sea level.

to Lago Guri to the west, form a hilly, undulating landscape that is covered by dense forests in the east and open shrub savannas in the west.

The second major lowland region is the middle Orinoco lowlands. The southern shore of the middle and lower Río Orinoco borders a large belt of lowland plains and low hill-lands stretching from Guri in the east to the mouth of the Río Meta in the west. This region exhibits considerable physiographic variation, since it is a transition zone between the Guayana mountains to the south and the flat plains of the Llanos to the north. The Río Caura acts roughly as a divide between a more hilly western section, with many granitic hills and a complex forest-savanna mosaic, and an undulating, mainly forested eastern section where extensive floodplains also occur.

The third lowland region, the Caura-Paragua peneplains, is mostly unexplored and extends along the middle course of the Caura and Paragua rivers. This area adjoins the middle Orinoco lowlands to the north and is limited in the west by the Sierra de Maigualida, Cerro Jaua, and Cerro Ichún, in the south by the Sierra Pakaraima, and in the east by Cerro Guaiquinima and Chaco-tepui. The landscape consists of gently undulating peneplains mixed with floodplains, both densely covered by forests.

The fourth large lowland region of the Venezuelan Guayana, the Ventuari-Orinoco-Casiquiare plains, is located in the central and southwestern portion of Amazonas state. It can be split into two subunits: the mainly forested Ventuari plains along the middle and lower course of the Río Ventuari, and the extensive Casiquiare peneplains, which are circumscribed by the Orinoco, Atabapo, Guainía, Negro, and Casiquiare rivers. These nearly level plains are covered by a complex mosaic of partly flooded forests, savannas, and other types of herbaceous vegetation.

Uplands (500–1500 m)

Uplands of the Guayana region are extensive hill-lands and low mountain ranges that occur mainly in the submesothermic altitudinal belt between 500 and 1500 m. The predominant topographic features of the uplands are rounded hills and lower mountain tops, undulating high plains, the lower slopes of high tepuis, and the summit areas of low elevation tepuis such as Cerro Guaiquinima, Cerro Ichún, and Cerro Yapacana. There are nine main upland regions in the Venezuelan Guayana.

The Gran Sabana region is located in the southeastern corner of Bolívar state. It consists of a high, undulating plain that covers close to 30,000 km^2, and it is inclined from the north, at 1450 m elevation, to the south, at 750 m (Schubert and Huber 1990; see Figure 1-18). Its southern edge, along the Brazilian border, is formed by the western branch of the Sierra Pakaraima, which here is considered to extend only as far west as the Serranía Marutaní. The Gran Sabana is covered mainly by treeless savannas that alternate with montane and gallery forests (Plates 18–21).

The Sierra de Lema and Sierra Senkopirén form a huge, mainly rocky,

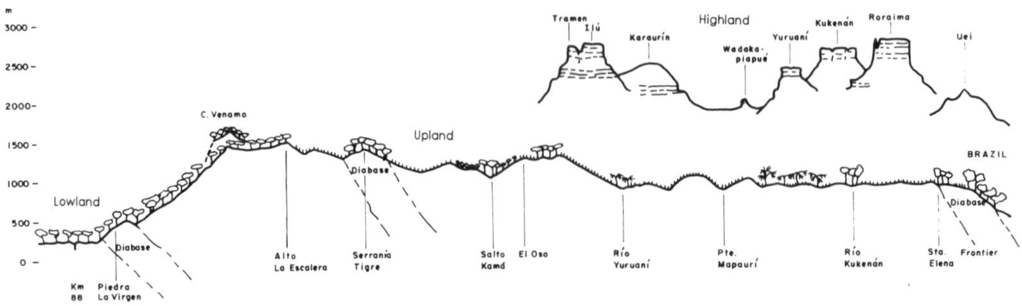

Figure 1-18. Schematic north-south cross sections through the Gran Sabana showing the three main altitudinal levels of the Venezuelan Guayana. The lower transect roughly follows the main road that traverses the Gran Sabana from Kilómetro 88 to the Brazilian border near Santa Elena de Uairén. The upper transect cuts through the eastern tepui chain along the southeastern border of Bolívar state.

upland plateau that extends irregularly from the northeastern edge of the Gran Sabana westward to the middle Río Caroní, southwest of Uaipán-tepui. It varies in elevation from 700 to 1650 m. The two mountains west of Los Testigos range known as Los Hermanos (Amaruay-tepui and Padapué-tepui, Figure 1-19) and the two small tepuis east of the tourist resort of Canaima, Cerro Venado and Kurún-tepui, belong to this unit. The entire area is predominantly covered by scrub and low forest, but it is still very poorly explored.

Chaco-tepui, a densely forested mountain range with rounded summits, lies to the west of the Icabarú and Caroní rivers in south-central Bolívar state. This area is virtually unexplored, but its average altitude is estimated at between 800 and 1000 m. Most of the summit of Cerro Guaiquinima, a large sandstone table mountain with a series of circular piedmont escarpments, ranges from only 730 m in the south to 1650 m in the northeast (Figure 1-20) and therefore belongs to the Guayana uplands rather than to the highlands. Its summit plains of 1096 km^2 are covered mainly by dense forests alternating with scrub and herbaceous vegetation (Steyermark and Dunsterville 1980).

Cerro Ichún lies close to the Brazilian border in southern Bolívar state and has the largest surface area of all the sandstone table mountains in the Guayana Shield, with a total area of 3260 km^2 (Figure 1-21). The densely forested upland summit forms a concave plateau of 2460 km^2, with its borders reaching 1100 m elevation on the eastern side and 1400 m on the western side, whereas the central depression varies from between 500 m elevation in the north to 700 m in the south.

The extensive northwestern uplands stretch in a wide arc from the Río Cuchivero (Bolívar state) in the northeast to the Río Guayapo (Amazonas state) in the southwest. Most of this area consists of dissected, granitic hill-lands with an average altitude between 500 and 1000 m. It is covered by tall forests, with numerous savannas and shrublands in isolated patches.

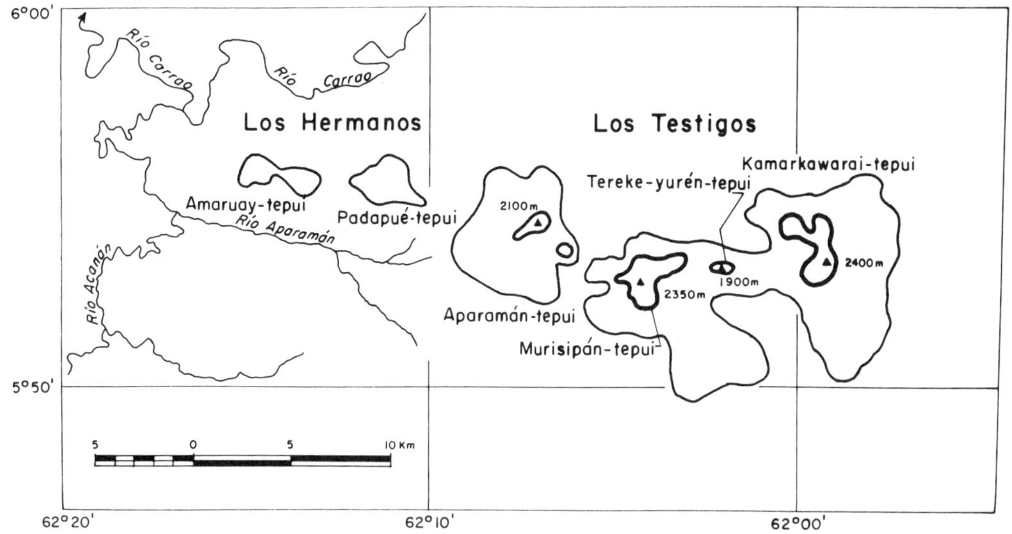

Figure 1-19. Los Hermanos (upland) and Los Testigos (highland) massifs. Light contour = slope area, dark contour = summit area. Based on side-looking airborne radar image NB-20-11.

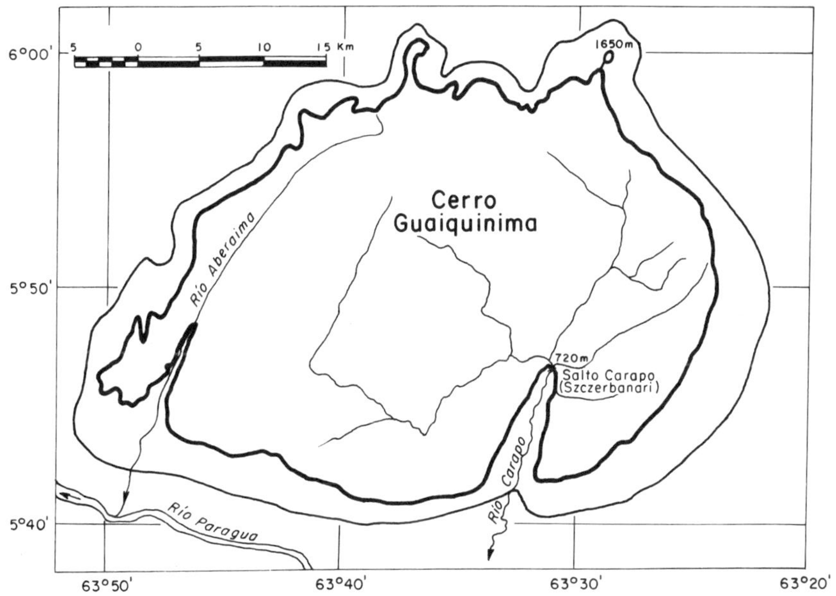

Figure 1-20. Cerro Guaiquinima. Light contour = slope area, dark contour = summit area. Based on side-looking airborne radar image NB-20-10.

The widely visible and isolated sandstone mountain of Cerro Yapacana rises abruptly from the surrounding lowlands 40 km southeast of the confluence of the Orinoco and Ventuari rivers in central Amazonas (Holt 1933; Figure 1-22). Its small, level summit plateau of only 10 km^2 reaches an altitude of approximately 1300 m. The summit is densely forested and is surrounded by a slope area of approximately 30 km^2.

The Sierra Parima, an extensive mountain range along the border of Brazil and eastern Amazonas state, extends for about 250 km from the upper Río Ventuari in the northwest to the Orinoco headwaters in the southeast. The summits of these mountains are mainly granitic and therefore mostly rounded or gently sloping. Together they form a continuous, undulating high plain, the Parima uplands, between 750 and 1300 m elevation. The Sierra Parima is covered mostly by forests, but natural and anthropogenic savannas also occur, especially in the southern section. This is part of the traditional homeland of the Yanomami Amerindians (Huber et al. 1984).

The southern uplands are located in southeastern Amazonas state and include a series of ill-defined and almost entirely unexplored uplands, ranging mainly between 600 and 1400 m elevation. Some isolated higher peaks are also found there, such as Cerro Aratitiyope (Plate 24) and Cerro Tamacuari.

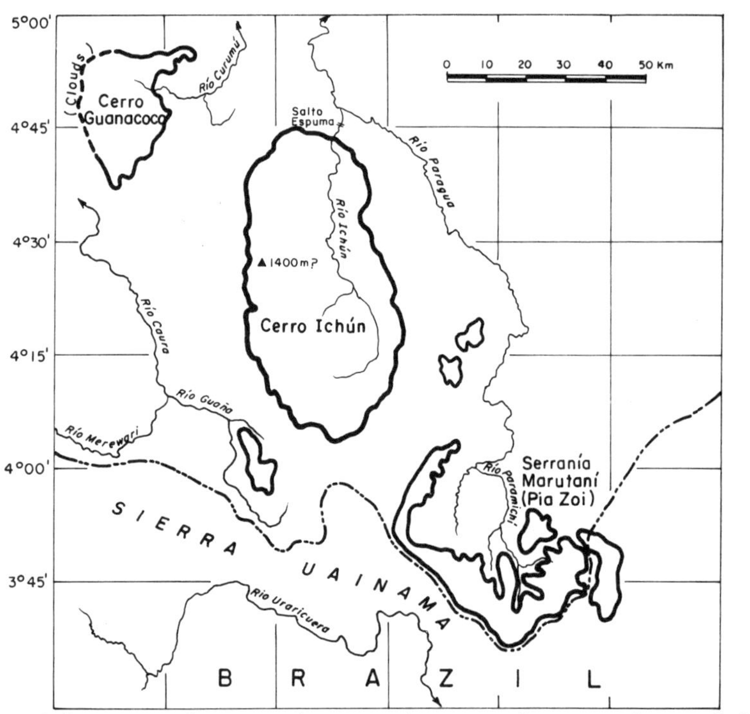

Figure 1-21. Cerro Ichún and Serranía Marutaní (Pia Zoi). Dark contour = slope and summit area. Based on LANDSAT 3, 251-057, 23 August 1980.

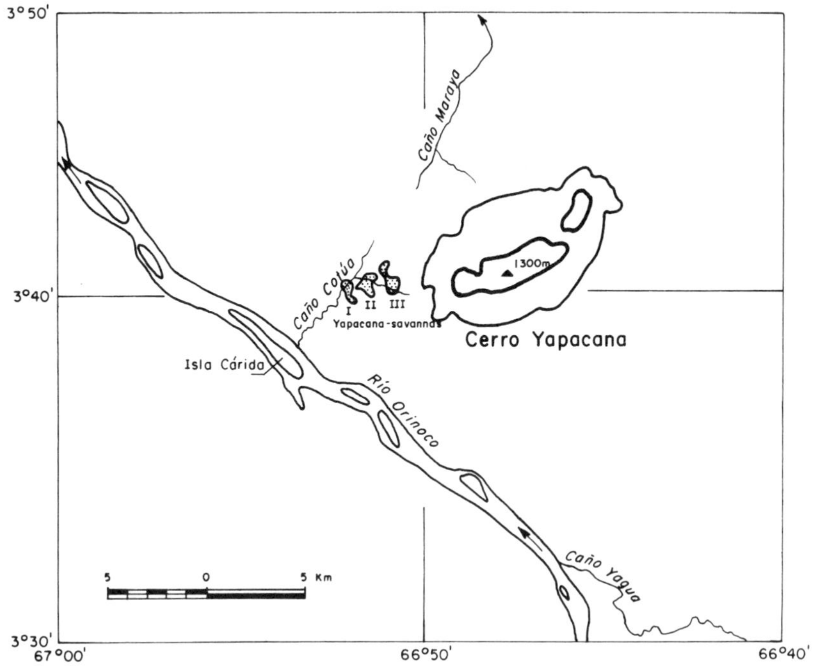

Figure 1-22. Cerro Yapacana. Light contour = slope area, dark contour = summit area. Based on side-looking airborne radar image NA-19-4.

The main mountain systems of this remote region are the Sierra Urucusiro, Sierra Curupira, and Sierra Tapirapecó along the southern border with Brazil, and Sierra Unturán and Cerro Vinilla toward the north of the Río Matapire and west of the Río Mavaca. Except for the sandstone Cerro Vinilla, the other mountains consist of granites and gneisses and are mostly covered by tall, dense forests.

Highlands (1500–3000 m)

The most characteristic physiographic feature of the Venezuelan Guayana is the occurrence of many sandstone table mountains, which often have spectacular sheer cliffs and flat-topped summits. The Pemón Amerindians of the Gran Sabana call this kind of mountain *tepui*. Although the original use of the term *tepui* is restricted to a small part of the Venezuelan Guayana, modern explorers use it to designate any table mountain in the Guayana area. Thus names like Huachamacari-tepui or Aracamuni-tepui have appeared in the literature (e.g., George 1988), although the indigenous populations living near these mountains are not closely related to the Pemón tribe and do not use the word.

As part of their comprehensive study of the bird fauna of the Guayana mountains, Mayr and Phelps (1955, 1967) coined the term *Pantepui* to designate the assemblage of sandstone table mountains in the Venezuelan Guayana and adjacent Brazil and Guyana. They offered no precise definition of the term, although they clearly meant to restrict it to high-elevation, meso- or microthermic life zones. The concept of Pantepui was used quite differently by subsequent authors; Steyermark (1979b, 1982), for instance, considered it very broadly to include all the intervening lowland areas as well.

The term *Guayana Highland* has also been applied indiscriminately, often to include both lowland and upland biota that have little in common with the high mountain biota found on the tepui summits. This is particularly evident in *The Botany of the Guayana Highland* series by Bassett Maguire and collaborators (see Buck 1990 for complete citations of component parts), which first appeared in 1953 and treats many lowland taxa, thus contradicting the basic aim and geographical range indicated in the title.

Based on the intensive study of the flora and vegetation of more than 50 mountain summits in the Venezuelan Guayana, it is now clear that the high mountain biota of this region forms a homogeneous biogeographic complex, the tepui life zone. This area corresponds in many ways to the páramo life zone in the upper Andes of tropical South America. Consequently, it is now possible to redefine Pantepui as a biogeographical province comprising all high-tepui ecosystems that are restricted to the Guayana Shield between 1500 and 3000 m elevation, but are best developed between 1800 and 2700 m (Huber 1987; see also Chapter 4). This province is composed not only of the typical quartzite and sandstone tepuis of the Roraima Group, but also includes the high granitic ranges such as the Sierra de Maigualida and part of the Sipapo Massif.

Table 1-3 (see end of chapter) presents data on the maximum elevation and surface area of the roughly 50 tepuis known from the Venezuelan Guayana. More detailed information on each tepui is provided in the following text. A few of the mountains in this table may not actually belong to the Pantepui zone according to the above criteria, but are included there either because their summits are insufficiently explored or because they have been treated as tepuis in earlier geographical or scientific literature.

The eastern tepui chain, or Roraima-Ilú range, is located in the southeastern corner of Bolívar state along the border with Brazil and Guyana; it is a visually impressive series of tepuis that extends some 60 km from southeast to northwest (Figure 1-23; Plates 26–28). Farthest to the south is the isolated Uei-tepui (Wei-tepui or Cerro El Sol), followed northwards by Roraima- (Roroima-) tepui, which at 2723 m elevation is the third highest mountain in the Guayana Shield. Next follows Kukenán- (Mataui-, Matawí-) tepui, Yuruaní-tepui, the smaller, tower-like Wadakapiapué-tepui, Karaurín- (Caraurín-) tepui, and finally Ilú- (Uru-) tepui (Plate 20), which includes the imposing Tramen-tepui. These tepuis all have open, rocky summits, except Uei-tepui, which is covered by dense, herbaceous vegetation, and Karaurín-tepui, with shrubby vegetation. Roraima-tepui is the only tepui in this group with a relatively easy terrestrial access, along a ledge on its southwestern face.

Cerro Venamo, a low mountain in eastern Bolívar along the border with

Guyana, marks the northeastern corner of the Gran Sabana. It is a moderately inclined sandstone mountain with a few open, rocky walls on the northern and eastern slopes. Since it is covered by dense cloud forests, it is barely distinguishable on radar or satellite images. Julian Steyermark, the only known scientist to ascend Cerro Venamo, recorded a summit altitude of 1600 m, which would place it in a transitional position between the upland and highland biotas.

The northern limit of the Gran Sabana is formed by the irregular mountain system of the Sierra de Lema, which extends westward from Cerro Venamo toward Auyán-tepui (Steyermark and Nilsson 1962). The higher peaks of this mountain range include both flat-topped sandstone summits reaching almost 1600 m elevation and rounded summits probably consisting of diabase. Near the northern limit of the Sierra de Lema is the Cordillera Epicara, including Cerro Pitón in the headwaters of the Río Chicanán. Since most of the range is covered by forest and is very poorly explored and toponymically confused, it is difficult to determine its precise extension and altitudinal limits. The middle elevations (uplands) of the Sierra de Lema extend in wide plateaus toward Auyán-tepui.

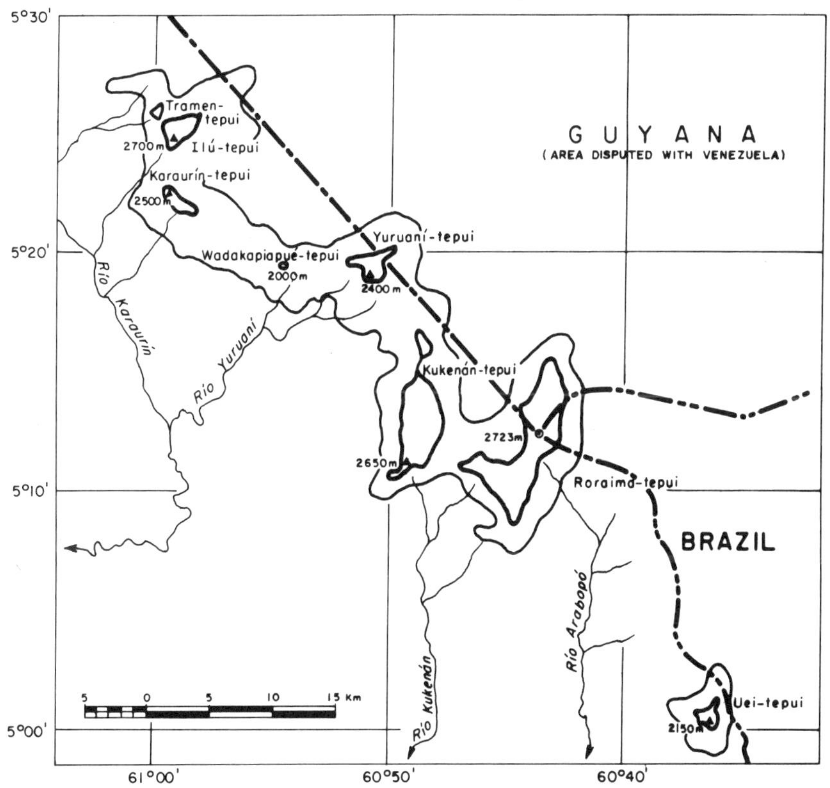

Figure 1-23. Eastern tepui chain, in southeastern Gran Sabana, Bolívar state. Light contour = slope area, dark contour = summit area. Based on side-looking airborne radar image NB-20-12.

The Ptari massif is a small mountain system consisting of three separate mountains, all located to the north of the Kavanayén mission in the northern Gran Sabana. The main mountain is Ptari- (Pu-tari-) tepui, or Cerro Budare, an almost perfectly symmetrical tepui of 2400 m elevation. The other two are Carrao- (Karrao-) tepui, to the northeast of Ptari-tepui, and Sororopán-tepui, which consists of a 10-km long ridge oriented northeast to southwest, with a steep wall on the southern face (Steyermark 1966; Figure 1-24). The summit of Ptari-tepui is dominated by open rock surfaces except for a more dissected area on the eastern edge (Plate 33), whereas Carrao-tepui and Sororopán-tepui are densely covered by forests.

Northeast of the Kamarata mission there is another visually impressive chain of tepuis that extends in an east–west orientation. This chain is generally called Los Testigos, or sometimes the Aparamán range. Four mountains between 1900 and 2400 m elevation are the principal members of this range (from east to west; see also Figure 1-19): Kamarkawarai- (Camarcai-barai-) tepui, Tereke-yurén-tepui, and Murisipán- (Murosipán-, Murochiopán-) tepui, all rising from a common basement, and Aparamán-tepui, which is separated from the others to the west. There is, however, confusion concerning the precise names of the two central mountains; according to Brewer-Carías (1978), Murochiopán-tepui is the name of a smaller lateral mountain of Aparamán, followed to the east by the high Tereke Yurén-tepui and the lower Tucuy-wo-cuyén-tepui, whereas members of the Terramar Foundation expeditions (Steyermark 1986a; Holst 1987; George 1988) apply the name Murisipán-tepui to Brewer's Tereke Yurén-tepui, and the name Tereke-yurén-tepui to Brewer's Tucuy-wo-cuyén-tepui, omitting the name of the smaller lateral mountain of Aparamán-tepui. All summits of the Los Testigos massif are mainly open rock surfaces similar to those of the eastern Roraima-tepui chain and of Ptari-tepui; the summit of Kamarkawarai-tepui has a large collapsed sinkhole open-

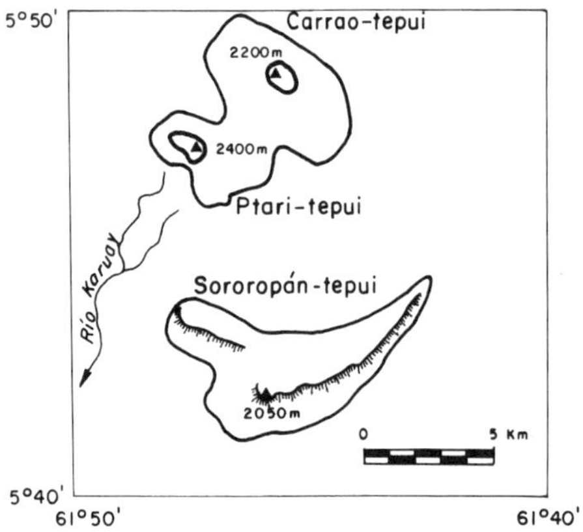

Figure 1-24. Ptari massif. Light contour = slope area, dark contour = summit area. Based on side-looking airborne radar image NB-20-11.

ing toward the west, whereas the summit of Murisipán-tepui has a small seasonal lagoon in the center. The summit of Aparamán-tepui is highly dissected, with innumerable small towers and large, irregular blocks which make traverses by foot or even helicopter landings extremely difficult.

The Auyán-tepui massif is a large mountain system east of the middle Río Caroní. Just southeast of the massif lies the Kamarata mission, where the only foot trail to the southern summit begins. The massif consists of the main Auyán-tepui (Plate 22); two smaller, tower-like tepuis at the end of a long forested ridge, Cerro La Luna and Cerro El Sol (Uei-tepui); and a southern satellite mountain, the mainly forested Uaipán- (Waipán-) tepui (Figure 1-25).

The large plateau of Auyán-tepui, whose summit is inclined from 2450 m elevation in the south to approximately 1600 m in the northwest, is roughly U-shaped. It has a mostly open, rocky eastern branch (Plate 31) and a larger, mainly forested western branch (Steyermark 1967). These two branches are separated by the deep valley of the Río Churún (locally called the Cañón del Diablo), which originates in the southern section of the plateau (Plate 30). The western branch has extensive diabase intrusions and is drained by a small river, which falls almost 1000 m from the edge of the mountain before joining the Río Churún farther downstream. This waterfall, most widely known as Angel Falls (Salto Angel), is the highest in the world. The indigenous name for the falls is Karepa-Kupai-merú, although it is often called Churún-merú, a waterfall farther southwest in the Cañón del Diablo.

Aprada-tepui is located 22 km northwest of the Chimantá massif and approximately 25 km east of the ancient mining settlement of Urimán along the middle Río Caroní. Its flat summit is mainly covered by open rock vegetation and small islands of tepui forests (Plate 29). To the east lies Araopán-tepui, which is connected to Aprada-tepui by a steep, forested ridge (Figure 1-26).

The extensive Chimantá mountain system is located on the western border of the Gran Sabana, approximately 50 km south-southeast of Auyán-tepui and about 35 km northwest of the Uonquén (Wonkén) mission in the Caruay (Karuai) valley. It consists of a series of separate mountains, divided by the Río Tírica into two sections (see Figure 1-27; Plate 34). The higher northern section includes Murey- (Eruoda-) tepui, with the highest point of the massif at 2650 m elevation, Tirepón-tepui, Apacará-tepui, Abacapá-tepui, Agparamán-tepui, Toronó-tepui, and Chimantá-tepui. The southern section is formed by Churí-tepui, Acopán-tepui (Plate 36), and Amurí-tepui (Huber 1992a; Plate 35). The species richness and ecological diversity of the plants on the Chimantá massif are among the richest of the entire Guayana Highlands. This is due at least in part to the great variety of summit landscapes, including wide plains, extensive broken rock surfaces, and large forested valleys and creeks (Plates 34–38).

Angasima-tepui and Upuigma-tepui are two spectacular, small tepuis just south of the Chimantá massif on the opposite side of the Río Aparurén valley (Figure 1-27). The triangular-shaped tower of Angasima-tepui is only 8 km away from Amurí-tepui. Its heavily windswept summit consists of a north-

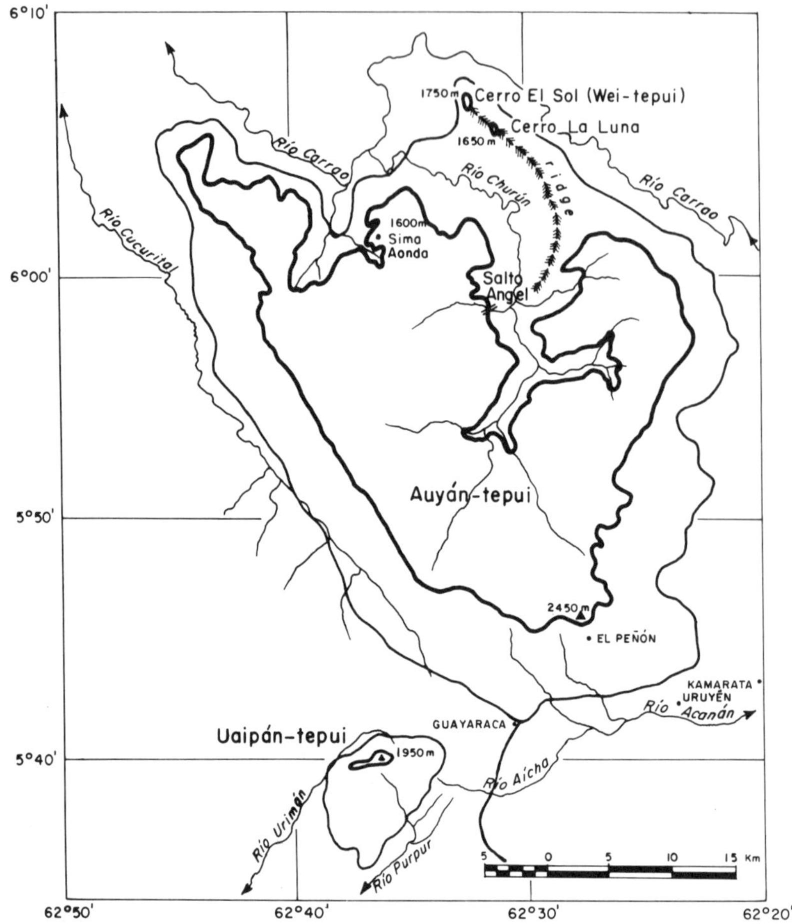

Figure 1-25. Auyán massif. Light contour = slope area, dark contour = summit area. Based on side-looking airborne radar images NB-20-7 and NB-20-11.

ern high plain covered by low herbaceous vegetation and a southern peak with dense tepui scrub. A small valley extends northeast from the peak to the lower slopes. The summit of Upuigma-tepui (also called El Castillo; Plate 35) consists of many small rock towers separated by deep crevasses, making it impossible to walk between them. The surface of the towers is covered by dense herbaceous and scrubby vegetation.

Although most of the large massif of Cerro Guaiquinima (Waikin-ima-tepui) belongs to the uplands, the northernmost rim reaches elevations of 1500 to 1650 m and can be considered true highlands (Figure 1-20). The high points of Guaiquinima are covered by broad-leaved, herbaceous meadows, mixed with sclerophyllous scrub and low tepui forests (Steyermark and Dunsterville 1980).

Serranía (Cerro, Sierra) Marutaní (Pia-Zoi) is a roughly U-shaped mountain range that belongs to the southeastern branch of the Sierra Uainama

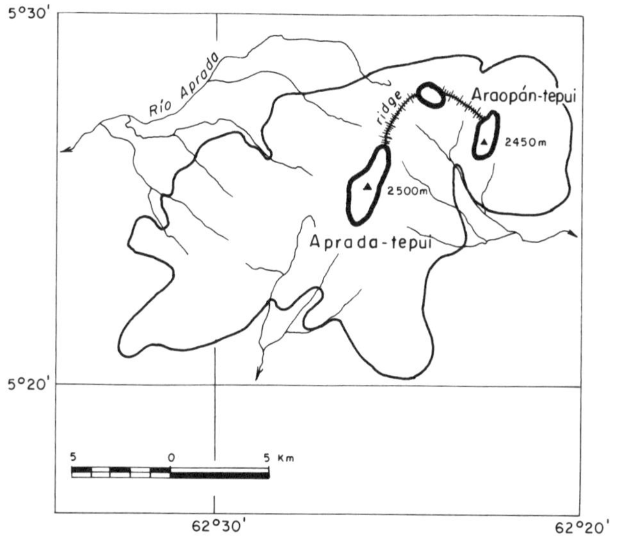

Figure 1-26. Aprada massif. Light contour = slope area, dark contour = summit area. Based on side-looking airborne radar image NB-20-11.

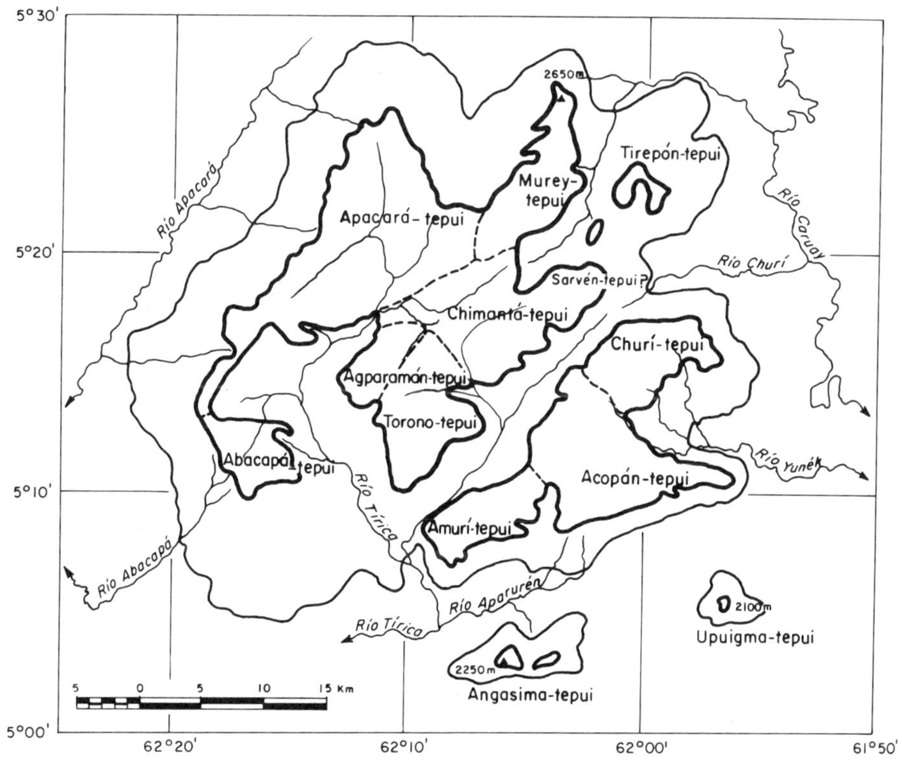

Figure 1-27. Chimantá massif. Light contour = slope area, dark contour = summit area. Based on side-looking airborne radar image NB-20-11.

(Figure 1-21). It occupies the southernmost point of Bolívar state and borders Brazil's northern Roraima state. The maximum elevation of the mountain is unknown, but probably is around 1500 m (Steyermark and Maguire 1984b). The summit plateau is drained by the headwaters of the Río Paramichi and descends from south to north. Most of the summit is covered by tall, dense, sclerophyllous scrub and low forests. The toponymy of this area is still unreliable, and the most common names for the mountain range are Marutaní, Pia-Zoi or Piazoi, Pia Soi, and Urutaní.

The relatively small Cerro Guanacoco (Wuanakoko; Figure 1-28) borders the right bank of the upper Río Caura, where it rises abruptly to about 1500 m elevation. Its summit forms a series of descending plateaus that are steeply inclined from west to east and drained by the Río Jumpetiri, a tributary of the Río Curutu. Most of the summit is covered by dense, moist tepui forests, occasionally interrupted by patches of swampy meadows. Cerro Guanacoco is loosely connected with the northwestern tip of Cerro Ichún by a series of ridges and low hills. The southwestern section of Cerro Guanacoco probably corresponds to the Cerro Ameha that Eugène André (1904) tried to climb unsuccessfully in 1901.

The Jaua-Sarisariñama massif consists of two separate mountains, Cerro Sarisariñama to the south and Cerro Jaua to the north (Steyermark and Brewer-Carías 1976), with a combined summit area of 1173 km^2 (Figure 1-29). The roughly circular Cerro Sarisariñama has a densely forested, level summit plateau with an average elevation of 1400 to 1600 m, and reaches its highest elevation of approximately 2350 m on Cerro Yacoto along the southwestern rim. Two large, circular sinkholes (*simas*) are located near the northeastern border, at approximately 1350 m elevation (Brewer-Carías 1976). Cerro Jaua is connected to Cerro Sarisariñama in the south by a lower mountain range; it covers more than 600 km^2 and has an inclined and dissected summit plateau, which descends from approximately 2250 m elevation in the southernmost tip to 1400 m in the north. Most of the summit is forested, but in the southern half, large extensions of tepui meadows alternate with gallery forests along the Río Marajano, which crosses the entire mountain from south to north.

Sierra de Maigualida is the largest continuous mountain system in the Guayana Shield, extending some 300 km from north to south and with an average width of 20 to 40 km (Figure 1-30). It forms the watershed between the Caura and the Ventuari river basins in the center and to the south, and between the Caura and the Cuchivero river basins in its northern section. It also forms the natural boundary between southwestern Bolívar and northeastern Amazonas states. Including Serranía Uasadi in the south and Serranía Nichare in the north, Sierra de Maigualida is the main mountain chain in the Venezuelan Guayana that consists of granitic rocks rather than the more common Roraima quartzites and sandstones. Several large rivers originate here, including the Yudi, Cuyuwí, Chajura, and Erebato rivers to the east, and the Iguana, Asita, and Ventuari rivers to the west and south. The entire mountain range is covered by dense, tall montane forests, except in the

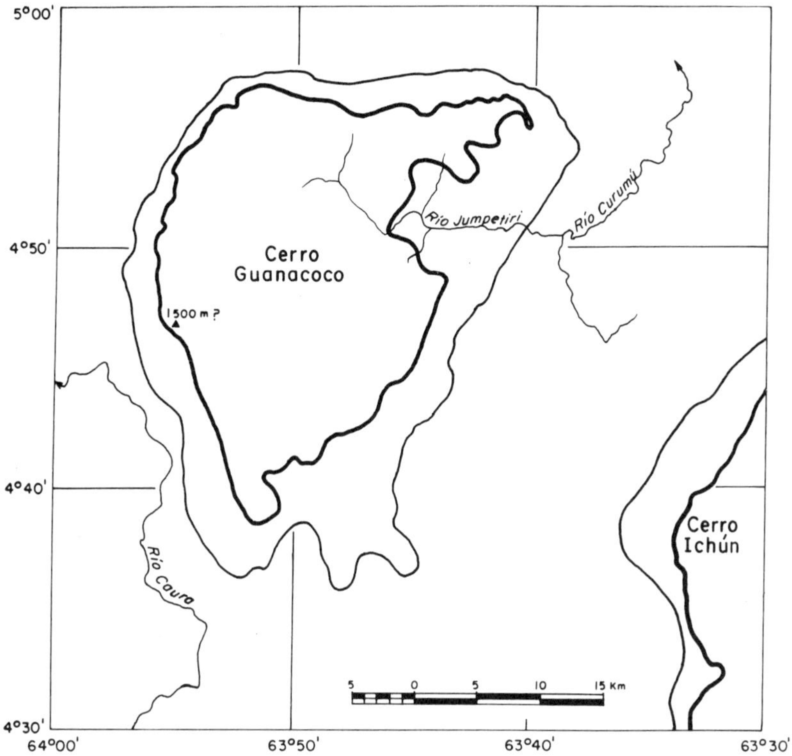

Figure 1-28. Cerro Guanacoco. Light contour = slope area, dark contour = summit area. Based on side-looking airborne radar image NB-20-14.

highest parts above 2000 m elevation (Cerro Yudi and Cerro Cuyuwí), where tepui-like meadows and scrub predominate.

Cerro Yaví is a small, imposing sandstone tepui that rises to a maximum elevation of 2300 m from the valley of the upper Río Parucito in northeastern Amazonas (Hitchcock 1947). Its horseshoe-shaped summit presents a very broken, rocky terrain covered by open scrub and tepui meadows mixed with low forest patches. There is also a small lake on the eastern side of the summit.

The large Yutajé-Coro Coro massif has a summit area of 275 km^2 along the northern border of Amazonas. It consists of two inclined summit plateaus drained by the Río Yutajé to the south and the Río Suapure to the north (Figure 1-31). The eastern section of this massif is called Cerro Yutajé, with a maximum elevation of 2140 m, whereas the western part is Cerro Coro Coro, with a maximum elevation of 2400 m. Most of the summit area consists of rough, rocky terrain covered by scrub and tepui meadows.

Like Cerros Coro Coro and Yutajé, Cerro Guanay is located on the border between Bolívar and Amazonas states (Maguire and Deery de Phelps 1951; Figure 1-31). Its inclined summit plateau descends from east to west and to a lesser degree from north to south, drained mainly by the headwaters of the Río

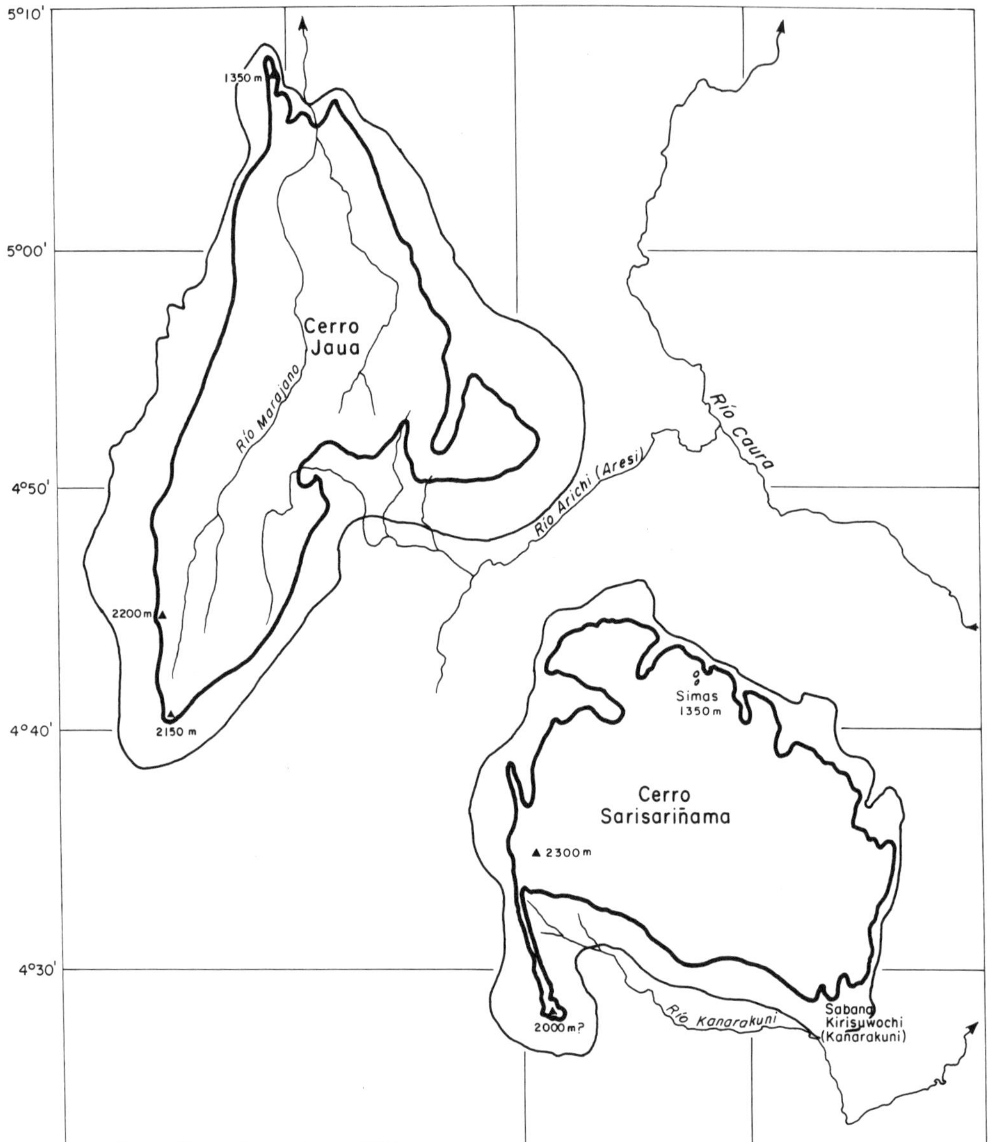

Figure 1-29. Jaua-Sarisariñama massif. Light contour = slope area, dark contour = summit area. Based on side-looking airborne radar images NB-20-9, NB-20-10, NB-20-13, and NB-20-14.

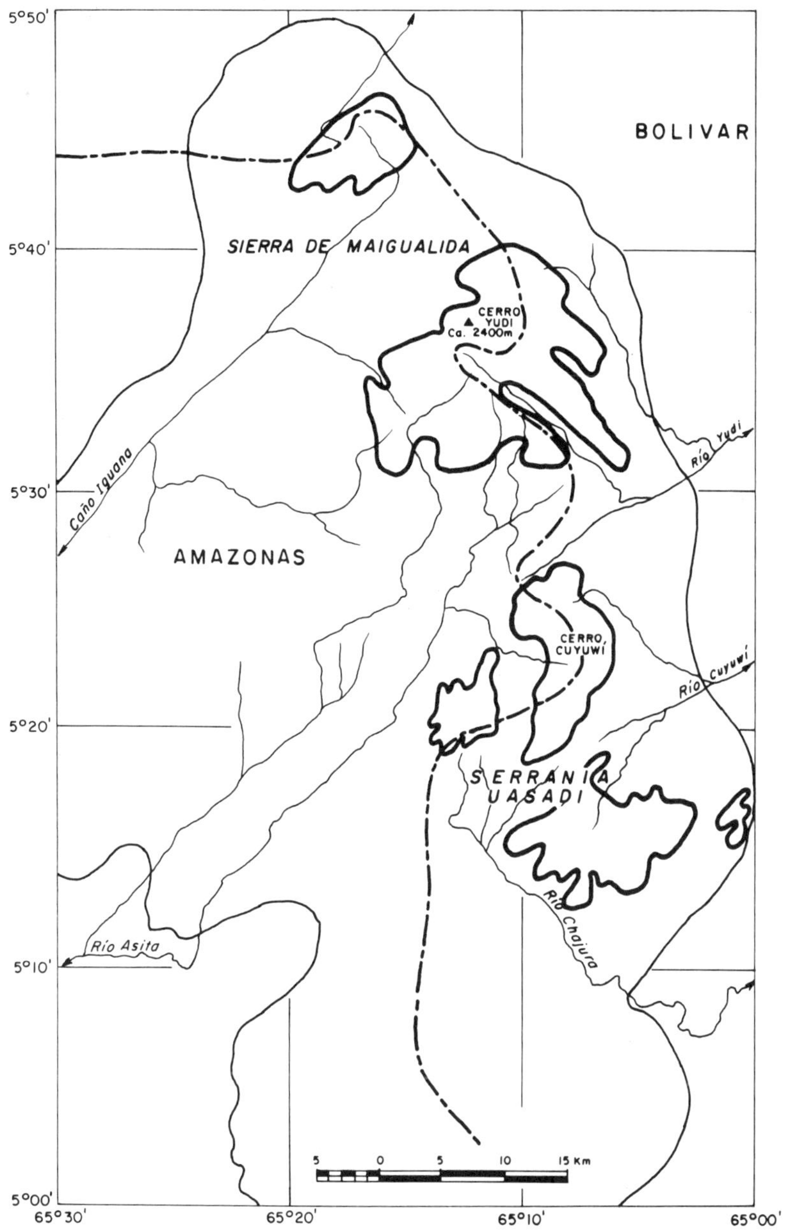

Figure 1-30. Sierra de Maigualida and Serranía Uasadi. Light contour = slope area, dark contour = summit area. Based on side-looking airborne radar image NB-20-9.

Parguaza. Its highest point occurs along the southeastern rim at 2080 m. Most of the summit is covered by dense tepui scrub growing on heavily broken, rocky terrain, with low gallery forests along the creeks.

The whitish, rounded cap of Cerro Camani is visible from far away in the Manapiare basin, rising from a large, low-lying sandstone mesa 30 km west of the town of San Juan de Manapiare (Figure 1-31). Its small summit reaches 1800 m elevation (erroneously cited as 2200 m in Maguire and Deery de Phelps 1951) and is covered by open herbaceous vegetation. At the southeastern edge of the mesa from which Cerro Camani arises, Cerro Morrocoy (800 m elevation) emerges above the adjacent Manapiare plains.

The extensive mountain system of the Cuao-Sipapo massif extends over about 1000 km^2 to the south and southeast of Puerto Ayacucho in Amazonas state (Figure 1-32). It consists of many separate mountain ranges, but the toponymy of the area is unreliable and sometimes contradictory. The Cuao-Sipapo massif forms the watershed between the Río Orinoco to the west and south, and the Río Ventuari to the east. The rivers draining into the Orinoco are, from north to south, the Cuao, Autana, Sipapo, and Guayapo, whereas the Río Marieta and Caño Guapuchí flow into the Río Ventuari.

Within this complex, three higher mountain ranges rising above the extensive Cuao-Marieta uplands can be distinguished: (1) the Cuao massif, to the northwest, consisting of sandstones and quartzites of the Roraima Group and with an estimated maximum elevation of 2000 m (the northern section of this massif probably corresponds to the Cerro Paraque visited by Phelps in 1946 and by Maguire in 1948 and 1949); (2) the Sipapo massif, in the center of the area, made up of granitic high plains and peaks, with an estimated maximum elevation around 1800 m (Plate 46); and (3) the lower Sierra Guayapo (often mistakenly spelled Guayabo) to the south, also granitic, with several peaks (Cerro Gallinero, Cerro Aracapo, and Cerro Ovana or Ouana) estimated between 1600 and 1800 m elevation. On the southeastern edge of this massif lies the semi-isolated Cerro Moriche (1250 m elevation), on the right bank of the Río Ventuari. The well-known Cerro Autana is a small isolated, towerlike sandstone mountain with a large cave traversing it from side to side. It emerges from a lowland area 5 km west of Cerro Cuao and reaches an elevation of 1300 m; its level summit plain is covered almost entirely by a dense tepui meadow (Steyermark 1974, 1975; Plate 16).

The large Parú massif is located in northeastern Amazonas (Figure 1-33). Its summit area covers an area of 930 km^2, of which 725 km^2 belongs to Cerro Parú and 205 km^2 to the steeply inclined Cerro Euaja, which is separated from Cerro Parú by the Río Hacha. Cerro Parú (Asisa or A'roko) consists of a series of extensive interior plateaus ranging in elevation between 500 and 1400 m, and a higher rim along the northwestern and western borders that reaches 2200 m elevation. The westernmost section of the mountain is also called Cerro Asisa or Cerro La Momia, which includes Laguna Asisa, a small circular lake at 1310 m (Hoyos 1973). Most of the interior plateaus of Cerro Parú are covered by savannas and tepui meadows, whereas the interior slopes of the higher borders are covered by dense shrublands.

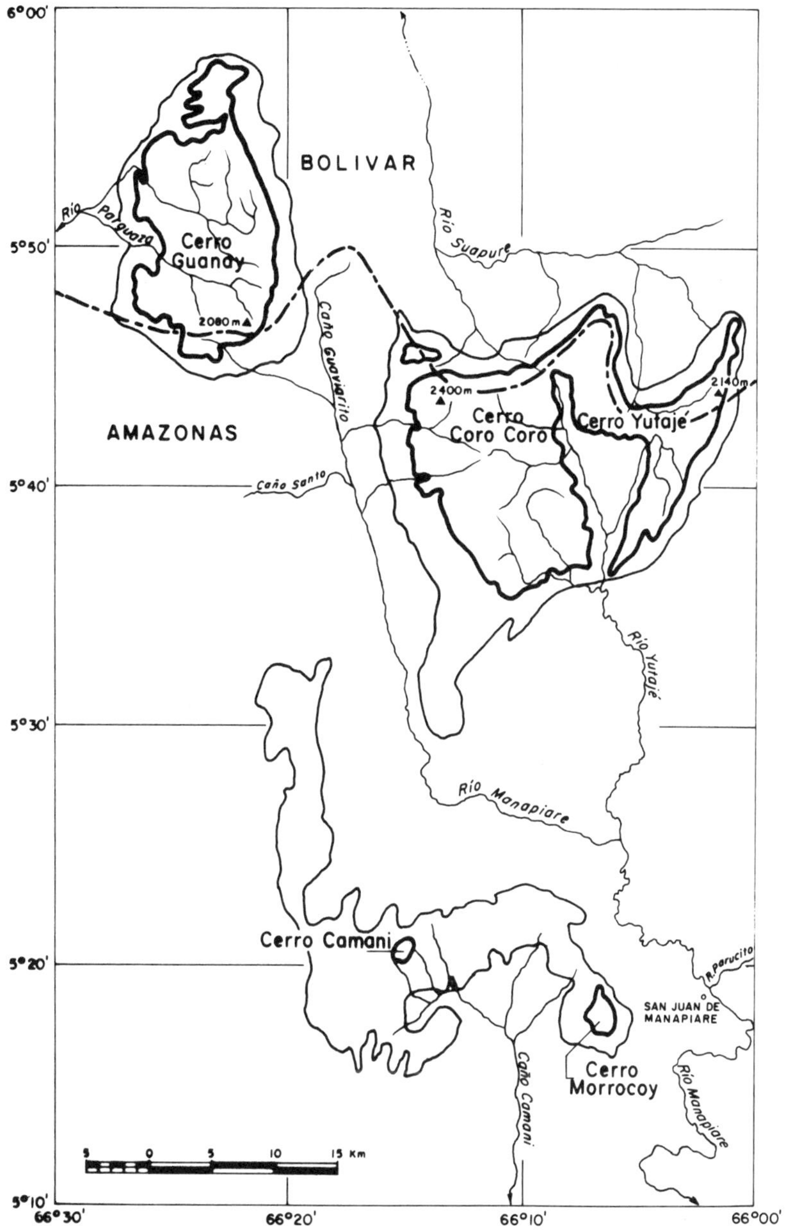

Figure 1-31. The upper Río Manapiare basin with Cerro Guanay, Yutajé-Coro Coro massif, Cerro Camani, and Cerro Morrocoy. Light contour = slope area, dark contour = summit area. Based on side-looking airborne radar image NB-19-12.

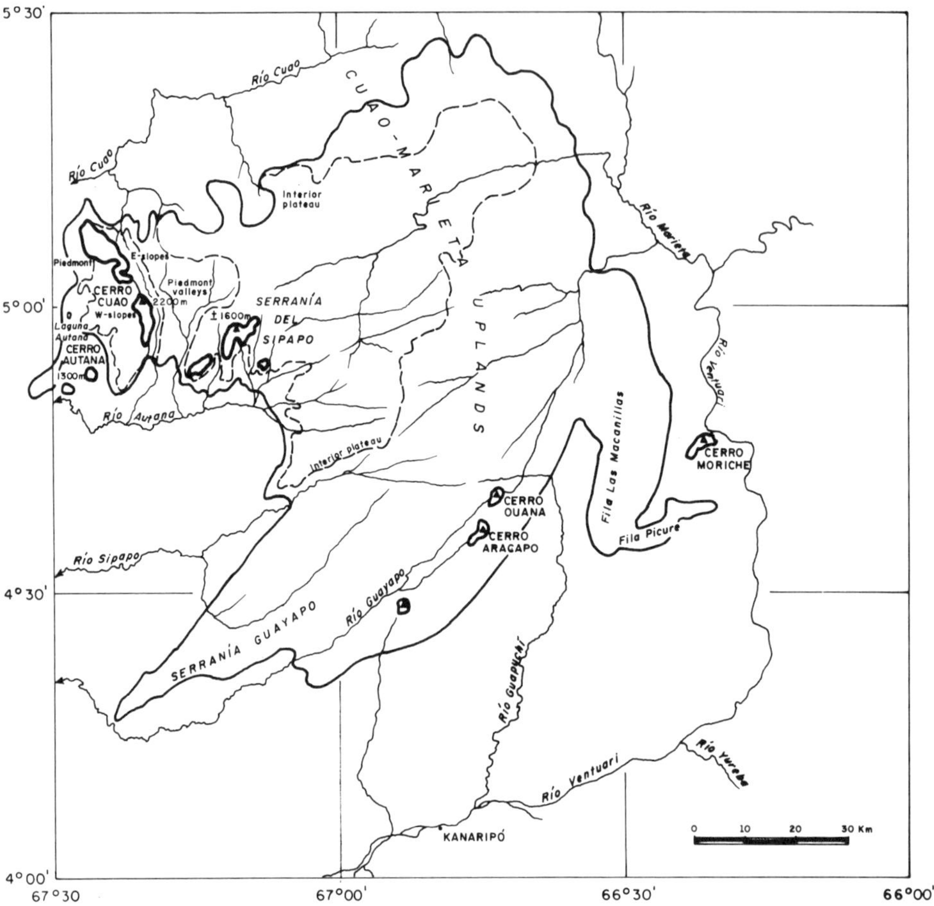

Figure 1-32. Cuao-Sipapo massif. Light contour = slope area, dark contour = summit area. Based on side-looking airborne radar images NB-19-12 and NB-19-16.

The Duida-Marahuaka massif is a large mountain complex in the center of Amazonas state formed by three well-separated ranges (Gleason 1931; Steyermark and Maguire 1984a): Cerro Duida (Yennamadi), with the highest point reaching 2358 m elevation on the southwestern rim; Cerro Marahuaka (Marahuaca), to the northeast of Duida and reaching 2800 m elevation (Plates 44 and 45); and Cerro Huachamacari (Huachamakari, Kushamakari), to the northwest of Duida, with a maximum elevation of 1900 m (Figure 1-34; Plate 42). The Marahuaka massif consists of two separate high plateaus. The first and larger one extends to the north and is known by the Yekwana Amerindians as Fufha (Huha). The second plateau, located to the southeast, has two local names, Fuif (Fhuif) at its northwestern edge, and Atahua'shiho (Atawa Shisho) for its southeastern portion. To the west of these two main plateaus, a huge, entirely forested ridge, Cerro Petaca, runs approximately north-south and reaches an elevation of at least 2700 m. Cerro Marahuaka is the second highest mountain in the Guayana Shield after Sierra de la Neblina.

Cerro Aratitiyope (Plate 24) is a spectacular, needle-like granite mountain of 1700 m elevation south of Cerro Vinilla, approximately 32 km southwest of the village of Mavaca (George 1988). Most of the mountain above 1100 m is bare rock, but there is considerable saxicolous vegetation and small patches of low forest. The lower shoulder of the mountain is mostly covered by dense forest.

Sierra Unturán separates the Mavaca and Siapa (Matapire) river basins in southern Amazonas and stretches for some 80 km from east to west and north, reaching an estimated elevation of 1600 m. The steeply inclined granitic slopes are predominantly covered by montane forests and shrublands.

The remote and still very poorly known Sierra Tapirapecó extends for about 100 km along the Venezuelan-Brazilian border in southern Amazonas state (FUDECI 1990). It includes another impressive conical granite mountain, Cerro Tamacuari, which reaches 2340 m elevation.

Finally, the Neblina massif is located in the southernmost part of Amazonas state along the border with Brazil (Brewer-Carías 1988; Figure 1-35). The massif comprises three mountains: Cerro Aracamuni (1600 m elevation)

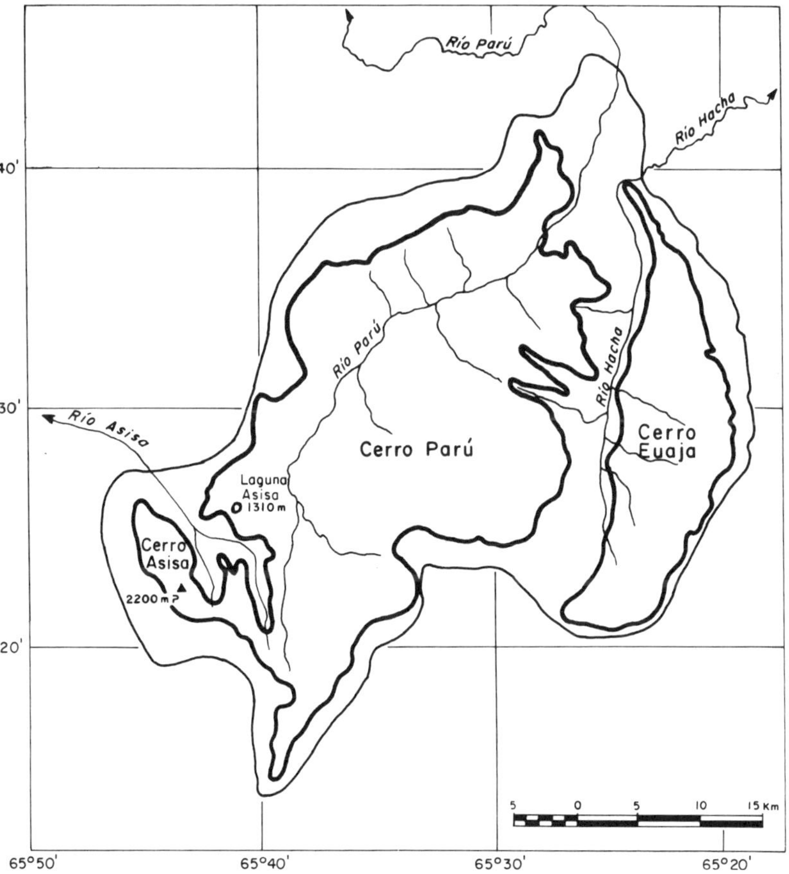

Figure 1-33. Parú massif. Light contour = slope area, dark contour = summit area. Based on side-looking airborne radar image NB-20-13.

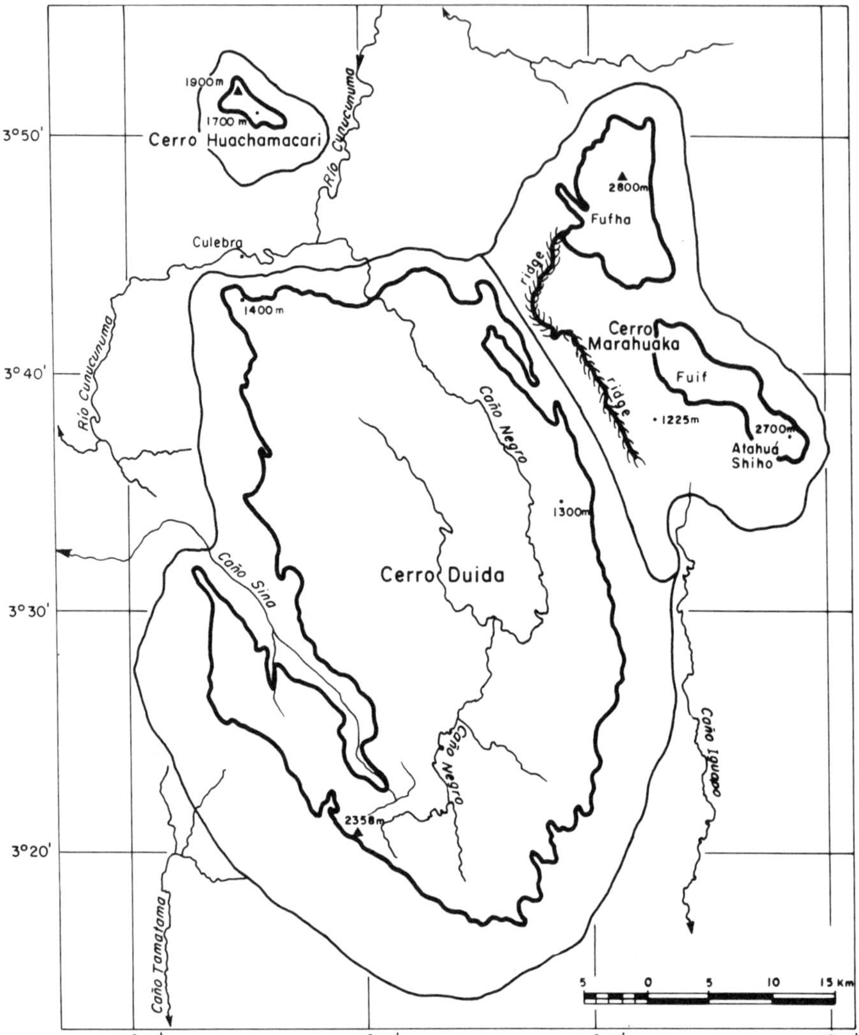

Figure 1-34. Duida-Marahuaka massif and Cerro Huachamacari. Light contour = slope area, dark contour = summit area. Based on side-looking airborne radar image NA-20-1.

to the north, Cerro Avispa in the center (also 1600 m maximum elevation), and the Sierra de la Neblina to the south (Plate 49). The latter mountain comprises extensive summit plateaus between 2000 and 2400 m elevation. It has a huge, deep valley called the Cañón Grande that crosses the entire mountain from northeast to southwest, drained by the headwaters of the Río Baría. Along the southernmost border, the twin peaks of Pico Phelps (2992 m elevation) and Pico da Neblina (3014 m elevation) emerge above the summit plateaus. Pico da Neblina lies just south of the Venezuelan border and is the highest peak in both Brazil and the Guayana Shield.

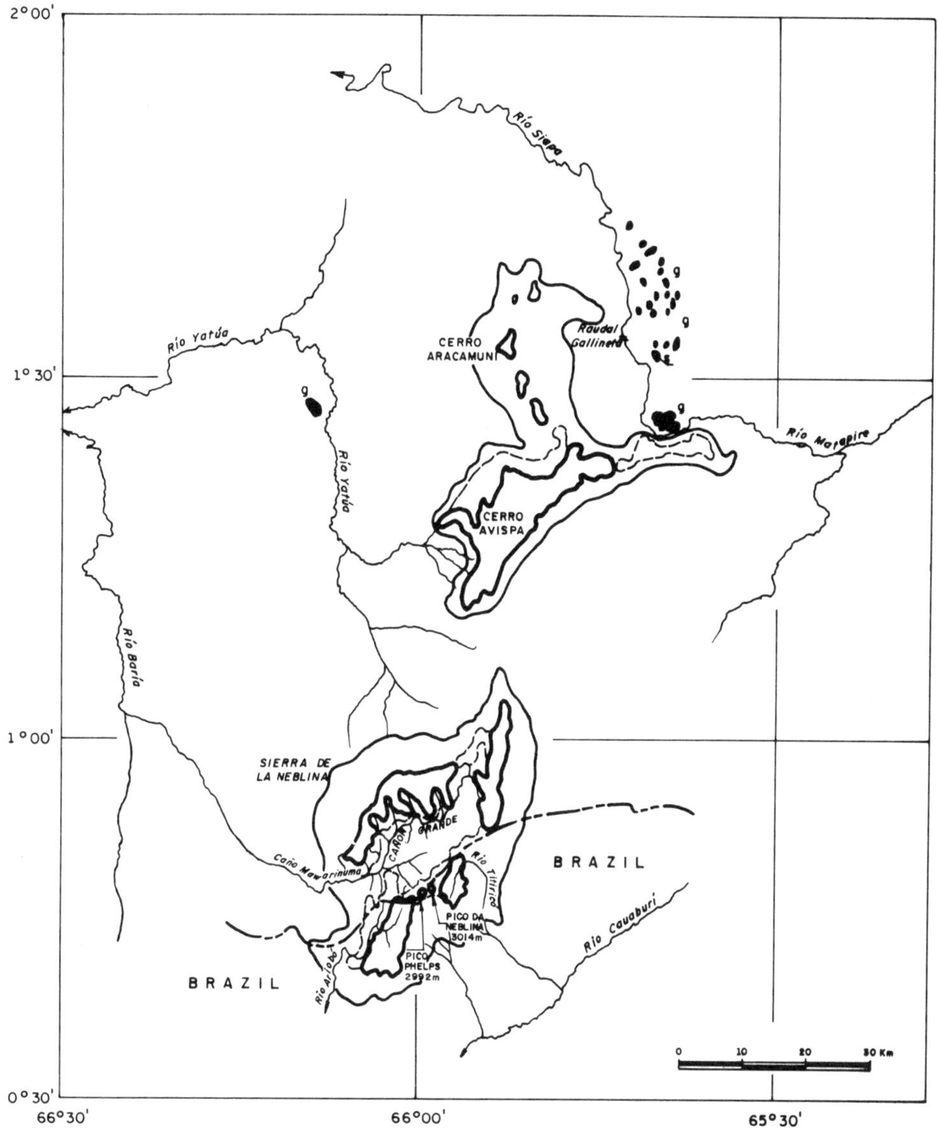

Figure 1-35. Neblina massif. Light contour = slope area, dark contour = summit area, g = granitic outcrops (*lajas*), s = sandstone outcrops, broken line = intermediate plateaus and escarpments. Based on NASA LANDSAT, E-21322, 5 September 1978; side-looking airborne radar images NA-19-16, NA-20-12, and NA-20-13; and Brewer-Carías (1988).

Human Populations

Although often cited as an example of empty space (e.g., Harris 1968), the Venezuelan Guayana is the homeland of a great variety of human races and ethnic communities. Despite the low population density compared to the northern half of the country, both the indigenous groups and later-arriving immigrants have played an important role in the ecology and history of the Venezuelan Guayana.

Indigenous Population

Except for the high mountain summits, all parts of the Venezuelan Guayana have long been inhabited by indigenous groups of different ethnic and linguistic affiliations. The presence of large indigenous populations in the Orinoco region was noted by the earliest European explorers in the 16th and 17th centuries, and some were portrayed as exotic creatures, for example, on the map of the Guiana published by Theodore De Bry in Frankfurt in 1599 (see also Plate 1). Since then, the area has attracted not only many missionaries, but also numerous anthropologists, ethnologists, and linguists, who have accumulated considerable knowledge about the cultural traditions and lifestyles of the Amerindians of northwestern South America. According to 1992 census data, 25 of the 28 indigenous tribes living in Venezuela are located in the Venezuelan Guayana, with a total of 100,614 persons, or approximately 32 percent of the indigenous population of Venezuela; this is only 0.55 percent of the total population of Venezuela and 8.8 percent of the population of the Venezuelan Guayana (OCEI 1993b).

Most of the indigenous populations of the Venezuelan Guayana have suffered dramatic declines in their number during the 19th and 20th centuries. They have also undergone many profound changes in their life habits, home ranges, and internal community structures. Nonetheless, a characteristic pattern of distribution of the principal Amerindian tribes living in the area can still be recognized (Figure 1-36).

The following account, edited by Professor Roberto Lizarralde of the Universidad Central de Venezuela, presents a brief overview of the main groups presently inhabiting the states of Delta Amacuro, Bolívar, and Amazonas. Anthropological data are based mainly on Coppens (1980, 1983, 1988), Lizarralde (1982), and Boadas (1983), with statistical figures from OCEI (1985, 1993b). Alternative tribal names frequently cited in the literature are listed in parentheses. Since labels of some botanical collections from the Venezuelan Guayana contain references to indigenous plant names, uses, and medicinal properties, the information provided here should help identify and locate the ethnic groups corresponding to particular collections.

Delta Amacuro state. According to the 1982 census, three tribes totaling 17,670 inhabitants were living in Delta Amacuro: the Arawak (Lokono), Kariña (Kari'ña), and Warao (Guaraúno); however, the 1992 census mentions only the Warao with a total population of 21,125. This important indigenous group is skilled in fishing, hunting, and gathering and has an independent linguistic affiliation (Heinen 1988). They live dispersed in an intricate labyrinth of river channels and creeks of the Orinoco Delta, where they have developed a specialized lifestyle adapted to the frequently inundated environment (Plate 72). They make extensive use of the abundant forest resources, particularly the palms *Mauritia flexuosa* (*moriche*) and *Euterpe* sp. (*manaca*). Today the Warao are rather acculturated to Venezuelan society, and their lifestyle has been affected strongly by the influences of outside settlers.

Bolívar state. Thirteen tribes totaling 34,028 inhabitants traditionally live in the extensive lowlands and uplands of Bolívar state. They are the Akawaio (Kapón), Arawak, Guahibo (Guajibo, Hiwi), Hoti (Chicano, Jodi), Kariña (Kari'ña), Mapoyo (Wanai), Panare (E'ñapa, E'ñepá), Pemón (including subgroups Arekuna, [Arecuna], Kamarakoto, Makushi (Makuxi), and Taulipáng [Taurepan]), Piaroa (De'áruwa, Uwotjuja, Wothuha, Wótuha), Sapé, Uruak (Arutaní), Wapishana (Pidian), Yanomami (including the subgroups Sanema [Sanuma] and Yanam [Ninam]), and Yekwana (Dekuana, Makiritare, Maquiritare, Mayonggong, Yecuana, Yekuana, Ye'kwana). In addition to the number cited above, some 950 immigrants from adjacent states have come from indigenous families such as the Bale, Baniva, Curripaco, Piapoco, Pumé, and Warao, and should be added (OCEI 1993b). The three largest indigenous groups are the Pemón, Yekwana, and Panare.

The Pemón (approximately 19,000 persons) belong to the Carib-speaking family and occupy the southeastern region of the state, especially the Gran Sabana, the Caroní basin, and the lower Paragua basin. They live on shifting

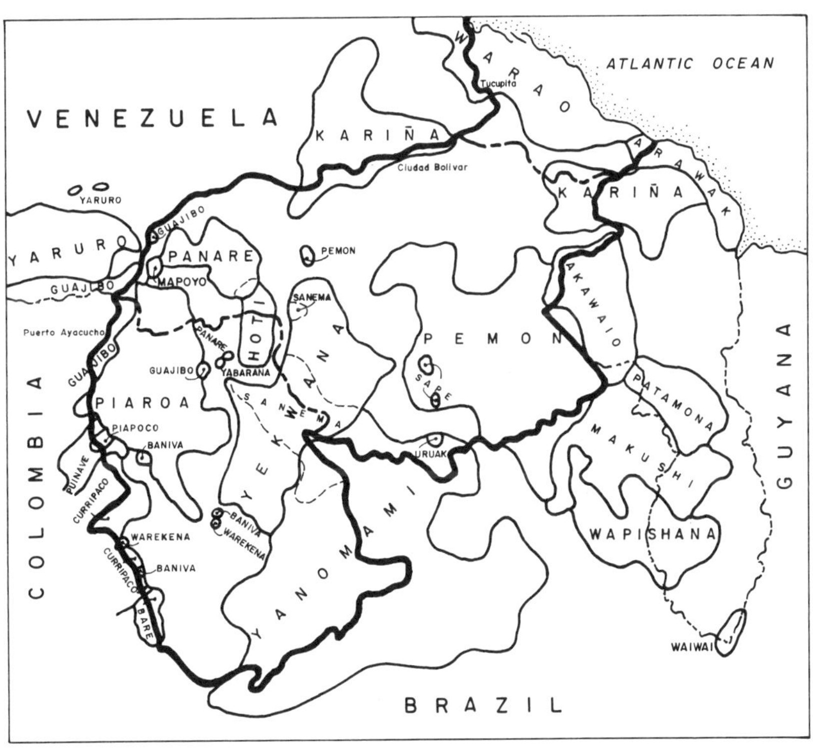

Figure 1-36. Distribution of the main indigenous groups (light lines) in the Venezuelan Guayana and adjoining areas. The heavy line is the outline of the flora area. Based on Lizarralde (1985).

agriculture, with a staple of bitter manioc (*yuca amarga* or *Manihot* sp.), and hunting. Their villages are small, usually in open savannas, and they have an interesting, nonauthoritative, egalitarian community structure (Thomas 1982). The largest Pemón communities are established around the Catholic missions of Kamarata, Wonkén (Uonquén), Kavanayén, and Santa Elena de Uairén, all established in the 1930s and 1940s in the Gran Sabana. Other large Pemón settlements are in the mining centers of Icabarú, Urimán, San Salvador de Paúl, Karún, and Las Claritas, as well as in the relatively newer established villages of San Francisco de Yuruaní, San Ignacio de Yuruaní, Perai-tepui, and San Rafael de Kamoirán.

The Pemón probably began occupying the Gran Sabana uplands 500 to 600 years ago, immigrating from the adjacent savannas of the Río Branco to the south. During their extensive hunting trips and migrations from one site to another, they often lit great fires in the savannas. The Pemón consider the summits of the large table mountains sacred sites belonging to their gods and, therefore, avoid setting foot in these upper montane regions.

The Yekwana, numbering approximately 4400, live in southwestern Bolívar (about 1800 persons) and northeastern Amazonas states (about 2600), mainly in the headwaters of the Caura and Ventuari rivers. They belong to the Carib-speaking linguistic family, and their traditional form of subsistence consists of slash-and-burn agriculture (with bitter manioc as the staple), hunting, and fishing. They are also skilled boat makers and renowned navigators, with an intimate knowledge of all the water courses in their homelands.

The Yekwana live today in small settlements of up to 20–30 houses loosely arranged around a larger circular communal meeting house. Santa María de Erebato, established in the forests on the left shore of the upper Río Erebato, is the main Yekwana settlement in Bolívar state, whereas Cacurí (Kakurí), located in a large savanna on the right shore of the upper Río Ventuari, is the largest Yekwana settlement in Amazonas. In the Yekwana language, the larger mountains are called *jidi* (or *hidi*), which roughly corresponds to the Pemón term *tepui*.

The Panare (E'ñepá) Amerindians (about 3100 persons) occupy the westernmost part of Bolívar state, where they live mainly in small villages established in semideciduous lowland forests of the southern Orinoco floodplains and adjacent hill-lands, but their home range also extends into the evergreen forests of the uplands of the Cuchivero and Suapure basins. Like the Yekwana and Pemón, they belong to the Carib-speaking family. Their subsistence is based on shifting cultivation, with yam (*ñame* or *Dioscorea alata*), sweet manioc (*yuca dulce* or *Manihot esculenta*), and plantain and banana (*plátano* and *topocho* or *Musa* spp.) as staples, together with fishing, hunting, and gathering of forest products. Important anthropological literature on the Panare was produced by Henley (1982, 1988), while Boom (1990b) published results of a detailed ethnobotanical study of the Panare culture.

The remaining Amerindian populations reported for Bolívar state are either very small, localized, and acculturated groups such as the Mapoyo (178 members), Sapé (28 members), Uruak (45 members), and the Wapishana (5 members in the 1982 census, none in the 1992 census), or belong to larger, peripheral groups living mainly in adjacent regions, such as the Akawaio and Makushi in Guyana; the Kariña in Anzoátegui and southern Delta Amacuro; the Guahibo in Colombia, Apure, and Amazonas; the Piaroa, Hoti, and Sanema in Amazonas; and the Yanam in Brazilian Roraima.

Amazonas state. Fourteen different ethnic groups, totaling 43,129 inhabitants are known to live in Amazonas state: Bale (Baré), Baniva (Baniwa, Baniba), Curripaco (Kurrim, Kurripako), Guahibo (Guajibo, Hiwi), Hoti (Chicano, Jodi), Panare (E'ñapa, E'ñepá), Piapoco (Tsase), Piaroa (De'áruwa, Maco, Macu, Mako, Uwotjuja, Wiru, Wóthuha, Wótuha), Puinave (Puinabe), Sáliva, Warekena (Guarekena, Guarequena, Uarekena, Walekhena), Yabarana (Yawarana), Yanomami (including Yanomam, Yanomamɨ, and Sanema [Sanimá, Sanuma] linguistic groups; additional names are Guaharibos, Guaica, Guaika, Parafiri, Shamathari, Shiriana, Shirishiana, Uaica, Waika, Xiriana, Yanoama, and Yanomama [see Lizot 1988 for further details]), Yavitero (Yavitano), and Yekwana (Dekuana, Makiritare, Maquiritare, Mayonggong, Yecuana, Yekuana, Ye'kwana). In addition, there is a mixed-ethnic, but distinct linguistic group of approximately 750 persons known as Yeral (Geral) present in southernmost Amazonas, as well as other minor families belonging to tribes from neighboring states which sum approximately 650 persons (OCEI 1993b).

Most of these groups inhabit the dense, basimontane forests of the lowlands and occasionally the submontane forests of the uplands that stretch in a wide arc from northwest to southeast. Another group of tribes lives along the axis of the Atabapo and Negro rivers. The only area not inhabited by Amerindians seems to be the Casiquiare, lower Siapa, and lower Pasimoni floodplains. The largest indigenous groups of Amazonas are the Yanomami (including the Sanema), Guahibo, Piaroa, and Yekwana.

The term *Yanomami* refers to a great Amerindian nation composed of at least four different linguistic subgroups: the Yanomamɨ, the Yanomam, the Yanam, and the Sanema (Lizot 1988). In spite of numerous studies made since the 1940s, Yanomami linguistic affiliations still remain unclear. The Yanomami constitute the largest surviving indigenous nation of the Guayana region. Their total population in Venezuela and Brazil is estimated at approximately 25,000 members, of which approximately 15,000 live in Venezuela, including about 2100 Sanema (OCEI 1993b).

The Yanomami living in the Venezuelan Amazon belong to the Yanomamɨ (in the Sierra Parima to Sierra Tapirapecó and Sierra de la Neblina), Yanomam (headwaters of the Orinoco), and Sanema (northern Sierra Parima, upper Río Ventuari) linguistic subgroups. Furthermore, some small and dispersed settlements of the Sanema and Yanam (Ninam) subgroups are also found in southern Bolívar state.

The Yanomami usually live in forest environments, where they practice shifting cultivation, hunting, gathering of forest products, and, to a minor degree, fishing. Interestingly, one of their main staples is the banana or plantain (*Musa* spp.), which was introduced centuries ago from the Old World tropics. They build large communal houses called *shapono* or *shabono* for 40–250 persons, consisting of a circular roof construction surrounding a wider open plaza (Plate 75). Since their life style, social organization, and religion are strongly centered on the upland forest habitat, they have not developed particular abilities for river navigation.

During the 1980s and 1990s, some sectors of the Venezuelan Yanomami suffered serious cultural and health impacts from a heavily increased number of non-Amerindian immigrants into their homelands, especially miners, missionaries, and tourists. In the southern part of the Sierra Parima, numerous Yanomami families have already begun to abandon their traditional forest habitat and settle in open, artificially created savannas (Smole 1976, Huber et al. 1984).

The Guahibo are the second largest indigenous group in Amazonas, numbering approximately 9400 members. Although their traditional homeland is located in the Colombian Llanos (Friel 1924), the Guahibo have become an important ethnic component of the indigenous population in adjacent northwestern Amazonas state, especially in and around the capital Puerto Ayacucho, and along the Río Orinoco as far south as its confluence with the Río Vichada. A small, isolated group has also lived in San Juan de Manapiare since 1970. Originally renowned for their interesting circular migration cycles (Conaway 1984), they now live in settlements located principally in savannas near the forest edge, practicing shifting cultivation (with bitter manioc as a staple), hunting, and fishing. Linguistically, the Guahibo belong to the group of tribes of independent affiliation (Metzger and Morey 1983).

The Piaroa number about 9370 members and are a socially well-organized nation of forest dwellers of the small (sometimes considered independent) Sálivan language family (Overing and Kaplan 1988). Their homeland extends over the region of low hills and mountains of the Cuao-Sipapo massif in the northwestern corner of Amazonas state, crossed by the Cataniapo, Cuao, Autana, Sipapo, and Guayapo rivers. A small part of their territory also lies in the Parguaza basin of Bolívar state and in the upper Manapiare valley and the lower Río Ventuari of Amazonas. Included here as well are the Mako (Maco, Maku, Wiru), a small group living at the edge of the forests (Plate 73) in the lower Ventuari, Marueta, and Yureba rivers.

For their subsistence, the Piaroa practice slash-and-burn agriculture, with bitter manioc as their staple, as well as hunting, fishing, and gathering of numerous plant and animal forest products. Traditionally, their villages are small and consist of a large, conical and circular communal house called *churuata* (Plate 74) where all the 15–50 village members live. The Piaroa have an excellent geographic knowledge of their homeland, and in their mythology, the summits of the various mountains of the Cuao-Sipapo massif, especially that of the spectacular, tower-like Cerro Autana, play an important role as sacred sites.

The Yekwana, who were already discussed for Bolívar state, are the fourth largest ethnic group in Amazonas, where their territory extends from the upper Río Ventuari to the Parú, Cunucunuma, Padamo, and Iguapó rivers around the Parú and the Duida-Marahuaka massifs. The Yekwana seem to be in the process of expanding their territory towards the lower Cunucunuma and adjacent Orinoco rivers.

There are a number of other indigenous tribes in Amazonas with much more reduced population sizes than those treated above. These include the Hoti, a small Carib-speaking group of less than 650 members living on both sides of the Sierra de Maigualida, in the upper Río Cuchivero basin (Bolívar state), and in the Asita, Iguana, and Parucito river valleys (Amazonas state). The Yabarana are another small Carib-speaking tribe with about 300 members in San Juan de Manapiare and in nearby villages mixed with Piaroas. There is also a complex of various tribes belonging to the Arawak family (except the Puinave, which are independent), all living along the black-water Atabapo, Guainía, and Negro rivers in southwestern Amazonas. These mostly acculturated tribes are the Puinave (774 members, Orinoco-Atabapo confluence), the Piapoco (about 1200 members, Orinoco-Guaviare confluence), the Baniva (1166 members, dispersed from the Río Guainía to the upper Orinoco), the Warekena (427 members, along the Río Guainía), the Bale (1225 members, along the Río Negro and the lower Casiquiare), and the Curripaco (2770 members, along the Guainía, Temi, and Atabapo rivers) (OCEI 1993b).

In summary, although clearly less numerous and varied than in the central and western Amazon basin (Lizarralde 1993), there is still an impressive ethnic diversity throughout the Venezuelan Guayana. This may be due in part to the high ecological diversity of the area, but also to the fact that this region is a crossroads of three main indigenous linguistic families, the Arawak, the Carib, and the Tupí (Mason 1950, cited in Lizot 1988), as well as the non-affiliated linguistic groups. Unfortunately, much of this ethnic diversity is vanishing rapidly due to the continual acculturation process, which is leading to the irreversible loss of a rich cultural (including ethnobotanical) heritage for humanity.

Immigrants and Settlers in the 19th and 20th Centuries

The process of European colonization of the Venezuelan Guayana started soon after the area's discovery during the 16th century. Originally, the Spanish, British, and Dutch Crowns were competing with each other to obtain permanent access and possession of the coveted wealth of the mythical El Dorado, an Amerindian kingdom believed to be somewhere in the "large, rich and beautiful Empire of Guiana" (Raleigh 1596; see Plate 1). For this purpose, several military expeditions were conducted to the area, eventually leading to the establishment of fortified stations along the lower and middle Orinoco by the Spaniards. Soon after Dutch and British ambitions in the region were abandoned in the 17th century, the area was colonized by the Spanish conquistadors and later by Catholic missionaries, who established permanent missions along the shores of the Río Orinoco and its main lower tributaries.

The colonization activities by missionaries during the 17th and 18th centuries, especially the Jesuits and the Capuchins, extended deeply into the Guayana and upper Orinoco regions, producing considerable new geographical knowledge about these remote territories and their inhabitants. The detailed accounts written by Gumilla (1741) and Gili (1780–1784) testify to the many discoveries made during this phase of exploration. Overall, though, the establishment of new settlers and immigrants in the interior of the Venezuelan Guayana proceeded slowly and was mainly concentrated along the lower and middle Orinoco (e.g., Misiones del Caroní, founding of El Dorado, Upata, and Tumeremo in the Río Cuyuní). During the first half of the 18th century, missionary activities were banned by the Venezuelan government, and the colonization process was practically stopped during the rest of that century.

As a result of Solano and Iturriaga's *Expedición de Límites* to the upper Orinoco and the Río Negro between 1755 and 1760, a new phase of occupation by Spanish settlers began, leading to the founding of many new villages, such as San Fernando de Atabapo in 1758, Santa Bárbara del Orinoco in 1759, Esmeralda in 1760, and San Carlos de Río Negro in 1760. At the time of Humboldt and Bonpland's visit to the upper Orinoco and Casiquiare in 1800, however, most of these settlements were nearly abandoned, with only a few people living under precarious conditions (Humboldt 1818–1829).

Towards the end of the 19th century, the general Amazonian rubber boom had invaded parts of the Venezuelan Amazon, leading to the establishment of numerous small settlements for processing and shipping the wild rubber collected in the surrounding forests, especially along the Casiquiare (Capihuara), upper Orinoco (San Antonio), and the lower Ventuari rivers (Yacurai or Las Carmelitas). During this phase, the European population increased steadily and expanded into the most remote corners of the Guayanan lowlands. At the same time, enslavement of Amerindians for labor forces by the European settlers soon caused the retreat of indigenous tribes towards more inaccessible parts of the Guayana. The regional center of the rubber boom period was San Fernando de Atabapo, the capital of Amazonas until 1924, when it was moved to the newly created town of Puerto Ayacucho. The rubber boom lasted about a century, from 1860 to 1960 (Perera 1990).

Besides rubber exploitation, gold and diamond mining activities also developed on a larger scale, especially in the eastern Venezuelan Guayana. The most important mining towns founded during the 19th century were Guasipati and El Callao in the Río Cuyuní basin. As a direct consequence of these increased economic activities, the larger towns located on the lower Orinoco, especially Ciudad Bolívar (then still called Angostura), San Félix, and Santo Tomé, received many new immigrants.

During the 20th century, the colonization of the Venezuelan Guayana increased dramatically after 1950, due to the discovery and massive industrial exploitation of several abundant natural resources in the region. The most important mineral resources are iron in the area of El Pao and Ciudad Piar (Cerro Bolívar, Cerro San Isidro, Cerro Altamira, and Cerro Arimagua, all in northeastern Bolívar), bauxite (in Los Pijiguaos, western Bolívar; Plate 79), and gold and diamonds (in the Cuyuní, the lower Caroní, and the middle Guaniamo river basins). The country's largest heavy industry factories for iron, steel, and aluminum production were built west of Ciudad Guayana (Puerto Ordaz), offering direct and indirect employment to approximately 500,000 people. The energy supply for this huge industrial complex is provided by one of the world's largest hydroelectric power plants, the dam at Guri. This complex was built on the lower Río Caroní over a period of almost 20 years. Another series of dams is under construction downstream from Guri (at Macagua, Tocoma, and Caruachi).

The intensive industrialization of the Venezuelan Guayana has focused on the region of the lower Río Orinoco in Bolívar state, where the principal mineral and energetic resources are concentrated. A special governmental agency, the Corporación Venezolana de Guayana, or CVG, was created in the 1950s to oversee and stimulate this development. Today, this corporation acts as a gigantic holding company with several subsidiary companies, each devoted to the industrial and commercial exploitation of a particular resource. These include CVG-FERROMINERA (iron); CVG-BAUXIVEN (aluminum); CVG-EDELCA (hydroelectricity); and CVG-TECMIN (mining of gold, diamonds, and other minerals). After the national petroleum industry, the CVG is the largest economic power concentration in modern Venezuela.

In 1969 the governmental development agency CODESUR (Comisión para el Desarrollo del Sur de Venezuela) was created to expand colonization and industrialization in the Venezuelan Amazon. This agency was originally designed to counterbalance Brazil's intensive colonization plans in the Amazon region from 1960 to 1970. In 1974 and 1975, however, it was transformed from an economically oriented development agency to an environmental management organization. In 1989, another governmental institution ascribed to the environmental ministry, SADA-Amazonas, became responsible for the design, implementation, and supervision of regional management plans for land use and population policy in the Venezuelan Amazon.

Although there have been no officially sponsored colonization processes in the hinterlands of the Venezuelan Guayana, illegal immigration has occurred in certain areas due to intensive mining activities. This is the case at the headwaters of the Río Orinoco, the Sierra Parima, the Sierra Pakaraima, and in the Río Cuyuní basin. The construction of paved roads in southeastern Bolívar state (El Dorado to Santa Elena de Uairén) and in northwestern Bolívar (Caicara to Puerto Ayacucho) has clearly stimulated the immigration and establishment of large human settlements and villages in these parts of the Venezuelan Guayana. Finally, besides traditional intensive land use activities such as agriculture, logging, ranching, and mining, tourism has begun to attract many people to the region.

Table 1-3. Maximum elevation, slope areas, and summit areas of the high mountains and tepuis of the Venezuelan Guayana. These are arranged in geographical subunits from southeast to north, west, and then south. When not otherwise indicated, altitudinal data were measured directly by Otto Huber with Thommen altimeters or GPS (global positioning system) instruments.

	Maximum elevation (m)	Summit area[1] (km²)	Slope area[1] (km²)
Estado Bolívar			
Eastern tepui chain		(70)[1]	(320)
Uei-tepui	2150	2.50	20
Roraima-tepui	2723[2]	34.38	
Kukenán(Mataui)-tepui	2650	20.63	
Yuruaní-tepui	2400	4.38	300
Wadakapiapué-tepui	2000	< 0.01	
Karaurín-tepui	2500	1.88	
Ilú(and Tramen)-tepui	2700	5.63	
Cerro Venamo[3]	± 1600[4]	N.A.[5]	N.A.
Sierra de Lema[3]	1650[6]	N.A.	N.A.
Ptari massif		(2.50)	(58)
Sororopán-tepui[3]	2050	N.A.	30[7]
Carrao-tepui	± 2200[6]	1.25	28
Ptari-tepui	2400	1.25	
Los Testigos massif		(12)	(116)
Kamarkawarai-tepui	2400	5.00	
Tereke-yurén-tepui	1900[8]	0.63	88
Murisipán-tepui	2350	5.00	
Aparamán-tepui	2100[9]	1.25	28
Auyán massif		(670)	(795)
Auyán-tepui	2450	666.90	
Cerro La Luna	1650	0.20	715
Cerro El Sol	1750	0.60	
Uaipán-tepui	1950[10]	2.50	60
Cerro Venado[3]	1320	0.60	17
Kurún-tepui[3]	1100	0.60	
Aprada massif		(6)	(210)
Aprada-tepui	2500	4.37	210
Araopán-tepui	2450	1.25	
Chimantá massif		(615)	(915)
Murey(Eruoda)-tepui	2650	51.25	
Tirepón-tepui	± 2600[6]	8.75	
Apacará-tepui	2450	173.12	
Abacapá-tepui	2400	28.13	
Agparamán-tepui	2400	22.50	
Toronó-tepui	2500	59.38	915
Chimantá-tepui	2550	93.75	
Churí-tepui	2500	47.50	
Acopán-tepui	2200	92.50	
Amurí-tepui	2200	36.88	
Angasima-tepui	2250	2.00	32
Upuigma-tepui	2100	0.63	13
Cerro Guaiquinima[3]	1650	1096.26	410
Sierra Marutaní (Pia-Zoi[3])	± 1500[11]	N.A.	± 740[7]
Cerro Ichún[3]	± 1400[6]	2460.00	798
Cerro Guanacoco	± 1500[6]	526.25	400
Jaua massif		(1170)	(770)
Cerro Sarisariñama	2350	546.88	286
Cerro Jaua	2250	625.62	482
Sierra de Maigualida	2400	± 440	N.A.

	Maximum elevation (m)	Summit area[1] (km^2)	Slope area[1] (km^2)
Estado Amazonas			
Cerro Yaví	2300	5.62	70
Yutajé massif		(275)	(143)
Serranía Yutajé	2140	95.63	143
Cerro Coro Coro	2400	179.38	
Cerro Guanay	2080	165.00	113
Cerro Camani	1800[6]	1.88	N.A.
Cuao-Sipapo massif		(± 300)	N.A.
Cerro Autana[3]	1300	1.88	282
Cerro Cuao	2000[6]	80	
Cerro Sipapo	± 1800[6]	56	N.A.
Cerro Ovana (Ouana)	± 1800[6]	N.A.	N.A.
Cerro Moriche	± 1250[6]	± 0.20	31
Parú massif		(930)	(580)
Cerro Parú (Asisa, A'roko)	± 2200[12]	724.38	580
Cerro Euaja	± 2000[6]	205.62	
Cerro Yapacana[3]	1300[13]	10.50	38
Duida-Marahuaka massif		(1219)	(1100)
Cerro Huachamacari	1900	8.75	60
Cerro Marahuaka	± 2800	121.00	325
Cerro Duida	2358	1089.00	715
Cerro Aratitiyope	1700[14]	< 0.01	N.A.
Sierra Unturán[2]	± 1600[6]	N.A.	N.A.
Cerro Tamacuari	2340[15]	< 0.01	N.A.
Serranía Tapirapecó	2000[15]	N.A.	N.A.
Neblina-Aracamuni massif		(473)	(1515)
Cerro Aracamuni-Avispa	1600[16]	238.00	658
Sierra de la Neblina	3014[17]	235.00	857
Total summit area		12,290	
Total estimated Pantepui area (above 1500 m)		5,000	

[1]Summit and slope areas were measured by an electronic planimeter based on maps produced from radar and satellite images (Figures 1-19 to 1-35). Slope areas are estimated; the sums of summit and slope areas in massifs are rounded off and placed in parentheses.
[2]Altitude provided by Cartografía Nacional (MARNR). This is the altitude of the highest point of the Venezuelan sector of Roraima; possibly, the northern rim in Guyana is higher.
[3]Either insufficiently explored tepuis or lower mountains traditionally treated as tepuis.
[4]From Steyermark and Nilsson (1962).
[5]N.A. = Not applicable, too diffuse or poorly mapped to measure with reasonable accuracy.
[6]Estimated from helicopter overflights.
[7]Global estimate for summit and slope area.
[8]From Steyermark (1986a) and Holst (1987).
[9]From Holst (1987).
[10]From Mayr and Phelps (1967).
[11]From Steyermark and Maguire (1984).
[12]From Cowan and Wurdack's botanical labels.
[13]From Steyermark and Bunting's botanical labels.
[14]From George (1988).
[15]From FUDECI (1990).
[16]From Steyermark, Holst and collaborators (1989).
[17]From Brewer-Carías (1988).

CHAPTER 2

History of Botanical Exploration

Otto Huber

The exploration of the Venezuelan Guayana with regard to its flora and vegetation can be divided into three major phases. The earliest phase began in the 18th century and consisted of river-based expeditions in the lowlands. Starting in the mid-19th century, a second phase began with larger fluvial and land-based explorations that covered parts of the lowlands, uplands, and highlands of the flora area. The third phase, which began in the late 1960s and continues until the present, has included helicopter explorations, concentrating mainly on highland areas of the Venezuelan Guayana.

For the following overview, many of the original publications from participants on explorations in the Guayana were consulted. Secondary sources included articles by Urban (1906), Pittier (1920), Knuth (1928), Arnal (1943), Prance (1971), Weibezahn et al. (1983, 1990), Huber and Wurdack (1984), Weibezahn and Janssen-Weibezahn (1990), and Texera (1991). A summary of the state of exploration of major mountains and lowland areas of the Venezuelan Guayana is presented in Table 2-1 (see end of chapter). Herbarium acronyms follow Holmgren et al. (1990).

Pioneer Explorations in the Guayana Lowlands (1754–1951)

The first phase of botanical exploration in the Venezuelan Guayana involved expeditions to previously unexplored areas. Most of these expeditions had broader objectives than obtaining botanical information and were often aimed at acquiring basic geographical, cartographical, military, geopolitical, and ethnographical information on newly conquered or unexplored lands. This phase lasted nearly 200 years, from the first Spanish expedition to the upper Orinoco in the mid-18th century to the discovery of the source of the Río Orinoco in 1951. Because of the primitive logistic conditions during much of this period, the early expeditions were almost exclusively river-based and therefore confined to the most important waterways and surrounding lowland areas.

Botanical explorations along the Orinoco began in 1754, when the Spanish government sent to the Guayana an official expedition called the *Comisión de Límites*. The main mission of this group was to delimit the border of the Spanish and Portuguese colonies in northern South America, following the treaty signed in Madrid by both countries in 1750. The expedition also had scientific objectives, and a young Swedish botanist and student of Carl Linnaeus, Pehr Loefling, was put in charge of a group of naturalists that included two doctors and the illustrators Bruno Salvador Carmona and Juan de Diós Castel. Besides obtaining general information on the natural history of the Guayana, the members had instructions to investigate earlier reports of varieties of cinnamon, cacao, and quinine in the area (Ramos 1946; Pelayo 1990; Pelayo and Puig-Samper 1992).

In April 1755, after several months of collecting plants along the coast near Cumaná, Loefling and two of his naturalists traveled overland to the lower Orinoco to meet the main part of the expedition that had traveled up the Orinoco from its mouth. In the latter half of 1755, Loefling collected in the lower Río Caroní area, but he fell ill with severe fever and died there in February 1756. The illustrators Carmona and Castel continued on the expedition until 1760, working in the upper Orinoco between 1758 and 1760. Although none of Loefling's plant collections from Venezuela are known to have ever reached Europe, some of his notes survived and were published as part of *Iter Hispanicum* (Loefling 1758). Also, many of the botanical illustrations survived, and some were published (Pelayo 1990).

One of the commanders of the Comisión de Límites, Eugenio de Alvarado, also made extensive notes on the vegetation and plant products of the lower Orinoco, including ones known locally at the time as *quina, algarrobo,* and *carapa.* His reports were deposited in the Archivo General de Simancas (Ramos 1946), and parts were first published by Cuervo (1893).

At the start of the 19th century, from April to July 1800, the German naturalist Alexander von Humboldt (Figure 2-1) and the French botanist Aimé Bonpland (Figure 2-2) made their landmark journey to the middle and upper Orinoco (Figure 2-3). After crossing the Venezuelan Llanos, they entered the Orinoco at the mouth of the Río Apure, then traveled upstream by boat (*piragua*) and passed through the treacherous rapids of Atures (Plate 7) and Maipures. Farther upstream, they followed the Río Atabapo to the small settlement of Yavita on the Río Temi. At that point, they had to drag their boat overland along a rudimentary trail through the dense forests of the Isthmus of Tuamini to the Caño Pimichín, a tributary of the Río Guainía. This historic portage is still used today, although it was modified in 1970 when the Venezuelan government made a new road that goes directly from Yavita to Maroa on the Río Guainía.

From Pimichín, Humboldt and Bonpland navigated downstream to the Río Guainía, past the confluence with the Río Casiquiare, and reached the village of San Carlos de Río Negro on May 7, 1800. At the time, this was the

Figure 2-1. Alexander von Humboldt. Courtesy of Hunt Institute for Botanical Documentation, Carnegie Mellon University, Pittsburgh, Pennsylvania.

southernmost Spanish settlement in Venezuela. On May 10, 1800, they paddled upstream on the Casiquiare Canal to the bifurcation of the Río Orinoco near Tamatama on May 21. Reaching this spot was a major highlight of their three-year journey in the New World tropics. From Tamatama, they continued upstream on the Orinoco to La Esmeralda, then a small Amerindian settlement near the base of Cerro Duida. The uppermost point they reached on the Río Orinoco was the mouth of the Caño Iguapo. They began their return back down the Orinoco from La Esmeralda on May 23, hardly stopping to collect or make observations. This is shown by the rapidity of their trip, reaching San Fernando de Atabapo by May 27. They passed the rapids of Maipures on May 30, the rapids of Atures the next day, and arrived at the town of Angostura (today's Ciudad Bolívar) on June 14, 1800. Once at Angostura, both Humboldt and Bonpland were victims of violent fevers and spent a month recovering their health before proceeding on to Barcelona and Cumaná (Sandwith 1925; Humboldt 1956).

Figure 2-2. Aimé Bonpland. Courtesy of Hunt Institute for Botanical Documentation, Carnegie Mellon University, Pittsburgh, Pennsylvania.

The amount and high quality of the data gathered by these two naturalists during their explorations in the Venezuelan Guayana was extraordinary. It included contributions on zoology, geography, ethnography, astronomy, and botany, constituting the true starting point for the scientific study of the region. Since Humboldt and Bonpland's plant collections were the first ones made in the Venezuelan Amazon, there were many species new to science. Their specimens formed an important floristic base for the botany of the upper Orinoco and Río Negro basins, and their accurate descriptions of the plant communities observed along their itinerary are still a valuable information source for botanists. Today, the main herbaria holding Humboldt and Bonpland's specimens are at Paris (P-Bonpl.) and Berlin (B-W).

The next major botanical explorers in the Guayana were two German brothers and geographers, M. Richard and Robert H. Schomburgk (Figure 2-4). They were commissioned by the British Crown to determine the border between the British colony of Guiana and the Republic of Venezuela. Both broth-

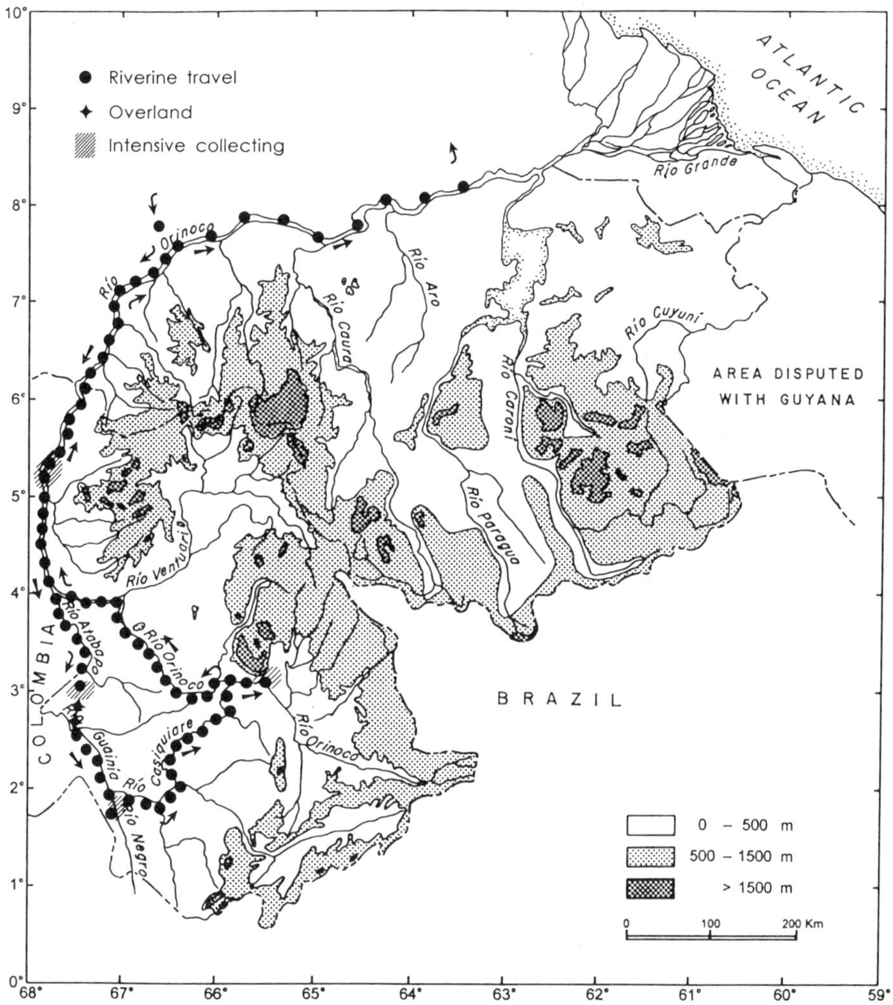

Figure 2-3. Route of Humboldt and Bonpland in the Venezuelan Guayana, 13 April–2 June 1800.

ers traveled extensively between 1835 and 1844 in the headwaters of the Essequibo, Rupununi, and Branco rivers. They were the first persons to make plant collections from the upper tepui slopes of Roraima-tepui. Robert Schomburgk then made a large and arduous journey from the base of Roraima-tepui west to La Esmeralda, crossing part of the Pakaraima and Parima mountains (Figure 2-5). From La Esmeralda he continued by river to the Río Casiquiare and then via the Río Negro to Brazil (R. H. Schomburgk 1840b, 1840c). The scientific results of these explorations were published in a monumental volume on the flora and fauna of the Guayana, the third volume in the work (M. R. Schomburgk 1847–1848). This publication also included the first description and subdivision of life zones in the former British Guiana. The main sets of Schomburgk collections are now principally held at four herbaria (B, BM, K, P).

Figure 2-4. Robert H. Schomburgk. Courtesy of Hunt Institute for Botanical Documentation, Carnegie Mellon University, Pittsburgh, Pennsylvania.

The famed British entomologist Alfred Russel Wallace and the botanist Richard Spruce also visited the Venezuelan Amazon. Both entered Venezuela from the south, Wallace in 1851, and Spruce in 1853, traveling up the Río Negro from Manaus (then called Barra). In February and March 1851, Wallace collected insects and some plants in Yavita. His botanical collections were lost during a shipwreck in the Atlantic in 1852 (Wurdack 1960). Some of his field notes, however, including those on palms, were rescued and later published in his beautifully illustrated monograph on Amazonian palms (Wallace 1853).

Richard Spruce (Figure 2-6) traveled the southern and northwestern part of the Venezuelan Amazon from April 1853 to November 1854. He explored along the Negro, Guainía, Atabapo, Orinoco, Casiquiare, Cunucunuma, and Pasimoni rivers (Spruce 1908, Figure 2-7). He made more than 800 botanical collections in the upper Orinoco and Río Negro region alone (Reichenbach 1873), many of them representing species new to science. The main set of

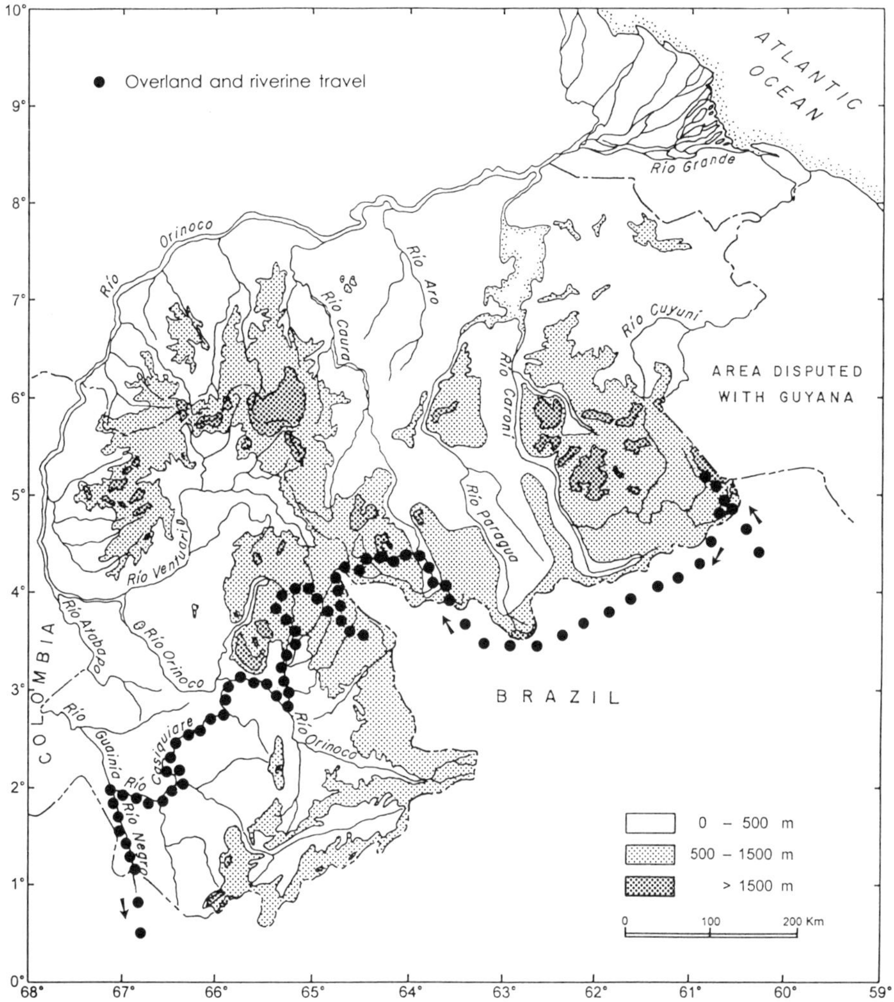

Figure 2-5. Route of Robert H. Schomburgk in the Venezuelan Guayana, October 1838–March 1839. Schomburgk collected extensively on the slopes of Roraima-tepui, then traveled up the Rio Uraricuera in northern Brazil before reentering the Venezuelan Guayana at the western end of the Sierra Pakaraima. He then crossed the Sierra Parima and descended the Río Padamo to reach the Río Orinoco and Río Casiquiare.

Spruce's collections is located at K, but duplicates are widely distributed in other major herbaria as well.

The French explorer Jean Chaffanjon made the first of three visits to Venezuela in 1885. He made geographical and anthropological studies along the Río Orinoco and also intended to discover the river's source, which was still in doubt after Robert Schomburgk's unsuccessful attempt to reach it in 1839. Chaffanjon's first trip was from the mouth of the Orinoco to the mouth of the Río Meta, and then from the mouth of the Río Caura to the mouth of the Río Erebato (from January to March 1885). On his second trip, from April 1886 to

Figure 2-6. Richard Spruce. Courtesy of Hunt Institute for Botanical Documentation, Carnegie Mellon University, Pittsburgh, Pennsylvania.

January 1887, he traveled farther up the Orinoco than any previous explorer, reaching the rapids of the Raudal de los Guaharibos at the base of Sierra Parima (which he erroneously described as the source of the Orinoco). His final trip was from 1888 to 1890, when he visited the Río Caroní and the Río Cuyuní on his way to the British, Dutch, and French Guianas. In an adventure-style book published in Paris in 1889, Chaffanjon offered a vivid description of his experiences during the first two expeditions along the Orinoco and the Caura (see Perera 1986 for a Spanish-language translation). He made 565 plant collections that were deposited at P. There, Paul Maury studied many of Chaffanjon's plants, later publishing his results (Maury 1889).

Another French explorer, Albert Gaillard, visited the Orinoco region between Atures and San Fernando de Atabapo from April to September 1887. Gaillard mostly collected fungi (Patouillard and Gaillard 1888); these and the earlier collections of mosses and liverworts by Spruce provided the initial taxonomic information base for the cryptogamic flora of the upper Orinoco region.

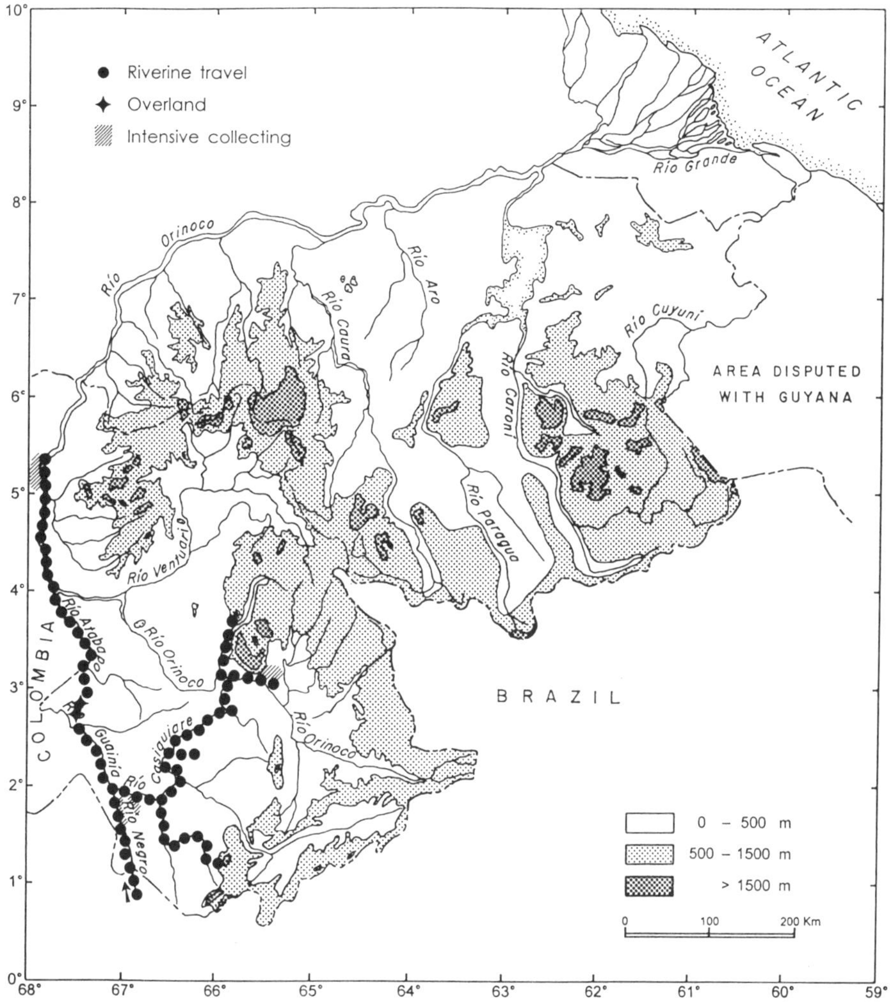

Figure 2-7. Route of Richard Spruce in the Venezuelan Guayana, 1853–1854. Spruce traveled up the Rio Negro from Brazil to reach Venezuela. On the Río Orinoco he traveled as far downriver as Maipures before turning back to return to Brazil.

In October 1887, Alfredo Jahn, Jr., became the first Venezuelan-born naturalist to visit the upper Orinoco, when he and the ethnologist and chemist Vicente Marcano reached San Fernando de Atabapo. Jahn became seriously ill there and had to end his trip, but he returned with a small botanical collection from the area around Atures. The specimens were studied and the results were published by Adolfo Ernst (1888).

In 1896, a pharmaceutical botanist from New York, Henry Hurd Rusby, and his companion, Roy W. Squires, became the first North American botanists to collect along the Río Orinoco. They made a large set of plant collections along the lower Orinoco, especially around Sacupana in the Orinoco Delta (Rusby 1896), that is deposited at NY.

Other botanical explorations were centered along the Caura and Cuchivero rivers, both southern tributaries of the lower Orinoco. In 1897 and 1898, the French naturalist Eugène André traveled along the lower Río Caura to La Prisión, where he made several trips looking primarily for birds and orchids. At the end of 1900, he made a second expedition along the Río Caura. This time he reached its upper part (the Río Merewari), where he tried unsuccessfully to climb the summit of the Cerro Ameha, part of the Cerro Guanacoco massif (André 1904).

In 1901, the German geographer Siegfried Passarge and W. M. Selwyn, a London merchant in charge of the astronomical determinations, made a geographical and botanical survey in the lower Río Cuchivero basin. They collected around 800 numbers of plants and sent them to B; these were the first specimens to be made from this area. The publications from this survey offer a good description of the predominant vegetation types in this area of the lower Orinoco (Passarge 1903, 1933).

The foremost explorer of the Venezuelan Guayana was Félix Cardona (Figure 2-8). He produced some of the first detailed maps and topographical measurements in the area and made important botanical and zoological collections (especially birds) between 1935 and 1950. His nearly 2500 numbers of plant collections are deposited at VEN and US.

In 1929–1930 and 1931, the National Geographic Society of Washington, D.C., organized two ornithological expeditions led by Ernest G. Holt. Their itinerary in Venezuela included travel along the Orinoco, Casiquiare, and Negro rivers. Holt and his collaborators, Wilhelm Gehriger and Emmet R. Blake, collected about 1000 botanical specimens, including the first ones from Sierra Imerí near Sierra de la Neblina (Holt 1931, 1933; Friedmann 1948). Their specimens were deposited at US.

In 1939, the Venezuelan government sponsored a multidisciplinary scientific exploration of the Gran Sabana in southern Bolívar state. Gaylord G. Simpson, a member of the commission, made a small botanical collection that he sent for determination to Henri Pittier at the Herbario Nacional de Venezuela (VEN) in Caracas (Aguerrevere et al. 1939). Prior to this study, only the southeastern section of the Gran Sabana had been initially characterized through the explorations of Robert Schomburgk in 1838, Ernst Ule in 1909 and 1910, and Theodor Koch-Grünberg in 1911.

One of the last traditional river expeditions that involved many support workers, technicians, and scientists, was the French-Venezuelan expedition to the headwaters of the Río Orinoco in 1951 (Anduze 1960; Rísquez-Iribarren 1962; Lichy 1978). The official botanist of this expedition was the biogeographer Léon Croizat, who collected close to 1200 numbers of botanical specimens before becoming seriously ill at the mouth of the Río Ugueto. Apparently, a smaller set of plant collections from this same expedition was made by José Maria Cruxent from above the Río Ugueto; these plants were discovered at the New York Botanical Garden (NY; Boom and Moestl 1990). Most of Croizat's collections are deposited at NY, with some duplicates at VEN (Holst and Todzia 1990).

Figure 2-8. Félix Cardona, pioneering explorer of numerous mountains in the Venezuelan Guayana. Photograph by Julian Steyermark.

Initial Explorations of Pantepui (1838–ca. 1960)

Since most tepuis are quite difficult to reach and are unsuited for agriculture or sustainable human occupation, they were hardly visited during the pioneer phase of the exploration of the Venezuelan Guayana. Besides, most of the native Amerindian communities have long considered the tepuis to be sacred places forbidden to humans.

The tepui summits and their upper slopes form a biogeographical entity called the Pantepui Province and include areas of the Guayana Shield at elevations above 1300–1500 m (Mayr and Phelps 1967; Huber 1987). The first botanical collections from Pantepui were made on the southern slopes of Roraima-tepui (or Mount Roraima) by Robert Schomburgk in 1838. In the following decades, there were numerous attempts to reach the high summit of Roraima-tepui, but a successful access route was not found until 1884, when the English naturalist Everard F. Im Thurn, curator at the Georgetown museum, and his companion Harry I. Perkins reached the summit on December

18. They returned with many unusual plants unknown to science (Im Thurn 1885a, 1885b; Perkins 1885), which were later deposited and studied at K (Im Thurn 1886, 1887). Their findings stimulated an intense interest among many scientists to learn more about the natural history of the tepuis.

Encouraged by these discoveries, two other English explorers, Frederick V. McConnell and John J. Quelch, climbed to the summit of Roraima-tepui in November 1894 and stayed for three days (Quelch 1895). They returned in October 1898 and spent nine more days on the summit. On both occasions they made a thorough inventory of the flora and fauna of the southern part of Roraima-tepui's summit; their botanical collections were deposited at K. From these explorations, a total of 239 spermatophytes, 88 pteridophytes, 63 bryophytes, and 11 thallophytes were recorded from Roraima-tepui above about 1650 m elevation (Brown 1901).

The imposing massif of Roraima-tepui continued to attract naturalists and explorers in the following decades. Ernst Ule, a German botanist, made extensive botanical collections on the slopes and summit from December 1909 to January 1910. He was one of the first visitors to offer a detailed description of the vegetation of Roraima-tepui (Ule 1915). The German ethnologist Theodor Koch-Grünberg also climbed to the summit of Roraima-tepui in 1911, but he apparently made no plant collections there (Koch-Grünberg 1917).

Prior to World War I, German and English naturalists were the main explorers of the Guayana mountains. Both countries had powerful scientific and governmental organizations that stimulated and supported their work, especially to increase the scientific collections of their museums. After 1920, however, U.S. scientists largely took over the exploration of the area. Most expeditions from this time on were large and institutional-sponsored ones that required careful planning and strong logistical support. During this phase, biological exploration in the Guayana was designed to systematically explore as many tepui summits as possible and to obtain large numbers of specimens for museums. Over a period of four decades, four institutions, one Venezuelan and three from the United States, were responsible for almost all the botanical exploration in the Venezuelan Guayana. The most important achievements of these institutions in the exploration of Pantepui are described below.

Within a single decade, the American Museum of Natural History carried out three large scientific expeditions that produced major advances in the knowledge of the Pantepui biota. The first expedition was to Roraima-tepui in 1927 and 1928 (Tate 1930). The second was the Tyler-Duida expedition to Cerro Duida in the Upper Orinoco in 1928 and 1929 (Tate and Hitchcock 1930), and the third was the joint expedition in 1937 with William H. Phelps, Sr., to the summit of Auyán-tepui in southeastern Bolívar state (Tate 1938a, 1938b). Although these expeditions were primarily zoological and were led by the zoologist George Tate, many botanical collections were gathered and subsequently deposited at NY (Gleason 1929, 1931; Gleason and Killip 1939).

The Caracas-based Phelps Ornithological Collection (Colección Ornitológica Phelps) organized many expeditions to the Venezuelan Guayana. These began in 1937 with a joint expedition with the American Museum of Natural History to Auyán-tepui. This famous group of ornithologists was first led by William H. Phelps, Sr., from 1937 to 1946, and later by William (Billy) H. Phelps, Jr., from 1946 to 1967. They explored many tepuis throughout the states of Bolívar and Amazonas and gathered the most complete collection of birds from the region (Mayr and Phelps 1967, 1971). On some expeditions, Kathleen Deery de Phelps, wife and field companion of William H. Phelps, Jr., made small but valuable botanical collections, which included the first plants from then-unknown mountains such as Cerro Sipapo (1946), Cerro Yaví (1947), and Cerro Parú (1949). These collections were sent to VEN, where Tobías Lasser studied them with the help of Bassett Maguire from the New York Botanical Garden (Lasser and Maguire 1950; Maguire and Deery de Phelps 1951).

In 1944, Julian Steyermark (Figure 2-9), then a curator at the Chicago Natural History Museum (now called the Field Museum of Natural History; F), made his first expeditions to the Venezuelan Guayana. He climbed to the summit of Cerro Duida in August 1944 and to the top of Roraima-tepui in September 1944. He then explored the upper slopes of Ptari-tepui and adjacent Sororopán-tepui and Carrao-tepui in the northern Gran Sabana between September and December 1944 (Steyermark 1966). In 1953, he returned to the region to explore the Chimantá Massif in Bolívar state. Two years later (January to March 1955), he returned to Chimantá on a joint expedition with John Wurdack of the New York Botanical Garden (Maguire 1957; Steyermark and Maguire 1967). During these five expeditions, Steyermark collected more than 6000 botanical specimens, which he deposited at F and partly at VEN (Steyermark 1951, 1952, 1953, 1957). Steyermark worked for the Chicago Natural History Museum until 1958; in 1959 he moved to Venezuela, where he joined the scientific staff at the Instituto Botánico in Caracas. His full itinerary in the Venezuelan Guayana appears in Figure 2-10.

The most important institution in the initial botanical exploration of Pantepui was undoubtedly the New York Botanical Garden. This institution developed the most ambitious and successful exploration programs in the Guayana Shield. Bassett Maguire (Figure 2-11) directed this program, with the collaboration of Richard S. Cowan, John J. Wurdack (Figure 2-12), and others (Maguire 1959, 1964). Their expeditions began in Suriname in 1944 and in Venezuela in 1948, where they visited many tepui summits: Cerro Cuao and Cerro Sipapo (1948–1949), Cerro Duida (1949, 1950), Cerro Huachamacari (1950), Cerro Yapacana (1951), Cerro Parú (1951), Cerro Moriche (1951), Cerro Camani (1951), Cerro Guanay (1951), Cerro Guaiquinima (1951–1952), Ilú-tepui (1952), Churí-tepui (1953), Cerro Yutajé (1953), and Sierra de la Neblina (1953–1954, 1957–1958). More than 10,000 numbers of botanical collections from the different tepui slopes and summits of the Venezuelan Guayana were gathered on this series of expeditions (see Figure 2-13). The specimens were deposited mainly at NY and partly at VEN; they are also the basis of a series of publications titled *The Botany of the Guayana High-*

Figure 2-9. Julian A. Steyermark, holding a branch of *Bonnetia steyermarkii* on Auyán-tepui, 1978. Pressing plants beside him is Victor Carreño, Steyermark's long-time field assistant. Photograph by Roy McDiarmid.

land, directed by Maguire and published between 1953 and 1989 in the *Memoirs of the New York Botanical Garden* (for citations of individual parts and a cumulative index to the taxa treated, see Buck 1990).

Besides these major institutional expeditions, there were several smaller trips during this period, such as the ascent of Roraima-tepui by Albert S. Pinkus in 1938 and 1939. Félix Cardona also made important collections on Auyán-tepui, Cerro Sarisariñama, Cerro Guaiquinima, Macizo del Chimantá, Upuigma-tepui, and other lesser-known mountains of the Venezuelan Guayana such as Cerro Arepuchi, Cerro Murú, and Cerro Arabayén.

Between 1927 and 1967, the Pantepui Province became much better known botanically, and probably even better explored than the surrounding lowlands. During this period, approximately 20,000 botanical specimens were collected on the summits of the Venezuelan tepuis. Many of these collections are still being studied by taxonomists around the world.

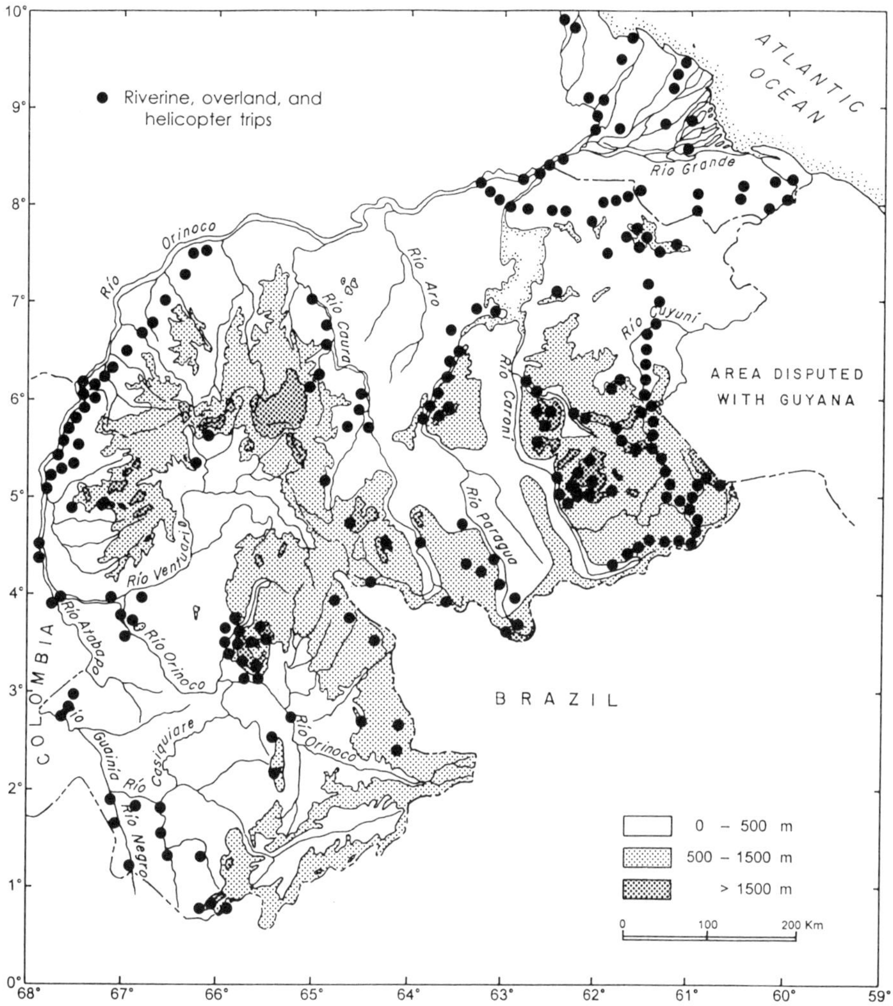

Figure 2-10. Collecting sites of Julian Steyermark in the Venezuelan Guayana, 1944–1986. Over this period, Steyermark made 27,939 different plant collections in the flora area. He visited some areas several times, such as the main road through the Gran Sabana, and he reached the summit of mountains such as Roraima-tepui and Cerro Duida both by foot and later by helicopter.

Modern Explorations in the Venezuelan Guayana (late 1990s)

Starting in the 1950s, the rate of botanical exploration in the Venezuelan Guayana increased dramatically. It became common for expeditions to be organized and run by two or more institutions, and there was a marked increase in the number of participating Venezuelan scientists. Also, many of the trips concentrated on specific scientific disciplines.

The main reason for these drastic changes in exploration style in the region was the use of helicopters. They were first used in remote areas of the

Figure 2-11. Bassett Maguire holding a plant of *Gongylolepis huachamacari* on the summit of Cerro Huachamacari, December 1950. Courtesy of Celia Maguire.

Venezuelan Guayana by Cándido Montoya Lirola on an exploratory trip to the headwaters of the Río Paragua in 1956 (Montoya 1958). The first use of helicopters to support scientific collecting on the tepui summits was in 1967, during the Phelps expedition to Cerro Jaua (Steyermark, personal communication). To justify the high operational costs and complex logistical support needed for helicopter missions in remote areas of the Guayana, multidisciplinary expeditions became more common, starting with one sponsored by the Asociación Venezolana para el Avance de la Ciencia (AsoVAC) in 1969 (Medina 1969). Although helicopters have been most useful to reach remote or otherwise inaccessible tepui summits, many helicopter expeditions were also made to remote lowland sites of the upper Orinoco. This allowed visits to many areas that are far from rivers and often ecologically quite distinct from more accessible localities.

The geopolitical importance and great economic potential of the Venezuelan Guayana has led to a much higher level of interest and activity in the

Figure 2-12. John J. Wurdack. Courtesy of Hunt Institute for Botanical Documentation, Carnegie Mellon University, Pittsburgh, Pennsylvania.

area by the Venezuelan government since the early 1960s, but particularly after 1970. As a result, several Venezuelan institutions began to gather information on the natural resources of the Venezuelan Guayana to develop more specific and detailed criteria for the design and execution of territorial management plans and economic development schemes. The remainder of this chapter outlines the activities of the main Venezuelan institutions that have been involved in botanical activities and related fields in the Guayana.

Ministerio de Agricultura y Cría

Before 1936, the Venezuelan government sponsored botanical investigation through a unit called the Sección Botánica of the Museo Comercial. In 1936, this unit was renamed Servicio Botánico and transferred to the newly created agriculture and livestock ministry under the name Ministerio de Agricultura y Cría (MAC). Henri Pittier was the director of this service until his death in

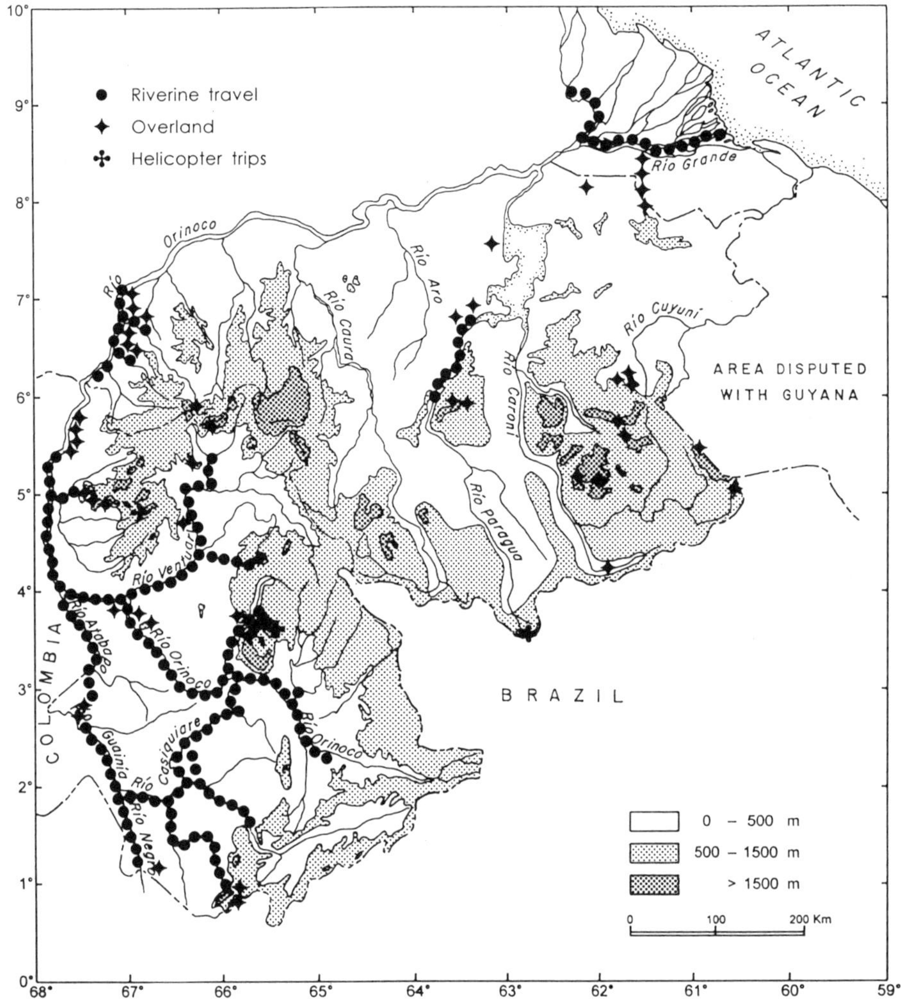

Figure 2-13. Routes and collecting sites of New York Botanical Garden expeditions in the Venezuelan Guayana, 1948–1981, as part of the *Botany of the Guayana Highland* Program.

1950. In 1939, the Servicio Botánico appointed U.S. Forest Service botanist Llewelyn Williams (Figure 2-14) to systematically explore the forest resources of the Venezuelan Guayana. From 1939 to 1944, Williams collected around 4000 numbers of excellent botanical specimens, the first ones from the lower and middle Río Caura (Williams 1940, 1941, 1942). Later he collected in the lowland forests of the Venezuelan Amazon as far south as the Brazilian border at Cocuy (Figure 2-15). He also made several thousand collections of wood samples from the area. Williams deposited his plant collections at F and VEN. He placed the full collection of wood samples at Y, but these were later transferred to MAD. Duplicates of this valuable collection were given to VEN, but parts of the collection were subsequently lost.

Figure 2-14. Llewelyn Williams demonstrating how blowgun darts are dipped in curare. Reprinted with permission from the *Chicago Sun-Times,* © 1992.

During World War II, the U.S. government began a program of botanical research in the New World tropics to locate new sources for natural pharmaceutical and industrial products. In 1941 and 1942, Williams served as a senior field technician in the Rubber Development Corporation, in charge of organizing and supervising the harvest of wild rubber in the upper Orinoco-Casiquiare drainage. As part of the same program, Julian Steyermark traveled to Venezuela in 1943 to search for natural quinine sources for antimalarial drugs (Steyermark and Meyer 1945–1946); besides exploring the Venezuelan Andes, he also collected on several tepuis in the Guayana region. The Servicio Botánico served as the Venezuelan counterpart of this program, and the Herbario Nacional de Venezuela (VEN) benefitted greatly from the intense botanical activities of Williams and Steyermark.

In March and April 1943, Ellsworth P. Killip of the U.S. National Herbarium (US) was invited by Henri Pittier and the MAC to make botanical collections in Venezuela. He botanized near Ciudad Bolívar and along the lower Río Paragua (Salto Uraima, Río Tonoro) to the lower western slopes of Cerro

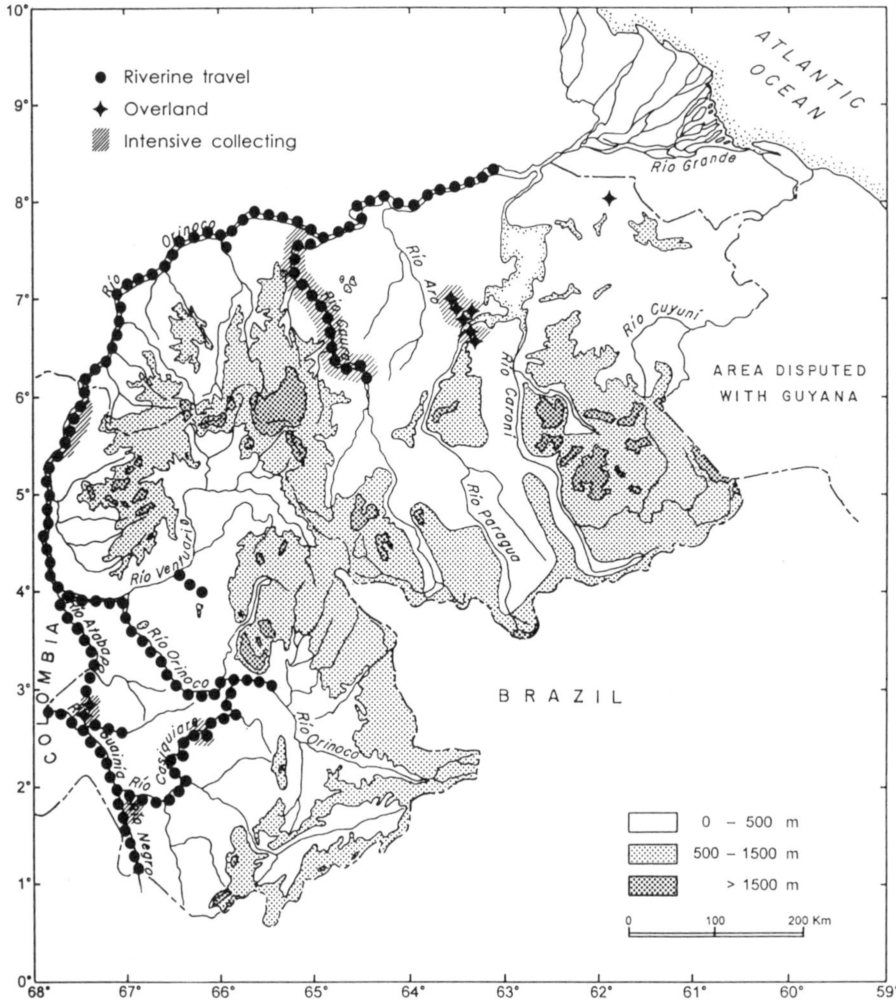

Figure 2-15. Routes and collecting sites of Llewelyn Williams in the Venezuelan Guayana, 1939–1944.

Guaiquinima. He deposited the approximately 500 specimens gathered on this trip at US and VEN.

In 1946, two botanists from MAC, Francisco Tamayo and Tobías Lasser, collected plants in the Gran Sabana (near Santa Elena de Uairén and Kavanayén), near Tumeremo, and between Ciudad Bolívar and Caicara (Tamayo 1961). They made more than 600 collections, adding significantly to the knowledge of the savanna flora of this area. Between 1959 and 1961, the MAC, the Corporación Venezolana de Guayana (CVG), and the Consejo de Bienestar Rural (CBR) carried out a large agroforestry survey of the eastern Venezuelan Guayana called the "Reconocimiento agropecuario forestal del Oriente de la Guayana Venezolana." This was the first comprehensive study of its kind south of the Río Orinoco. The survey made extensive inventories in the forests of the Serranía de Imataca and in the savannas of northeastern Bolívar. Leandro Aristeguieta, Zoraida Luces de Febres, Mauricio Ramia, and Julian A. Steyermark collaborated in the botanical part of this project (CBR 1961).

Between 1964 and 1968, the Programa Forestal de Guayana generated many botanical collections and much forestry information from the Venezuelan Guayana. This joint program between the MAC and the United Nations Food and Agriculture Organization (FAO) centered on the forests of the Serranía de Imataca in northeastern Bolívar and southwestern Delta Amacuro states. The most important botanical collectors involved in this program were Carlos Blanco of the Herbario Nacional de Venezuela (VEN), Luis Marcano-Berti of the Universidad de los Andes (MER), and Bernard Rollet of the FAO.

In 1959, Julian Steyermark moved from the Chicago Natural History Museum to the Instituto Botánico in Caracas (the former Servicio Botánico), where he worked until 1984. He spent much of this time conducting expeditions in the Venezuelan Guayana. Steyermark's number of plant collections in this area, including the ones made in 1944 and 1945, totals 27,939; this represents the most complete single botanical collection ever made in the region and is a scientific heritage of immense value (see Figure 2-10). Steyermark's specimens are deposited mainly at VEN, MO, NY, F, and US.

Two other botanists on the staff of the Instituto Botánico, Getulio Agostini and Gilberto Morillo, also made significant botanical collections in the Venezuelan Guayana. Agostini collected around Canaima and on Uaipántepui with Tetsuo Koyama (then at the New York Botanical Garden), and Morillo collected around Puerto Ayacucho, along the Río Guayapo, near Santa Bárbara del Orinoco, around San Carlos de Río Negro, Cerro Asisa, Cerro Duida, the middle Río Caura, and the Gran Sabana. Their collections are deposited mainly at VEN.

Ministerio de Obras Públicas

In 1969, the Venezuelan Government created a regional development agency called the Comisión para el Desarrollo del Sur (CODESUR), which was part of the public works ministry, Ministerio de Obras Públicas (MOP). Although CODESUR was created primarily to promote development and colonization in what was then designated as the "Región Sur" of Venezuela and included the Territorio Federal Amazonas and Distrito Cedeño of Bolívar state, one of its most important achievements was the completion of an extensive radar survey of the region at a scale of 1:250,000. This survey produced the area's first reliable cartographical record (Aero-Service Corporation 1972). In 1972, CODESUR completed a series of thematic maps, including the first vegetation map of Venezuela's Amazonas territory. The following year it published an atlas of the entire southern region (CODESUR 1973).

CODESUR commissioned several forest inventories along the lower and middle Río Suapure in Bolívar state (OTEHA 1971; Finol 1973), and it provided important logistic support to several expeditions in the region. These included the first explorations of the summits of Cerro Autana by Julian Steyermark in 1971 (Steyermark 1974, 1975), Cerro Avispa by Félix Cardona and G. C. K. Dunsterville in 1972, and Cerro Marahuaka by Stephen Tillett in 1975 (Tillett and Steyermark 1982). CODESUR also assisted Luis Marcano-Berti and Carlos Blanco in 1971 in their survey of the lower Cuao and Sipapo rivers. In 1975 and 1976, Paul Berry was hired by CODESUR to make ethnobotanical collections and detailed surveys of the biology and distribution of the *seje* palms (species of *Oenocarpus* and *Jessenia*) in the northern and central sections of Venezuelan Amazonas (Berry 1976). After 1974, CODESUR changed from a developmental orientation to concentrate on basic research of natural resources in the Venezuelan Amazon, and in 1981, it was officially disbanded.

Ministerio del Ambiente y de los Recursos Naturales Renovables

In April 1977, the first environmental and natural resources ministry in Latin America was created in Venezuela under the name Ministerio del Ambiente y de los Recursos Naturales Renovables (MARNR). Since its creation, it has sponsored and carried out important botanical and ecological activities in the Venezuelan Guayana. The Instituto Botánico was transferred here from MAC and continued its botanical expeditions in the Guayana. The most important trips were made by Julian Steyermark, Gilberto Morillo, and Francisco Delascio.

When the new environmental ministry was created, CODESUR was transferred to it. The organization devoted particular attention to ecological studies in Amazonas state. Between 1977 and 1981, CODESUR and the Consejo Nacional de Investigaciones Científicas y Tecnológicas (CONICIT), the Venezuelan national research council, sponsored a botanical-ecological inventory of the savannas, meadows, and shrublands of the lowlands of Venezuelan Amazonas. As part of this program, Otto Huber organized a dozen expeditions, during which he made about 6000 collections of botanical specimens which were deposited at VEN, NY, and K (Huber 1982, 1985a, 1985b; Huber et al. 1984). A map of Huber's collecting sites in the Venezuelan Guayana appears in Figure 2-16.

The MARNR also made several botanical surveys with personnel from its regional offices (then called Zona 10 in Amazonas and Zona 11 in Bolívar). These included forest inventories in the Reserva Forestal del Sipapo during 1977 and 1978, with the botanical collaboration of Gilberto Morillo (Catalán 1980); in Limón de Parhueña in 1979 (Canales and Catalán 1981); and in the Río Cataniapo valley (all three sites in Amazonas state). In 1977 and 1978, Charles Brewer-Carías, then an expedition commissioner for the MARNR, organized the first helicopter expeditions to Roraima-tepui, Kukenán-tepui, Aprada-tepui, Auyán-tepui, Ptari-tepui, and Chimantá-tepui (Brewer-Carías 1978). The MARNR also made soil and vegetation inventories around Santa Elena de Uairén in the Gran Sabana in 1981 and in the lower Río Paragua in

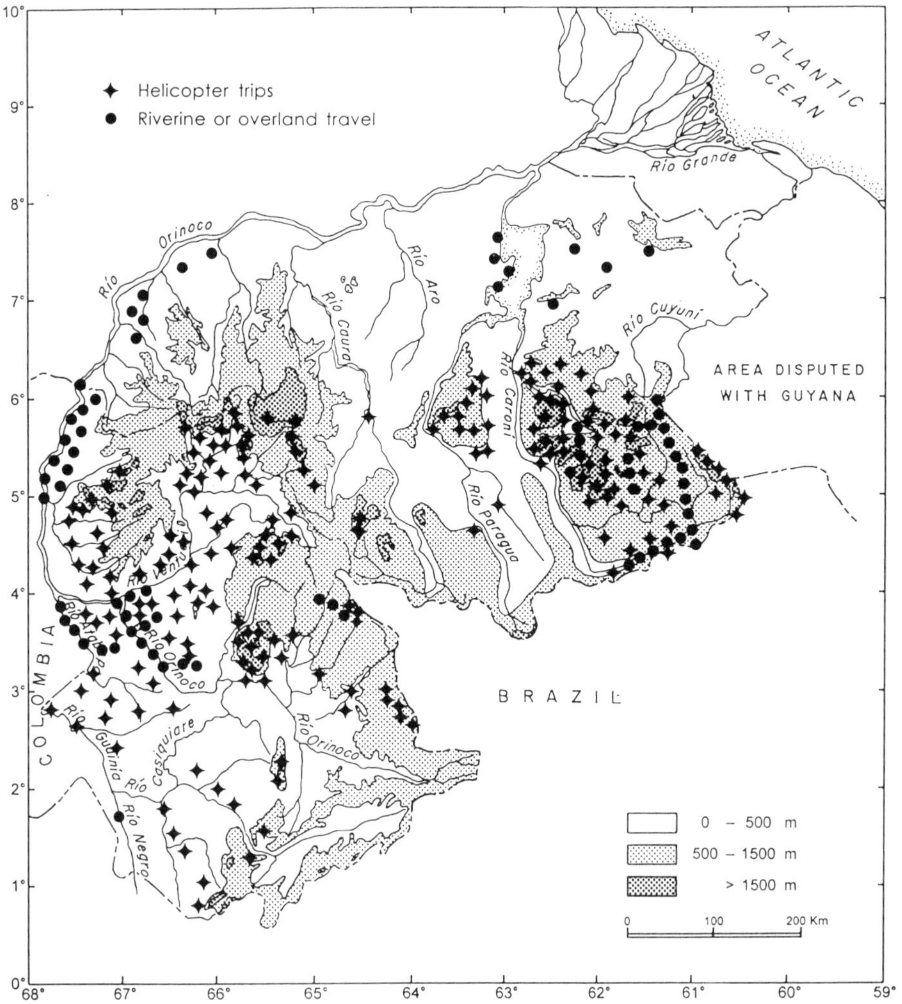

Figure 2-16. Collecting sites of Otto Huber in the Venezuelan Guayana, 1977–1993. Over this period, Huber made 12,817 different plant collections in the flora area.

1982, both in Bolívar state (unpublished reports). A second edition of the atlas by CODESUR (1973) was published six years later (MARNR 1979a), covering Amazonas but not Distrito Cedeño of Bolívar state.

Besides sponsoring field work, the MARNR established a regional herbarium in Puerto Ayacucho (TFAV) in March 1980. Francisco Guánchez was the founder and curator of this herbarium until 1987, to which he added 6000 botanical specimens (partly his own collections from the vicinity of Puerto Ayacucho and other areas of Amazonas, and partly duplicates deposited by visiting botanists). TFAV now contains almost 10,000 specimens from the Venezuelan Amazon and has been assigned to the Centro Amazónico de Investigaciones Ambientales Alejandro Humboldt (CAIAH), a branch of the government agency Servicio Autónomo para el Desarrollo Ambiental del Estado Amazonas (SADA-Amazonas).

Corporación Venezolana de Guayana

The Corporación Venezolana de Guayana (CVG) is a regional development corporation created in 1960 to construct and oversee the largest industrial center of Venezuela. It was not a major sponsor of botanical activities during its first decades of existence, although it did support agroforestry research in northeastern Bolívar with the MAC in 1960 and 1961.

Starting in 1982, however, the CVG began to promote intensive botanical, ecological, and paleoecological studies in the Río Caroní and the Río Paragua basins. This was done mainly through its subsidiary company, Electrificación del Caroní, C.A. (EDELCA), to obtain basic information for the sound management of the large catchment area feeding the hydroelectric power plants at Guri. Between 1982 and 1984, researchers from the school of forestry engineering at the Universidad de los Andes (ULA) made a series of forest inventories for EDELCA. They made most of their botanical collections in forests and shrublands that were later flooded during the third and final phase of Lago Guri. Overall, the ULA produced 16 theses and 12 internal reports on these forests. A final report was made by Pernía (1985), with 33 thematic maps and vegetation maps. Most of the botanical collections were made by Henry Rodríguez and are deposited at MER.

Between 1983 and 1989, EDELCA sponsored two projects in the Venezuelan Guayana. The first one, a botanical-ecological inventory of the savannas of Bolívar state, was directed by Otto Huber and co-funded by CONICIT. Carlos Schubert led the second project, a paleoenvironmental study of the Guayana region (Schubert et al. 1986). From 1983 to 1986, Huber's program was co-sponsored by the New York Botanical Garden, and several botanists from that institution were invited to participate (Ghillean Prance on Auyán-tepui; James Luteyn and John Pipoly on Macizo del Chimantá). About 25 multidisciplinary expeditions were carried out in these programs, covering parts of the lowlands, uplands, and highlands of Bolívar state, as well as most of the mountain summits (Grupo Científico Chimantá 1986; Huber 1986, 1992a). Approximately 8000 botanical collections were deposited at MYF, VEN, NY, K, MO, and US.

From 1985 to 1990, Horst Fölster of the University of Göttingen (Germany) directed a research project on the forest-savanna dynamics in the eastern Gran Sabana region (Fölster 1986; Hernández 1987, 1992; Dezzeo 1990). During this project, Lionel Hernández, Nelda Dezzeo, and collaborators of EDELCA made about 500 collections of botanical specimens mainly in the montane forests of the eastern tepuis such as Roraima-tepui and Kukenán-tepui. Their specimens are deposited at VEN and MYF, and partly at a small reference herbarium established in San Ignacio de Yuruaní by EDELCA in 1987.

In 1989 and 1990, EDELCA made a detailed multidisciplinary inventory of the natural resources in the area surrounding Lago Guri on the lower Río Caroní, including botanical and vegetation studies. Judith Rosales, Elio Briceño, and Gabriel Picón collected approximately 400 numbers there, now deposited at VEN and MYF. Starting in 1988, EDELCA began explorations of the previously unknown Sierra de la Maigualida, under the direction of Otto Huber.

In 1985, another subsidiary of CVG, Técnica Minera (TECMIN), started the ambitious Proyecto Inventario de los Recursos Naturales de la Región Guayana (PIRNRG). This group produced a series of detailed environmental reports and thematic maps on the different natural resources of the region between 1985 and 1991, including both flora and vegetation (Zinck 1986; CVG-TECMIN 1991a, 1991b, 1991c, 1991d, 1991e). By 1993, participants in this program had made more than 10,000 collections of botanical specimens in widely scattered localities in Bolívar, Delta Amacuro, and Amazonas states. Most came from forests and were made by Angel Fernández, Elio Sanoja, Yajaira Fernández, Jorge Velazco, Luz Delgado, Silvino Elcoro, and Euler Marín of the PIRNRG group from 1985 to 1990, with botanists from UNELLEZ's herbarium (PORT) in Portuguesa state (Basil Stergios, Gerardo Aymard, and Nidia Cuello). Many of these collections are particularly valuable as they come from previously unexplored areas, such as the slopes of Cerro Cuao visited by helicopter by Angel Fernández in 1989. Most specimens from this program now reside at PORT, VEN, NY, and MO.

In 1988, CVG-TECMIN started a project to establish rubber (*Hevea* spp.) plantations in Amazonas state, particularly in an area south and east of San Fernando de Atabapo. It sponsored botanists from the Guanare herbarium (Stergios, Aymard, and Cuello) to make a detailed botanical inventory of the nonflooded forests of that region, with about 700 collections deposited at PORT, VEN, and MO (Aymard et al. 1989).

Universities and Research Institutes

The Universidad Central de Venezuela (UCV) with campuses in both Caracas and Maracay) made its first scientific expedition to the Venezuelan Guayana to the summit of Auyán-tepui in 1956. The main participants were botanists and ecologists Ernesto Foldats, Ludwig Schnee, and Volkmar Vareschi. Until 1982, more than 30 professionals from the university's faculties of science, agronomy, and pharmacy made botanical collections in Venezuelan Amazonas (Huber and Wurdack 1984). Aníbal Castillo of the science faculty compiled a florula (unpublished) of the Río Cataniapo region southeast of Puerto Ayacucho. This study was part of the interdisciplinary research program called Proyecto Amazonas, which was begun by the university in the early 1980s. In 1983 and 1984, Nelson Ramírez and students from the science faculty made observations on the floristics, physiognomy, floral biology, and phenology of plant communities of the northern Gran Sabana (Ramírez et al. 1988). Most of the botanical specimens collected by university staff members are deposited at VEN, MY, and MYF.

UCV's herbarium at Caracas (MYF) houses many important ethnobotanical collections made by ethnologists and anthropologists in the native communities of the Venezuelan Guayana, especially from Amazonas. The most significant ethnobotanical collections housed there are the following (in alphabetical order):

- *Brian Boom and Margot Grillo,* and *B. Boom and A. Eisenberg,* 1985–1986 (373 numbers from the lower Maniapure region, among Panare Amerindians; Boom 1990a, 1990b)
- *Marcus Colchester and John Lister,* 1975–1976 (about 1350 numbers from the upper Ventuari and Manapiare basins, among Yekwana, Guahibo, and Piaroa Amerindians)
- *Marcus Colchester,* 1979–1980 (approximately 300 numbers from the upper Ventuari and Erebato basins, among Sanema Amerindians)
- *Emilio Fuentes,* 1978–1980 (approximately 370 numbers from the lower and middle Río Ocamo, among Yanomami Amerindians; Fuentes 1980)
- *Shirley Hoffmann,* 1990 (approximately 200 numbers from the vicinity of San Carlos de Río Negro, among Curripaco Amerindians)
- *Johannes Wilbert,* 1983–1984 (approximately 120 numbers from the Río Guiniquina, Delta Amacuro state, among Warao Amerindians)
- *Stanford Zent,* 1984–1987 (approximately 1000 numbers from the upper Cuao basin, among Piaroa Amerindians)

The Universidad de los Andes (ULA) through its herbaria of forestry science (MER) and pharmacy (MERF) carried out inventories made by Luis Marcano-Berti in the Imataca Forest Reserve, La Paragua Forest Reserve, and in the Orinoco Delta. Between 1954 and 1956, Alessandro L. Bernardi explored the forests of the area of Urimán and Icabarú (Bernardi 1956, 1957). Between 1982 and 1984, the university made several forest and botanical inventories in the lower Río Caroní in areas that were later flooded by the dam at Guri. These and many other botanical collections made by university personnel in the Guayana are mainly deposited in MER and VEN. The collections from Amazonas and the Gran Sabana by Luis Ruiz-Terán are housed at MERF.

The herbarium (PORT) of the Universidad Nacional Experimental de los Llanos Ezequiel Zamora (UNELLEZ, Guanare-Mesa de Cavacas) was founded in 1981 and has been the host institution for a series of botanical expeditions to the Venezuelan Guayana since its inception. The main botanists from this herbarium who have collected in the Venezuela Guayana are Basil Stergios,

Gerardo Aymard, and Nidia Cuello. Together they have made more than 30 trips throughout the flora area, especially to the upper Río Cuyuní (Delta Amacuro and Bolívar states), the area around Lago Guri and different sections of the Río Caura (Bolívar state), and the many tributaries of the Río Casiquiare (Amazonas).

Since 1973, the Instituto Venezolano de Investigaciones Científicas (IVIC), the Max Planck Institut of Germany, and the Institute of Ecology of the University of Georgia in the United States jointly directed an intensive ecological research program on the forest ecosystems (Rio Negro [Amazon] *caatinga*) and shrublands (*banas*) of the upper Río Negro area. This program was most active between 1974 and 1982; it formed part of the UNESCO Man and the Biosphere Program (MAB) but also received additional support from CONICIT. The project was based in San Carlos de Río Negro and was directed by Ernesto Medina and Rafael Herrera of IVIC's ecology center (Klinge et al. 1977). In addition to the ecological, ecophysiological, and pedological research activities, several botanists made collections in the *caatinga* and *bana* vegetation of the area. The largest collections were made by Howard L. Clark, the resident project manager in San Carlos for several years, and Ronald Liesner, a botanist at the Missouri Botanical Garden in St. Louis. They produced a list of flowering plant species of San Carlos de Río Negro, based on their own collections and others made by Eberhard Brünig, Hans Klinge, Ernesto Medina, and Christopher Uhl (Jordan 1989, and unpublished). Most of the botanical specimens of Clark and Liesner are deposited at VEN, MO, and NY. The area between San Carlos de Río Negro and Solano is now one of the botanically best-explored parts of the Venezuelan Amazon.

Independently of the IVIC–MAB San Carlos Project, Gudrun M. Christenson and C. Rose Broome of the U.S. National Cancer Institute, and Francisco Delascio of the Herbario Nacional de Venezuela (VEN) collected about 70 numbers of bulk plant samples near San Carlos in 1981. The bulk specimens went to the Anti-Cancer Screening Program of the Cancer Chemotherapy National Service Center at Beltsville, Maryland, and the vouchers were deposited at VEN and US.

Between 1985 and 1988, IVIC sponsored another botanical research program, the Proyecto Mapire, just north of the limits of the Venezuelan Guayana in Anzoátegui state. This study concentrated on the plant ecology and vegetation dynamics of riparian forests near Mapire, located on the northern bank of the middle Río Orinoco. The principal botanical collections of the riparian flora were made by Judith Rosales and are deposited in MYF and VEN.

Venezuelan Non-Governmental Organizations

Since its creation in 1940, the Sociedad de Ciencias Naturales La Salle has led several explorations to the Orinoco Delta, eastern and southeastern Bolívar state, and Venezuelan Amazonas. In 1980, the society's natural history museum began an ecological and phytosociological study of the riparian forests of the lower Orinoco near Mamo, on the northern bank of the Río Orinoco (Colonnello et al. 1986, 1988). Other trips yielded important botanical results from places such as San Juan de Manapiare in 1958 (Jam Lander 1958); Laguna Asisa on Cerro Asisa (or Cerro Parú) in 1973 (Hoyos 1973); and the area of El Abismo in the Sierra Pakaraima in 1984 (Colonnello 1986). The botanical specimens are deposited at CAR and VEN.

In April and May of 1968, a *Geographical Magazine (London)* expedition traveled along the Negro, Casiquiare, and Orinoco rivers from Manaus (Brazil) to Port of Spain (Trinidad) by hovercraft. Among the participating scientists were Michael J. Eden and Ernesto Medina, who made botanical collections between San Carlos de Río Negro and Puerto Ayacucho (Eden 1968; Medina 1971).

In January and February of 1969, the Asociación Venezolana para el Avance de la Ciencia (AsoVAC) and the Universidad Central de Venezuela (UCV) jointly sponsored the first helicopter expedition in Amazonas state. Starting from the base camp in La Esmeralda, Mario Fariñas, Ernesto Medina, and Justiniano Velásquez made general botanical collections in the upper Orinoco and on the summit of Cerro Duida (Medina 1969). Their specimens were deposited at VEN.

In February and March of 1974, the Sociedad Venezolana de Ciencias Naturales (SVCN) organized a large multidisciplinary expedition directed by Charles Brewer-Carías, visiting the tepuis of Sarisariñama, Jaua, and Guanacoco in southwestern Bolívar (Orejas and Quesada 1976). Julian Steyermark and his field assistant Victor Carreño Espinoza were the principal botanists on these expeditions, with Galfried C. K. and Ellinor Dunsterville the orchid specialists. Steyermark's main set of collections is at VEN, with duplicates at NY, MO, and US (Steyermark and Brewer-Carías 1976).

Fundación Terramar, a private foundation begun by Fabián and Armando Michelangeli in 1983, has sponsored a series of major trips to tepui summits in the Venezuelan Guayana. The most important trips were to Cerro Marahuaka, Cerro Huachamacari, and Cerro Duida in 1983 and 1984 (*Corpovoz* 1984), again in 1985 and 1988 (Delascio and Steyermark 1989; Michelangeli 1989); Cerro Aratitiyope in 1984; the Los Testigos Massif in 1986 (Steyermark 1986a) and 1987 (Holst 1987); the Río Siapa and Cerro Aracamuni in 1987 (Steyermark and Holst 1989); the eastern tepuis (Roraima, Kukenán, Ilú) in 1989; and Auyán-tepui in 1990. Participating botanists in these expeditions were Julian A. Steyermark, Ronald Liesner, and Bruce Holst (Missouri Botanical Garden, St. Louis); Roy Halling (New York Botanical Garden), Francisco Delascio and Germán Carnevali (Herbario Nacional de Venezuela), and Paul Berry and Teresa Iturriaga (Universidad Simón Bolívar, Caracas). Many of the botanical results were published in Steyermark's series of articles on the flora of the Venezuelan Guayana, published in the *Annals of the Missouri Botanical Garden* between 1984 and 1989. Duplicates of these collections are deposited at VEN, MO, and NY.

The Fundación para el Desarrollo de las Ciencias Físicas, Matemáticas y

Naturales (FUDECI) is a foundation affiliated with the Venezuelan National Academy of Sciences. It has sponsored three large, multidisciplinary expeditions to remote areas of the Venezuelan Guayana. The largest one was actually a series of eight expeditions to the Sierra de la Neblina between 1983 and 1987; the trips were coordinated by Charles Brewer-Carías and enabled 144 scientists to participate (Brewer-Carías 1988). This effort intensively sampled the flora and fauna of the southwestern base (Río Mawarinuma) of the Sierra de la Neblina; it also sampled 12 localities ranging in elevation between 770 and 2400 m over various sections of the Neblina massif. There were 27 botanists who made a total of 13,986 collections (Brewer-Carías 1988). Most of the participating botanists belonged to institutions from the United States, such as the Missouri Botanical Garden (St. Louis), the New York Botanical Garden, The U.S. National Herbarium (Washington, D.C.), the Field Museum of Natural History (Chicago), the University of Michigan (Ann Arbor), and Vanderbilt University (Nashville, Tennessee). The only Venezuelan botanists who participated were Francisco Delascio and Teresa Iturriaga of the Herbario Nacional de Venezuela (Caracas). The full set of botanical specimens is deposited at VEN, with duplicates distributed by the collector's home institution or by VEN.

FUDECI's second expedition was to the Serranía Tapirapecó, along the border of Brazil and Venezuela in southeastern Amazonas state, with two trips in 1988 and 1989 directed by Eugenio de Bellard. Overall, 115 scientists from Venezuela, the United States, and Europe participated (FUDECI 1990). The main botanical collections were made by Libia Laskowski and Ivón Ramírez for the Herbario Nacional de Venezuela and by Andrew Henderson and Hans Beck for the New York Botanical Garden; Henderson also made collections of live plant material for the biological experiment Biosphere II.

The third expedition of FUDECI was to Cerro Guaiquinima in 1990, led again by Eugenio de Bellard. The main botanical collections were made by Francisco Delascio and Rafael Ortíz of the Herbario Nacional de Venezuela and Brian Boom of the New York Botanical Garden.

Although other groups and individuals have also contributed to our botanical knowledge of the Venezuelan Guayana, coverage in this volume is restricted to the main organizations that have worked in the area.

Summary

After nearly 250 years of botanical exploration in the Venezuelan Guayana, there is a wealth of information available about the plants that inhabit this region. Consolidating this information has allowed us to produce the current flora, which contains nearly 9400 species of vascular plants. Nonetheless, there are still large areas of the Venezuelan Guayana that are poorly known floristically. Most in need of basic floristic study are the many forest types found on the lower, middle, and upper slopes of the mountains. Lowland forests are also extremely diverse and very poorly sampled.

Although nearly all the tepuis and many other areas of the Venezuelan Guayana have been designated as protected areas (see Chapter 5), very few of these areas have any kind of management plan, due mainly to the lack of sufficient base-line information. Consequently, the production of sound vegetation and ecological maps is one of the most urgent needs to allow the zoning of the natural environments in each protected area. Because of the increasing pressure exerted on almost all areas of the Venezuelan Guayana by human activities such as mining, logging, road building, shifting cultivation, and tourism, future botanical research activities need to be broadened beyond general collecting to include (1) the study of the impacts of these human activities on the natural vegetation and (2) more widely applicable ecological research.

Table 2-1. Relative level of botanical exploration of tepuis and other lowland and upland areas of the Venezuelan Guayana.

Tepui or other area	Relative level of exploration[1]	Number of expeditions	Year of first botanical exploration	Main plant collectors (in chronological order)
Tepuis and Adjacent Uplands				
Uei-tepui	++	2	1954	Maguire, Huber
Roraima (slopes)	+++	> 5	1838	Schomburgk, Appun, others
Roraima (summit)	+++	> 20	1884	Im Thurn, McConnell & Quelch, Ule, Tate, Steyermark, Delascio
Kukenán (summit)	++	4	1977	Delascio, Huber, Liesner
Yuruaní (summit)	++	4	1983	Hernández, Huber, Liesner
Karaurín (summit)	+	1	1987	Huber
Ilú (slopes)	++	1	1952	Maguire
Ilú, Tramén (summit)	++	4	1984	Huber, Liesner, Delascio
Ptari (slopes)	+++	> 5	1944	Steyermark, Maguire, Lasser
Ptari (summit)	+++	4	1978	Steyermark, Huber, Holst, Weitzman & Kral
Los Testigos (slopes and summits)[2]	++	3	1986	Huber, Steyermark, Holst, Liesner
Auyán-tepui (slopes and summit)	+++	> 20	1937	Cardona, Tate, Vareschi, Steyermark, Foldats, Huber, Prance, Delascio
Uaipán (slopes and summit)	+	2	1946	Cardona, Koyama & Agostini
Aprada (slopes)	+	2	1946	Cardona, Bernardi
Aprada (summit)	++	2	1978	Steyermark, Huber
Chimantá (slopes and summits)[3]	+++	> 10	1946	Cardona, Wurdack, Steyermark, Bernardi, Huber, Luteyn, Ahti, Pipoly, Pruski, Cleef
Angasima (summit)	++	2	1986	Huber
Upuigma (slopes)	+	1	1947	Cardona
Upuigma (summit)	+	1	1986	Huber
Guaiquinima (slopes)	++	6	1943	Cardona, Maguire, Huber, Stergios, Aymard, Delascio
Guaiquinima (summit)	++	> 10	1943	Cardona, Maguire, Steyermark, Huber, Delascio, Boom
Ichún	+	1	1961	Steyermark
Guanacoco	+	2	1974	Steyermark, Stergios
Marutaní (Pia-Zoi)	++	2	1961	Steyermark, Maguire
Jaua (slopes and summit)	++	> 5	1937	Cardona, Steyermark, Huber
Sarisariñama	++	3	1942	Cardona, Dunsterville, Steyermark
Maigualida (north slopes)	+	1	1989	Aymard, Fernández
Maigualida (summit)	++	4	1988	Huber, Berry
Yaví	+	2	1947	Phelps, Huber
Yutajé and Coro Coro	++	4	1953	Maguire, Liesner & Holst, Huber
Guanay	+	2	1951	Maguire, Huber
Camani	+	1	1951	Maguire

(continued)

Table 2-1. Continued.

Tepui or other area	Relative level of exploration[1]	Number of expeditions	Year of first botanical exploration	Main plant collectors (in chronological order)
Cuao-Sipapo (slopes and summits)	++	4	1946	Phelps, Maguire, Steyermark & Liesner, Fernández, Huber
Autana (summit)	+++	1	1971	Steyermark
Moriche	++	1	1951	Maguire
Yapacana (slopes and summit)	+	3	1951	Maguire et al., Steyermark & Bunting, Givnish
Parú (summit)	++	6	1949	Phelps, Cowan & Wurdack, Hoyos & Morillo, Huber, Berry
Huachamacari	+++	> 5	1950	Maguire, Cowan & Wurdack, Steyermark, Guariglia, Delascio, Fernández
Duida (slopes and summit)	+++	> 10	1928	Tate, Steyermark, Maguire, Cowan & Wurdack, Fariñas et al., Tillett et al., Guariglia, Delascio, Fuertes & Cardiel, Huber
Marahuaka (summit)	+++	> 10	1975	Tillett et al., Steyermark et al., Guariglia, Holst, Delascio, Huber
Parima	++	± 10	1839	Schomburgk, Cardona, Vareschi, Steyermark, Colchester, Huber, Guánchez
Vinilla	++	4	1981	Huber, Steyermark & Berry
Aratitiyope	++	1	1984	Steyermark et al.
Tapirapecó	+	2	1988	Laskowski & Ramírez, Henderson et al.
Unturán	+	1	1989	Henderson et al.
Aracamuni (slopes)	++	3	1959	Wurdack & Adderley, Huber, Liesner
Aracamuni (summit)	++	1	1987	Liesner, Delascio, Carnevali
Avispa (summit)	+	1	1972	Cardona, Dunsterville
Neblina (slopes and summits)	++	> 10	1953	Maguire et al., Ewel, Steyermark, Luteyn, Liesner, Gentry & Stein, Davidse, Kral, Plowman, Funk, Buck, Nee, Boom, Renner, and others
Lowland Areas				
Upper Orinoco (above Río Ocamo)	++	5	1887	Chaffanjon, Croizat, Cruxent, Guánchez, Berry
Casiquiare basin (including southern tributaries)	++	> 10	1800	Humboldt, Schomburgk, Spruce, Holt, Williams, Maguire et al., Wurdack & Adderley, Vareschi, Stergios, Kubitzki, Aymard
San Carlos de Río Negro	+++	> 20	1800	Humboldt, Spruce, Holt, Maguire et al., Williams, Schultes, Steyermark, Clark, Liesner, and others

Tepui or other area	Relative level of exploration[1]	Number of expeditions	Year of first botanical exploration	Main plant collectors (in chronological order)
Piedra Cocuy	++	3	1853	Spruce, Maguire, Schultes?
Atabapo/Guainía basin	++	> 10	1800	Humboldt, Spruce, Williams, Maguire et al., Wurdack & Adderley, Vareschi & Foldats, Steyermark & Bunting, Huber, Tillett, Davidse, Guánchez
Cariche	++	1	1959	Wurdack & Adderley
Yapacana (meadows)	+++	> 10	1931	Holt, Maguire et al., Steyermark & Bunting, Huber, Kral
Upper Ventuari	+	3–5	1838	Schomburgk, Cardona, Huber
Manapiare/Parucito basin	+	± 10	1951	Maguire, Matos, Berry, Colchester & Lister, Huber
Atures	+++	> 50	1800	Humboldt, many others
Middle Orinoco (between Atures and mouth of Río Apure)	++	> 30	1800	Humboldt, Chaffanjon, Holt, Williams, Wurdack & Monachino, Marcano-Berti, Stergios, Steyermark & Holst, Guánchez & Huber, Boom
Lower and middle Caura	++	± 6	1885	Chaffanjon, Passarge, Williams, Liesner & Morillo, Fernández, Dezzeo
Paragua basin	++	± 6	1930	Cardona, Williams, Killip, Steyermark, Blanco, Stergios, Aymard
Lower and middle Caroní (including Río Icabarú)	+++	> 10	1754	Loefling, Cardona, Maguire, Bernardi, Wurdack & Guppy, Liesner, Delascio, Huber, ULA, Fernández, Rosales & Briceño, Picón
Gran Sabana	+++	> 50	1838	Schomburgk, Ule, Steyermark, Tamayo, Lasser, Maguire, Davidse, Huber, Kral, Fernández, Liesner, Ramírez, and others
Serranía de Imataca, Río Cuyuní	+++	> 10	1955	Wurdack, Steyermark, Bernardi, Blanco, Marcano-Berti, Rollet, Delascio
Orinoco Delta	++	> 10?	1896	Rusby & Squires, Wurdack, Steyermark, Breteler, Marcano-Berti, Davidse & González, Delascio, Fernández, Liesner

[1]+ very little, ++ moderately, +++ well explored.
[2]Includes the separate tepuis Aparamán, Kamarkawarai, Murisipán, and Tereke-yurén.
[3]Includes the separate tepui summits of Abacapá, Acopán, Agparamán, Amurí, Apacará, Chimantá, Churí, Murey (Eruoda), Sarvén, Tirepón, and Toronó.

CHAPTER 3

Vegetation

Otto Huber

The great floristic richness of the Venezuelan Guayana was quickly recognized by early scientific explorers of the 18th century such as Humboldt and Bonpland, the Schomburgk brothers, and Spruce. On into the 20th century, numerous botanical expeditions to Guayana were organized mainly to document the high diversity and endemism of plant species, but very little attention was ever devoted to the characteristics of the plant communities in which these taxa occurred. This fact is exemplified by *The Botany of the Guayana Highland* series published in the *Memoirs of the New York Botanical Garden* (see Buck 1990 for full citations), in which there is little discussion of vegetation types in more than 3500 published pages.

A series of explorations to the Venezuelan Guayana by Otto Huber during the 1970s and 1980s focused on vegetation mapping and analysis. These trips have shown that the area's very high floristic diversity is paralleled by a similarly high ecological diversity, both in the physiognomical variations of the plant communities and their geographical distribution patterns.

The vegetation map of the Venezuelan Guayana accompanying this volume illustrates both the geographical distribution and diversity of vegetation types in the flora area. Given the small scale of the map (1:2,000,000) and our incomplete knowledge of the Venezuelan Guayana, the vegetation categories treated are, by necessity, relatively broad. Even with these restraints, developments in remote sensing techniques such as side-looking radar (SLAR) and satellite imagery, coupled with the use of helicopters for field surveys, have improved data collection and made it possible to apply more precise criteria for vegetation classification, especially for previously inaccessible parts of the flora area. This chapter supplements the map by providing explicit definitions and an ecological interpretation of each vegetation unit.

Previous Vegetation Studies and Maps

The first vegetation descriptions of the Venezuelan Guayana were of limited scope and appeared in the travel accounts of early explorers such as Humboldt (1816–1831, 1818–1829, 1956), Schomburgk (1847–1848), Spruce (1908), and Chaffanjon (1889). Henri Pittier (1920) published the first vegetation map of Venezuela, called *Mapa Ecológico de Venezuela.* At a scale of approximately 1:2,000,000, this map recognized just five natural vegetation types in the Venezuelan Guayana: savannas (*sabanas*), xerophilous forests (*selvas xerófilas*) in the western Orinoco Delta region, seasonal forests (*selvas veraneras*), rain forests (*selvas pluviales*), and temperate forests (*selvas templadas*). The extent and distribution of these vegetation types on the map were clearly schematic and tentative, since there was very little information available at the time.

Both Tamayo (1958) and Hueck (1960) later published vegetation maps at a scale of 1:2,000,000, with more detailed units and better explanations than those of Pittier. However, since neither author had any field experience in the Venezuelan Guayana nor access to aerial photographs, their vegetation units in the flora area had to be inferred by second-hand information.

The first set of reliable, cloud-free images of the entire Venezuelan Guayana was produced in the early 1970s from a detailed airborne radar survey commissioned by the Venezuelan development agency CODESUR. The Aero-Service Corporation (1972), a subsidiary of the Goodyear Company, used north- to south-oriented bands of radar images to assemble accurate topographic, hydrographic, geologic, and geomorphological maps at a scale of 1:250,000. At that same scale, they also made a vegetation map consisting of 19 sheets. Five main vegetation categories and seven subcategories were designated, as follows:

1. Lowland forests
   - Riparian or periodically flooded forest (*várzea*)
   - Moist or permanently flooded forest (*igapó*)
2. Upland forests
   - Riparian upland forest complex
   - Montane upland forest complex
3. Summit and tepui vegetation
   - Montane savannas
4. Savannas
   - Dry savannas
   - Humid savannas
5. Aquatic and swamp vegetation

This set of maps was significant because it was the first time that the boundaries of the main vegetation formations in the Venezuelan Guayana were precisely mapped. Still, the vegetation units were not precisely defined, which could lead to misinterpretations of the actual vegetation cover. The lack of adequate field control and floristic reconnaissance in the different vegetation types also limited the reliability of these maps.

Several other vegetation maps were published since the Aero-Service map was done. Both editions of the atlas published by CODESUR (1973; MARNR 1979a) covered Amazonas state and contained slightly modified versions of Tamayo's (1958) phytogeographic map and Ewel and Madriz' (1968) *Zonas de Vida de Venezuela* (life zone map), both at a scale of approximately 1:4,000,000. Because they were based on rather theoretical classifications, however, these maps still did not reflect well the actual distribution of the major vegetation types in the area. In 1982, the Venezuelan environmental ministry (MARNR) published its first vegetation map of Venezuela at a scale of 1:250,000, with 75 blueprint sheets. These were based on the interpretation of SLAR inventories, which were by then completed for the entire country. The maps employed a more accurate classification of vegetation types that combined strictly vegetational data with climatic, geomorphologic, and land use information. For the 41 sheets that cover the Venezuelan Guayana, this map series is the most complete and authoritative source of information available for the region today; however, some inaccurate definitions of vegetation types and a conspicuous lack of field control and floristic data significantly limit its usefulness.

A detailed inventory of natural resources of the Venezuelan Guayana (known as PIRNRG) began in 1985 (see Zinck 1986). This has been carried out by an interdisciplinary team from CVG-TECMIN, an affiliate of the regional governmental development agency, Corporación Venezolana de Guayana (CVG). The group developed a detailed ecological classification scheme based on geomorphological reference units (García 1987), and they have begun to produce extensive descriptions and thematic maps of the Venezuelan Guayana at a working scale of 1:250,000. For each of the main vegetation types that can be recognized by radar and satellite imagery, a team makes floristic surveys and later prepares the vegetation maps. Until early 1994, the project had produced a limited publication of sixteen detailed vegetation maps of the southeastern section of Bolívar state, each at a scale of 1:500,000 (CVG-TECMIN 1987, 1989, 1991a, 1991b, 1991c, 1991d, 1991e).

Huber (1982) produced a small colored vegetation map of Amazonas state at a rough scale of 1:3,000,000. An expanded colored vegetation map of the entire country was later made at a scale of 1:2,000,000 by Huber and Alarcón (1988). The latter introduced a novel ecological classification of the main vegetation types, including more detailed floristic criteria. This same ecological classification has been adapted and slightly modified for the vegetation map of the Venezuelan Guayana accompanying this volume.

System of Vegetation Classification

This section explains the methods and criteria used to delimit and describe the different vegetation units recognized in this chapter and on the accompanying vegetation map. The following paragraphs discuss the individual plant formations and specific classification problems.

Methods and Criteria for Vegetation Classification

Because of the very uneven degree of exploration of the Venezuelan Guayana, especially in the extensive forested areas, it is not yet possible to make a detailed classification and description of all vegetation types found in the region. However, a general outline that accounts for the most distinctive features of the different vegetation types can be attempted and is the main objective of this chapter.

In broad terms, any natural (nonanthropogenic), terrestrial *vegetation type* found in the Venezuelan Guayana belongs to one of the following four major *plant formations,* which are indicated on the accompanying vegetation map by different color groups:

1. Forest formation .... Various shades of green
2. Scrub formation .... Light and dark brown
3. Herbaceous formation .... Light and dark orange
4. Pioneer (mainly saxicolous) formation .... Black

As seen by the preponderance of green-colored units on the vegetation map, the forest formation covers by far the largest area of the Venezuelan Guayana. The herbaceous formation has the next largest coverage, followed by the scrub and pioneer formations.

To make clear distinctions between different vegetation types in the Venezuelan Guayana, thereby allowing comparison with those of more distant regions, an explicit set of criteria and methods are established in the following paragraphs. Natural vegetation types are generally defined by a combination of intrinsic (or biotic) and extrinsic (or abiotic) parameters that are applied to their predominant plant communities. Although biotic and physical parameters are clearly interconnected, their distinction is useful in defining broad vegetation types in large areas such as the Venezuelan Guayana.

The intrinsic or biotic parameters include physiognomy, phenology, and floristic composition. *Physiognomy* (appearance) is defined by the predominant growth form, sometimes indicating its size and density. The four categories of growth forms recognized are *trees, shrubs, herbs,* and *thallophytes* (mostly algae, fungi, and lichens). The corresponding physiognomic plant communities are *forest, shrubland* (or scrub), *meadow,* and *pioneer vegetation.* Although palms are not true woody plants, they are treated as trees when they consist of a tall solitary trunk and a clearly recognizable crown. In the case of small multistemmed palms, their growth form corresponds better to shrubs, since they both occupy similar niches in the ecosystem. Climbers are considered trees or shrubs in the case of (woody) lianas, or herbs in the case of vines. However, since climbers and epiphytes require the presence of other supporting plants, they play only a subordinate role in the classification of vegetation types.

Phenology refers to the periodic succession of vegetative and reproductive phases in the plant community, especially regarding the foliage (deciduousness), and *floristic composition* includes the degree of species diversity and endemicity. Ideally, data on floristic composition should be available in the form of ranked species lists from phytosociological or forestry surveys accompanied by botanical collections. However, there have been few such studies in the flora area, especially in vegetation types such as forests and palm swamps. For extensive areas of the Venezuelan Guayana, the only floristic information available are observations from an airplane or helicopter, often consisting of the recognition of a few dominant taxa at the genus or even family level.

The extrinsic or abiotic parameters include climate, substrate, hydrology and orographic position. *Climate* refers especially to temperature, rainfall, evapotranspiration, and seasonality (for terminology, see Chapter 1). *Substrate* includes underlying parent rock material, geomorphologic position, and soils (especially nutrient contents, acidity, and texture; see Chapter 1). *Hydrology* refers especially to the intensity and temporal oscillations of the soil moisture content or water table, and the inundation regime, if present. *Orographic position* applies in the case of montane vegetation types (see below). In some cases fire represents another important physical parameter, although in the Guayana, it is usually associated with human interference.

There are cases when it is difficult to classify particular vegetation types. Considerable field experience may be necessary to recognize and correctly evaluate the ecological complexity of a particular vegetation type. For example, an herbaceous field often contains many shrubs whose total biomass in the ecosystem may well approach or even exceed that of the underlying herbaceous layer. In such a case, it is not immediately clear if the vegetation should be classified as a meadow or as a shrubland. If the herbaceous layer is continuous, probably making a major energetic input into the global ecosystem, it should be classified as a shrubby meadow. On the other hand, if the herbaceous layer is discontinuous and nonhomogeneous, the vegetation type should be classified as a shrubland. When the predominant growth form is intermediate between tall shrubs and low trees, the floristic composition is usually the best way to distinguish one vegetation type from another.

Altitudinal (or Orographic) Zonation of Vegetation

The general physiographic division of the Venezuelan Guayana into lowlands, uplands, and highlands (see Chapter 1) is correlated with different vegetation types that result from a clear altitudinal zonation in climate (particularly in temperature). This section deals specifically with montane vegetation types and is designed to provide a level of distinction that is finer than merely referring to upland or highland vegetation types. In the past, the altitudinal zonation on the tepuis has been misinterpreted because of the presence of sheer vertical walls up to 1000 m or more high, and many botanists concluded that there was an insular or azonal biota living on the isolated summits. After many comparative field studies, however, the mountains of the Venezuelan Guayana can be shown to have a well-developed altitudinal sequence of life zones similar to that found in the Andes or in the coastal cordilleras; they lack only the much higher elevation glacial and periglacial zones of the Andes. Sometimes, as in Mayr and Phelps (1967), the different altitudinal life zones were called "tropical," "subtropical," and "temperate" zones, but the misapplication of these terms has already been pointed out in Chapter 1. Instead, the following classification of montane life zones in the Venezuelan Guayana is proposed (see Figure 3-1).

Piedmont	
Premontane	
Basimontane	50–400 m elevation
Slope	
Lower montane (submontane)	400–800 m elevation
Montane	800–1500 m elevation
Upper montane	1500–2000 m elevation
Summit	
High-tepui	2000–3000 m elevation

A more complete definition of the altitudinal vegetation zones in the Venezuelan Guayana is given below. It should be emphasized that there are no rigid upper and lower limits for the different altitudinal zones in this area,

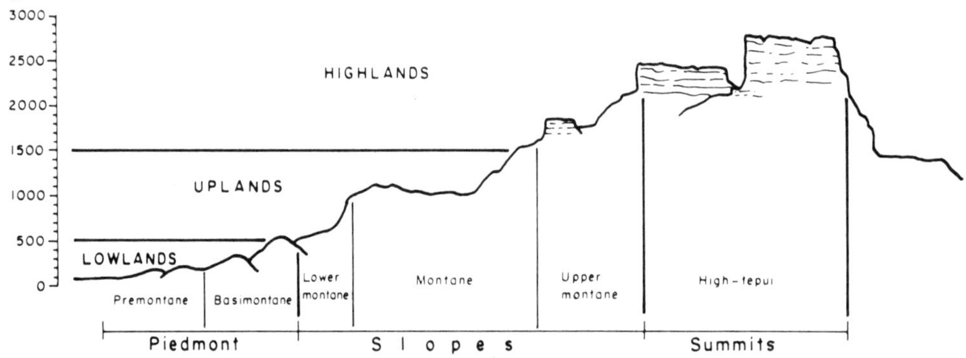

Figure 3-1. General physiography (lowlands, uplands, and highlands) of the Venezuelan Guayana, including a more detailed altitudinal sequence of life zones in the mountains of this region. Elevations above sea level.

since they vary considerably in relation to the overall elevation and mass of each mountain system. For this reason, the values given above for each zone are approximate.

Piedmont. The term *basimontane* indicates the lowermost vegetation zone along the base of a mountain, in the macrothermic belt. In this zone, most of the adjacent lowland floristic elements are still present, but grow under different ecological conditions due to the inclined terrain receiving significant inputs of ground water and nutrients from the upper slopes. This zone generally ranges between 50 and 400 m elevation in the Venezuelan Guayana.

An important distinction should be made for the piedmont category, which consists of the premontane and the basimontane zones (Figure 3-1). As defined here, the *pre*montane zone refers exclusively to the terrain that lies immediately *before* the mountain, thus extending from the base of a mountain slope into the adjacent level or hilly lowland unit. This transition zone is influenced by a series of external factors from the adjacent mountain slopes that significantly affect its ecology, such as subsoil water flow, nutrient runoff, and wind patterns. As used here, the term *premontane* differs fundamentally from Holdridge's concept of a premontane life zone (Holdridge 1979), which is located on the lower mountain slope and actually belongs to the montane environment.

Slope. The lower montane vegetation zone, in which the floristic elements of the adjacent lower belt are gradually replaced by typically montane elements, is a transitional vegetation zone in the submesothermic belt, usually between 400 and 800 m elevation. In this zone, there are often semideciduous vegetation types with large, membranous leaves and drip tips (Vareschi 1992a, pp. 90–92), especially in areas lying within a rain shadow.

The montane zone occurs mainly in the mesothermic belt and is characterized by a distinct montane flora and by physiological adaptations such as evergreen, coriaceous leaves, abundant epiphytes, and trees with buttresses. In the Venezuelan Guayana, this zone usually occurs between 800 and 1500 m elevation.

The upper montane zone is located entirely in the mesothermic belt and is characterized by the predominance of cloud forests and other perhumid vegetation types; in the Venezuelan Guayana, the upper montane zone is best developed near and along the base of vertical walls of tepuis between 1500 and 2000 m elevation. The cloud forests are characterized by short trees with gnarled stems and branches, flattened crowns, abundant vascular and nonvascular epiphytes, and small, thick, coriaceous, dark green leaves.

Summit. The high-tepui vegetation zone includes the typical tepui summit ecosystems that occur in the upper mesothermic and submicrothermic belts, mainly between 2000 and 3000 m elevation. This includes low-growing tepui forests, tepui scrub, and high mountain meadows and grasslands. Sclerophyllous, mainly microphyll leaves prevail among the trees and shrubs, while thick, coriaceous leaves characterize most herbs.

Approximately 80 percent of the tepuis of the Venezuelan Guayana are sufficiently high (above 2000 m elevation) to have the full range of altitudinal zonation described above, including the high-tepui ecosystems on their summits. In parts of the Eastern tepui chain and on Ptari-tepui, some of the lower elevation zones are missing. It is important to point out that if comparative evaluations are made between high-tepui ecosystems of different tepuis, care must be taken to compare only those belonging to the same elevational belt. Authors such as Steyermark and Dunsterville (1980) have asserted that there are "lowland taxa" on the "tepui summits" of Cerro Autana and Cerro Guaiquinima, but the parts of the summits they considered are only 1300 m and 760 m elevation respectively; these clearly do not correspond to the ecologically defined high-tepui belt of the Venezuelan Guayana.

As a major explorer of many tepui summits, Maguire (1970) concluded that high-tepui ecosystems had developed in isolation from the lowlands, separated mainly by their steep upper walls. This view has been echoed by numerous other visitors to the tepuis, who have described them as "islands in time" (see Brewer-Carías 1978; George 1988). In many tepuis, however, especially on the larger massifs of the Venezuelan Guayana, the summits are not completely isolated from the surrounding lower vegetation belts by vertical walls. Rather, there are usually large valleys and ridges that make a continuous connection between the lowlands, uplands, and highlands of a tepui massif. Total physical isolation of tepui summits occurs only in a few towerlike mountains, such as Roraima-tepui, Ilú-tepui, and the Los Testigos massif, where quite impoverished plant communities prevail. Rather than physical isolation from the lowlands, extremely adverse environmental conditions such as lack of soil, strong winds, and high radiation are the main cause of these depauperate communities.

Another refutation of the isolation theory of high-tepui ecosystems came from the first ecological explorations in the Sierra de Maigualida massif in the late 1980s. This large mountain range, which reaches 2400 m elevation and is situated near the center of the Venezuelan Guayana, has continuously ascending slopes instead of vertical walls because of its granitic, instead of sandstone, geology. Still, it has an altitudinal zonation very similar to that of the sandstone tepuis, with meadows and other nonforest high-tepui ecosystems on the rounded summit peaks. In fact, Huber (1988a) has shown that all high-tepui ecosystems (Pantepui) correspond to the same uppermost montane life zone that is present in all tropical and extra-tropical mountains above tree line, such as the páramos in the Andes and the alpine meadows of the Alps and the Rocky Mountains.

Forest Formations

Forests cover approximately 83 percent of the total surface area of the Venezuelan Guayana. The accompanying vegetation map contains 54 different forest units. This figure is highly tentative, however, and clearly reflects our incomplete knowledge of this key tropical biome in the region.

For a general classification of the forest types in the Venezuelan Guayana, a major division between lowland and montane forest types needs to be made. Lowland forests show the typical pattern of a horizontally structured mosaic, whereas montane forests have a more regular, vertical distribution pattern arranged in altitudinal belts. Although the boundaries between altitudinal belts vary across the flora area and usually intergrade quite gradually, the dichotomy of lowland and montane forests is very useful to account for the complexities of the Venezuelan Guayana. To distinguish the vertical distribution pattern of the montane forests, the altitudinal zones discussed above will be used.

The main criteria used to distinguish the various forest types are summarized as follows.

Size of the principal tree layer
- Tall (>25 m)
- Medium (15–25 m)
- Low (5–15 m)

Phenology
- Evergreen (>75 percent of trees are evergreen)
- Semideciduous (25–75 percent of trees are evergreen)
- Deciduous (<25 percent of trees are evergreen)

Soil moisture conditions
- Nonflooded (*tierra firme*)
- Flooded
  - Permanently
  - Periodically or seasonally
- Riparian
- Coastal and estuarine

Orographic position (see "Altitudinal Zonation of Vegetation" above)

Floristic composition

Many of the terms and definitions used here differ from those traditionally employed in forestry because they focus primarily on biological aspects of different forest types rather than on economic parameters such as size classes. Since extensive areas covered with dense forests are still poorly studied or inadequately inventoried, many of the following descriptions reflect significant gaps in our knowledge, especially regarding their floristic composition.

Lowland Forests (0–500 m elevation)

Lowland forests cover large extensions in the Orinoco Delta, the Caura basin, and the wide Ventuari-Casiquiare peneplains of the upper Río Orinoco. Most lowland forests are classical evergreen tropical rain forests as described by Richards (1952) or Beard (1944, 1955), but there are also semideciduous and deciduous forests in the drier regions of the northeastern Venezuelan Guayana. Other important lowland forest types are the coastal and estuarine forests of the Orinoco Delta, the riparian or gallery forests, and the peculiar sclerophyllous forests (Rio Negro caatinga) of the Río Negro drainage in southwestern Amazonas. Finally, the premontane and some basimontane forest types in the piedmont area of the main mountain systems of the Venezuelan Guayana are also included in the lowland forest category.

In contrast to their apparent uniformity when seen from the air, all these often vast lowland forests have a surprisingly high diversity in structure and floristic composition when analyzed more closely on the ground. Physical factors such as the locally underlying geology, the meso- and microclimate, the length and intensity of inundation, and the nutrient content of the soils, all play key roles in the distribution of lowland forest ecosystems. So do biotic factors such as intra- and interspecific competition, historical evolution, and fire.

Besides the biotic and abiotic factors that determine the present-day spatial variability of forest types in any given area, it is also important to understand historical changes in the forest communities. These include late Quaternary climatic changes as well as different degrees of human intervention during the past centuries. Advances in the study of tropical forest dynamics in several parts of the Amazon basin and other tropical regions have shown that many forest types are undergoing successional phases, each stage characterized by a different set of dominant tree species (see Gómez-Pompa et al. 1991 for an excellent and updated overview). The concept of mature forest, traditionally applied to many tropical lowland forests, must now be seen with respect to a particular time scale rather than as a rigid, timeless category in the forest history.

Since many of these spatial and temporal factors vary greatly from one microsite to another, the recognition of clearly distinguishable forest types in the species-rich tropical lowlands is usually very difficult. Instead, it is more realistic to consider the forest cover of the tropical lowlands as a mosaic composed of numerous tree communities without definite limits and steadily intergrading from one site to another. In this light, the present classification of the lowland forests of the Venezuelan Guayana is a first approximation of the actual distribution of forest types in the area.

The accompanying vegetation map of the Venezuelan Guayana contains 18 lowland forest units. These were compiled both from existing literature and from Otto Huber's notes during many years of field work in the area. More detailed information on the lowland forests is available in several volumes on vegetation published by CVG-TECMIN, including maps at a scale of 1:500,000 (CVG-TECMIN 1987, 1989, 1991a, 1991b, 1991c, 1991d, 1991e).

Plate 1. Early map of the Guayana (Guiana) region credited to John Ogilby in 1671. The map reflects the poor knowledge of the Guayana interior at the time, showing a fictitious Lake Parima on the shores of which Manoa, or El Dorado, should lie. Reproduced with permission of the Instituto Autónomo Biblioteca Nacional y de Servicios de Bibliotecas (Venezuela), Dirección de Servicios Audiovisuales, from the collection in the División de Cartografía e Iconografía.

Plate 2. LANDSAT image taken 15 January 1991 of a section of northwestern Amazonas state, Venezuela, approximately 184 km × 185 km. The junction of the Río Atabapo and the Río Orinoco appears in the lower left, and in the lower center is the junction of the Río Ventuari (continuing to the upper right) and the Río Orinoco. The large mountainous area in the upper left center is the Cuao-Marieta uplands, with the Cuao-Sipapo massif appearing in the extreme upper left. The small mountain in the bottom center is Cerro Yapacana. Areas in pink are herbaceous formations (mostly Amazonian savannas along the rivers), and areas in green are forested.

Plate 3. Riparian forest of the middle Orinoco Delta (unit 3 on vegetation map), Delta Amacuro state. The river edge is lined with floating colonies of *Eichhornia crassipes* (Pontederiaceae) and rooted plants of *Montrichardia arborescens* (Araceae). Two common palms of these forests are visible, *Mauritia flexuosa* and *Euterpe oleracea* (Arecaceae). Photograph by Ronald Liesner.

Plate 4. Lowland savanna (unit 71 on vegetation map) dominated by *Curatella americana* (Dilleniaceae) in the shrub layer and by *Trachypogon plumosus* (Poaceae) among the herbs, northeastern Bolívar state. Semideciduous forest (unit 7 on vegetation map) can be seen on the hills in the background. Photograph by Otto Huber.

Plate 5. A granitic outcrop or *laja* (unit 99 on vegetation map) at Los Pijiguaos, Bolívar state. *Curatella americana* (Dilleniaceae) trees appear in the savanna in the foreground (unit 75 on vegetation map). Photograph by Bruce Holst.

Plate 6. Open vegetation on steep slopes of granitic outcrops (unit 99 on vegetation map) at Cerro Guacamaya, Serranía de los Pijiguaos, Bolívar state. At the lower left is the branched rosette shrub *Vellozia tubiflora* (Velloziaceae), and in the center are rosettes of *Pitcairnia armata* (Bromeliaceae). The light green mats are *Selaginella* (Selaginellaceae), the palmately compound-leaved shrub is *Pseudobombax croizatii* (Bombacaceae), and the shrub with simple light green leaves clustered at the branch apices is *Plumeria inodora* (Apocynaceae). Evergreen upland forests on bauxite soils (unit 33 on vegetation map) cover the mountain summit in the background. Photograph by Andreas Gröger.

Plate 7. Raudales de Atures, a long series of rapids above Puerto Ayacucho, Amazonas state. Numerous granitic outcrops occur in this region, part of the igneous-metamorphic basement of the Guayana Shield. Photograph by Gustavo Romero.

Plate 8. Aerial view of San Fernando de Atabapo, Amazonas state. Three rivers converge near this site: the Río Atabapo, a black-water river with few suspended sediments (center left); the Río Guaviare (top left), originating in the Colombian Andes and with a high load of suspended sediments; and the Río Orinoco (just off the photograph to the lower right), a white-water river with a moderate sediment load that drains much of Amazonas state. Photograph by Paul Berry.

Plate 9. The Río Ventuari in central Amazonas state. This river is the main tributary of the upper Río Orinoco and is shown flowing through a large expanse of lowland evergreen forest (unit 14 on vegetation map). Photograph by Otto Huber.

Plate 10. The upper reaches of the Río Orinoco at ca. 350 m elevation, as it descends from the Sierra Parima above the Río Ugueto, Amazonas state. Rapids and waterfalls are common here as the river flows through evergreen submontane forests (unit 48 on vegetation map). Photograph by Paul Berry.

Plate 11. Raudal Pereza on the Río Autana, northwestern Amazonas state. This black-water river periodically floods adjacent forests (unit 15 on vegetation map). Cerro Autana, ca. 1300 m elevation, is visible in the distance to the right. Photograph by Gustavo Romero.

Plate 12. Banks of the Río Atabapo, westernmost Amazonas state. Lining this black-water river is a peculiar plant community (unit 66 on vegetation map) characterized by trees and shrubs with crooked trunks, sparse crowns, and low-density, balsalike woods known as *palo de boya*. The trees shown here are mainly *Malouetia glandulifera* (Apocynaceae). Photograph by Gustavo Romero.

Plate 13. Aerial view of the Caño Pimichín, a tributary of the Río Guianía, Amazonas state. This black-water stream is shown here in the dry season (February), when much of the white-sand substrate is exposed. The trees with purple flowers are *Eperua purpurea* (Caesalpiniaceae), characteristic of Rio Negro caatinga forests (unit 16 on vegetation map). Taller forest patches intergrade with shrubby formations (*bana,* unit 65 on vegetation map) or more open areas (white-sand meadows, unit 87 on vegetation map). Photograph by Otto Huber.

Plate 14. Transitional vegetation between lowland scrub (unit 64 on vegetation map) and lowland meadows on white sand (unit 87 on vegetation map), north of the lower Río Atacavi, Amazonas state. Here the scrub is dominated by the tall, slender-leaved *Humiria wurdackii* (Humiriaceae) and smaller bushes of *Archytaea angustifolia* and *Bonnetia crassa* (Theaceae). Photograph by Paul Berry.

Plate 25. Sandstone outcrop on southern Cerro Coro Coro, Amazonas state, ca. 1000 m elevation. Shown are the gray rosettes of *Lindmania thyrsoidea* (Bromeliaceae), the gray-leaved shrub *Graffenrieda sessilifolia* subsp. *occidentalis* (Melastomataceae), *Philodendron* sp. (Araceae), and in front of the larger rock, a tall-stemmed plant of *Vellozia tubiflora* (Velloziaceae). Photograph by Bruce Holst.

Plate 26. Northern tip of Roraima-tepui, Bolívar state, ca. 2700 m elevation. Shown are the vertical cliffs and rocky, dissected part of the summit called the "labyrinth." Photograph by Charles Brewer-Carías.

Plate 27. Lago Gladys at the northern extreme of the summit of Roraima-tepui, ca. 2700 m elevation. This rare tepui lake is approximately 2 m deep and covers 3–4 hectares. The borders of Venezuela, Brazil, and Guyana converge on the summit of Roraima-tepui, with Lago Gladys in Guyana. Photograph by Charles Brewer-Carías.

Plate 28. Valley on the summit of Roraima-tepui, Bolívar state, ca. 2650 m elevation. Small patches of low tepui forest appear on the mostly open terrain. Photograph by Charles Brewer-Carías.

Plate 29. Aprada-tepui, Bolívar state, ca. 2500 m elevation. Shown here after a rain shower, the summit has sparse vegetation with small islands of tepui forests. Photograph by Antonio Ahogado.

Plate 30. Salto La Cortina coming off the eastern sector of the summit of Auyán-tepui in the Cañon del Diablo, Bolívar state, ca. 1700 m elevation. Photograph by Charles Brewer-Carías.

Plate 31. Dissected rocky surface on the summit of Auyán-tepui, Bolívar state, ca. 2100 m elevation. Photograph by Charles Brewer-Carías.

Plate 32. Depression on the summit of Kamarkawarai-tepui, Bolívar state, ca. 2400 m elevation, showing the typical high-tepui forest vegetation (unit 23 on vegetation map). Two of the plants most characteristic of tepuis can be seen, the yellow-green rosettes of *Brocchinia tatei* (Bromeliaceae) in the foreground, and the dominant shrub *Bonnetia roraimae* (Theaceae). Photograph by Bruce Holst.

Plate 33. Distinctive erosion pattern of rocks on the edge of Ptari-tepui, Bolívar state, ca. 2400 m elevation. The protected crevices harbor lithophytic vegetation. Photograph by Bruce Holst.

Plate 34. Evergreen, upper montane forests (unit 22 on vegetation map) of the Río Tírica valley on the southern side of the Chimantá massif, Bolívar state, 1600–1800 m elevation. Photograph by Otto Huber.

Plate 35. Summit of Amurí-tepui (Chimantá massif), Bolívar state, ca. 1900 m elevation, with Upuigma-tepui in the background. The foreground and center show shrub islands of *Bonnetia roraimae* (Theaceae), and a high-tepui meadow can be seen on the left (unit 59 on vegetation map). Photograph by Otto Huber.

Plate 36. Dense high-tepui forest (unit 23 on vegetation map) on a diabase intrusion, Chimantá massif, Bolívar state, ca. 2400 m elevation. Photograph by Otto Huber.

Plate 37. Tall paramoid scrub (unit 59 on vegetation map) on Apacará-tepui (Chimantá massif), Bolívar state, ca. 2300 m elevation. This unique vegetation type is dominated by the endemic species *Chimantaea mirabilis* (Asteraceae). Photograph by Antoine Cleef.

Plate 38. Dense high-tepui forest (unit 23 on vegetation map) on the Chimantá massif, Bolívar state, ca. 2500 m elevation. The carnivorous, pitcher-shaped bromeliad, *Brocchinia reducta* (Bromeliaceae), is in the foreground, and the forest in the background is dominated by *Bonnetia roraimae* (Theaceae). Photograph by Otto Huber.

Plate 39. Wet, highland meadow (unit 59 on vegetation map) on Kamarkawarai-tepui, Bolívar state, ca. 2400 m elevation. The reddish pitcher plants are *Heliamphora heterodoxa* (Sarraceniaceae), the yellow-green vase-shaped plants are *Brocchinia tatei* (Bromeliaceae), and the plants with glossy, strap-shaped leaves are *Stegolepis humilis* (Rapateaceae). Photograph by Bruce Holst.

Plate 40. Pioneer vegetation (unit 102 on vegetation map) on the Chimantá massif, Bolívar state, ca. 2400 m elevation. The white-flowered plant is *Lindmania subsimplex* (Bromeliaceae), and the pitcher-shaped plant is *Brocchinia reducta,* a carnivorous bromeliad. Photograph by Otto Huber.

Plate 41. Cerro Huachamacari, Amazonas state, ca. 1900 m elevation, viewed from the savanna at Culebra. A Yekwana village is at the far end of the savanna. Photograph by Bruce Holst.

Plate 42. The summit plateau of Cerro Huachamacari, Amazonas state, 1700–1900 m elevation, covered by high-tepui shrubby meadows (unit 95 on vegetation map) and shrublands (unit 68 on vegetation map). Photograph by Otto Huber.

Plate 43. Undulating summit plateau of Cerro Duida, Amazonas state, ca. 2200 m elevation, showing extensive shrublands (unit 68 on vegetation map). Photograph by Otto Huber.

Plate 44. Tepui meadow (unit 94 on vegetation map) on Cerro Marahuaka, Amazonas state, ca. 2800 m elevation, dominated by *Marahuacaea schomburgkii* (Rapateaceae) and smaller rosettes of *Lindmania marahuacae* (Bromeliaceae). Photograph by Julian Steyermark.

Plate 45. Deep crevice on the summit of Cerro Marahuaka, Amazonas state, ca. 2700 m elevation. The shrub on the right is a species of *Weinmannia* (Cunoniaceae). Photograph by Ronald Liesner.

Plate 46. Cerro Sipapo, Amazonas state, ca. 1600 m elevation, showing large granitic domes and ridges with many rock outcrops. Photograph by Otto Huber.

Plate 47. Highland area of the igneous Sierra de Maigualida, Amazonas state, ca. 2000–2300 m elevation. The vegetation is a mosaic of high-tepui meadows (unit 91 on vegetation map) and montane and upper montane forest (unit 30 on vegetation map). Photograph by Otto Huber.

Plate 48. Highland meadow (unit 91 on vegetation map) on the Sierra de Maigualida, Amazonas state, ca. 2000 m elevation. The red-flowered herb is *Kunhardtia rhodantha* (Rapateaceae), and the small shrub in the center and right foreground is a species of *Spathelia* (Rutaceae). Photograph by Paul Berry.

Plate 49. Sierra de la Neblina seen from the Maturacá area of northern Brazil. On the right is Pico da Neblina, 3014 m elevation, which is the highest point in the Guayana Shield and in South America outside of the Andes. It lies just a few hundred meters south of the border with Venezuela. Photograph by Julian Steyermark.

Plate 50. *Stegolepis guianensis* (Rapateaceae; dark green strap-shaped leaves), *Orectanthe sceptrum* (Xyridaceae; gray rosettes), and *Nietneria paniculata* (Liliaceae; yellow-flowered herb), on the summit of Roraima-tepui, Bolívar state, ca. 2700 m elevation. These plants are characteristic of many of the eastern tepuis. Photograph by Charles Brewer-Carías.

Plate 51. *Chimantaea mirabilis* (Asteraceae, tribe Mutisieae) on the summit of Apacará-tepui (Chimantá massif), Bolívar state, ca. 2400 m elevation. Of the ten species of *Chimantaea,* nine are endemic to the Chimantá massif, and the tenth extends only to neighboring tepuis. Photograph by John Pruski.

Plate 52. Flowering branch of *Chimantaea mirabilis* (Asteraceae, tribe Mutisiae), Apacará-tepui (Chimantá massif), Bolívar state, ca. 2400 m elevation. Photograph by Otto Huber.

Plate 53. *Pterozonium spectabile* (Pteridaceae) on rock faces on the summit of Aparamán-tepui, Bolívar state, ca. 2100 m elevation. The genus *Pterozonium* is largely endemic to the Guayana Shield. Photograph by Bruce Holst.

Plate 54. *Kunhardtia rhodantha* (Rapateaceae) on the Sierra de Maigualida, Amazonas state, ca. 2000 m elevation. This mainly highland species ranges in a broad arc from Cerro Sipapo to Cerro Guanay and Cerro Coro Coro and then to Sierra de Maigualida. Photograph by Paul Berry.

Plate 55. *Saxofridericia grandis* (Rapateaceae) on Cerro Parú, Amazonas state, ca. 1100 m elevation. Photograph by Paul Berry.

Plate 56. *Gongylolepis jauaensis* (Asteraceae, tribe Mutisiae), on Cerro Coro Coro, Amazonas state, ca. 2200 m elevation. Photograph by Otto Huber.

Plate 57. *Pleurostima celiae* (Velloziaceae) on rocky outcrops on the summit of Cerro Yaví, Amazonas state, ca. 2300 m elevation. Photograph by Otto Huber.

Plate 58. *Bonnetia crassa* (Theaceae), with a distinctive sympodial branching pattern, on the summit of Cerro Guanay, Amazonas state, ca. 1750 m elevation. Photograph by Otto Huber.

Plate 59. *Duidaea rubriceps* (Asteraceae, tribe Mutisieae), on the northern tip of Cerro Duida, Amazonas state, ca. 1250 m elevation. The genus *Duidaea* has four species, three endemic to Cerro Duida and one endemic to neighboring Cerro Marahuaka. Photograph by Ronald Liesner.

Plate 60. *Heliamphora tatei* var. *tatei* (Sarraceniaceae), showing its unusual branched habit on the summit of Cerro Huachamacari, Amazonas state, ca. 1700 m elevation. The genus *Heliamphora,* with five species, is endemic to upland and highland areas of the Guayana Shield. Photograph by Charles Brewer-Carías.

Plate 61. *Navia aloifolia* (Bromeliaceae) on Sierra de la Neblina, Amazonas state, ca. 1900 m elevation. The genus *Navia* has approximately 100 predominantly terrestrial and rock-inhabiting species, all endemic to the Guayana Shield. Photograph by Charles Brewer-Carías.

Plate 62. *Hymenophyllopsis superba* (Hymenophyllopsidaceae) on Sierra de la Neblina, Amazonas state, ca. 2100 m elevation. The Hymenophyllopsidaceae, with one genus and eight species, is endemic to upland and highland areas of the Guayana Shield. Photograph by Joseph Beitel.

Plate 63. *Nautilocalyx pemphidius* (Gesneriaceae) is endemic to the Sierra de la Neblina but grows over a wide altitudinal range from 140 to 1250 m elevation. Photograph by Charles Brewer-Carías.

Plate 64. *Bonnetia maguireorum* (Theaceae), a gregarious, thick-stemmed shrub restricted to plateaus of the Sierra de la Neblina, ca. 1750–2000 m elevation. This species was previously known as *Neblinaria celiae.* Photograph by Charles Brewer-Carías.

Plate 65. *Brocchinia hechtioides* (Bromeliaceae), on the flanks of Cerro Sipapo, Amazonas state, looking toward Cerro Autana in the distance. The genus *Brocchinia* has 19 species, all endemic to the Guayana Shield, and is one of the most characteristic herbs of upland and highland areas. Photograph by Charles Brewer-Carías.

Plate 66. Colonies of *Rhyncholacis penicillata* (Podostemaceae) at Salto Hacha on the Río Carrao, just above Canaima, Bolívar state, ca. 400 m elevation. Photograph by Paul Berry.

Plate 67. *Utricularia neottioides* (Lentibulariaceae), growing in seeps on granitic outcrops (*lajas*) in northwestern Bolívar state, ca. 100 m elevation. Photograph by Bruno Manara.

Plate 68. *Tepuianthus savannensis* (Tepuianthaceae), in lowland scrub on white sand at Cucurital de Yagua, Amazonas state, ca. 100 m elevation. The Tepuianthaceae, with one genus and six species, is endemic to the Guayana Shield. Photograph by Otto Huber.

Plate 69. *Pentamerista neotropica* (Tetrameristaceae) in lowland savannas on white sand in the Río Atabapo basin, Amazonas state, ca. 120 m elevation. The monotypic genus *Pentamerista* is endemic to this area along the border of Colombia and Venezuela. The only other genus in the family, *Tetramerista,* occurs in Malaya, Sumatra, and Borneo. Photograph by Otto Huber.

Plate 70. *Aechmea brevicollis* (Bromeliaceae), a trunk epiphyte on the edge of a shrubland on white sand in western Amazonas state, ca. 100 m elevation. Photograph by Gustavo Romero.

Plate 71. *Thurnia polycephala* (Thurniaceae), an herb inhabiting the riverbeds of blackwater rivers, seen here at Yavita on the Río Temi, Amazonas state, ca. 130 m elevation. Most of the year the plants are under water, but during the dry season as shown here they are exposed and begin flowering. Photograph by Paul Berry.

Plate 72. Stilt houses at the mouth of one of the branches of the Orinoco Delta, Delta Amacuro state. These are used by fishermen and by Warao Amerindians as temporary or permanent residences. Photograph by Paul Berry.

Plate 73. Indigenous dwelling of a Mako Amerindian family in the Río Ventuari basin, Amazonas state. Photograph by Otto Huber.

Plate 74. Characteristic house of the Piaroa Amerindians, Río Manapiare valley, Amazonas state. Photograph by Paul Berry.

Plate 75. Yanomami Amerindian communal dwelling in the upper Río Orinoco region, Amazonas state. Photograph by Otto Huber.

Plate 76. A local inhabitant of western Amazonas state bundling fibers of *chiqui-chiqui* from the palm *Leopoldinia piassaba* for sale to broom manufacturers. This palm grows in areas drained by black-water rivers. Photograph by Paul Berry.

Plate 77. Young boy along the lower Río Atacavi, Amazonas state, holding bundles of inflorescences of *Schoenocephalium cucullatum* (Rapateaceae). These flowers are given as gifts or are sold locally as mementos from the white-sand regions of the upper Río Negro, Atabapo, and Inírida basins of Colombia and Venezuela. Photograph by Paul Berry.

Plate 78. Lago Guri, an artificial lake covering more than 4250 km^2 as a result of the construction of a dam at Guri, on the lower Río Caroní, Bolívar state. The dam supplies about 70 percent of the electricity needs of Venezuela, and its watershed is entirely within the Venezuelan Guayana. Photograph by Otto Huber.

Plate 79. Clearing of forest for bauxite mines in the Serranía de Los Pijiguaos, northwestern Bolívar state. Large-scale state-run mining concessions of bauxite in the region began in the 1980s. Photograph by Otto Huber.

Plate 80. Gold mining operation at Kilómetro 88, south of El Dorado, Bolívar state. Photograph by Nelda Dezzeo.

Plate 81. Fires likely set by gold miners in the white-sand savannas surrounding Cerro Yapacana, Yapacana National Park, Amazonas state. Photograph by Otto Huber.

Plate 82. Fires set by local Yanomami inhabitants in the southern Sierra Parima, Amazonas state. As a result of human-induced fires in this region, forest areas are being replaced by savannas. Photograph by Otto Huber.

Plate 83. View of part of the Sierra de la Neblina, southernmost Amazonas state, showing possible signs of fires (left) near the summit. Photograph by Charles Brewer-Carías.

Plate 84. Commercial logging in the San Pedro Forest Lot, Bolívar state. Photograph by Gabriel Picón.

Coastal and estuarine forests. Large colonies of dense mangrove forests form the outermost vegetation type growing along the extensive shoreline of Delta Amacuro state. These forests also stretch considerably inland in the Orinoco Delta, following the banks of numerous interconnected channels. According to Pannier and Fraino de Pannier (1989), mangroves cover approximately 4600 km^2 of the Orinoco Delta region. These medium-sized to tall forests (***unit 1*** on map) are composed of just a few highly specialized tree species, namely *Rhizophora mangle* (Rhizophoraceae) in waters with higher salinity, and *Avicennia schaueriana* (Verbenaceae) and *Laguncularia racemosa* (Combretaceae) in more brackish waters. In slightly more inland positions, mature mangrove stands are often mixed with palms (*Euterpe* spp.) and the buttress-rooted legume *Pterocarpus officinalis* (Danielo 1976; Canales 1985; Delascio 1985). Some of the mangrove forests of the Orinoco Delta have been increasingly exploited for tannins and other minor timber products, causing serious impact on the shoreline vegetation.

Inland forests of the Orinoco Delta and Amacuro plains. The zonation of forest vegetation in the Orinoco Delta depends on the different regimes of flooding that occur there. Locally, the outer sector with permanent flooding is called the Lower Delta (*Delta Inferior*). The next, more inland sector, with temporary but prolonged flooding is called the Middle Delta (*Delta Medio*). The innermost, only occasionally flooded sector is called the Upper Delta (*Delta Superior*). According to Canales (1985), the principal forest types associated with each of these sections are as follows:

1. *Lower Delta:* low (10–20 m), evergreen, permanently flooded swamp and palm forests (***unit 2*** on map), dominated by *Pterocarpus officinalis* (Fabaceae), *Symphonia globulifera* (Clusiaceae), *Euterpe oleracea* (Arecaceae), and *Tabebuia fluviatilis* (Bignoniaceae).

2. *Middle Delta:* medium/low (up to 25 m), evergreen, seasonally flooded marsh and palm forests (***unit 3*** on map; Plate 3), dominated by *Symphonia globulifera* (Clusiaceae), *Virola surinamensis* (Myristicaceae), *Carapa guianensis* (Meliaceae), *Pterocarpus officinalis* (Fabaceae), *Mora excelsa* (Caesalpiniaceae), *Pachira aquatica* (Bombacaceae), *Mauritia flexuosa, Manicaria saccifera, Euterpe oleracea* and *Bactris* sp. (Arecaceae), and *Phenakospermum guyannense* (Strelitziaceae), on periodically flooded ground.

3. *Upper Delta:* medium (up to 25 m), evergreen, unflooded or only briefly flooded forests (***unit 4*** on map). These are usually three-layered, dense forests with a few deciduous species, dominated by *Ceiba pentandra* (Bombacaceae), *Ocotea* sp. (Lauraceae), *Mora excelsa* (Caesalpiniaceae), *Erythrina* sp. (Fabaceae), *Tabebuia capitata* (Bignoniaceae), *Spondias mombin* (Anacardiaceae), *Triplaris surinamensis* (Polygonaceae), *Gustavia augusta* (Lecythidaceae), and *Licania densiflora* (Chrysobalanaceae).

In the southern alluvial plains and interior hill-lands of Delta Amacuro state, a mosaic of dense and tall forests extends from the Río Grande to the Serranía de Imataca and farther into the Amacuro-Cuyuní river basins. These forests, which are poorly explored botanically, are evergreen, two- or three-layered, between 25 and 30 m tall, and grow mostly on nonflooded higher ground (***unit 5*** on map). According to Steyermark (1968), their most important species are *Licania densiflora* (Chrysobalanaceae), *Eschweilera decolorans* and *Gustavia poeppigiana* (Lecythidaceae), *Tabebuia capitata* (Bignoniaceae), *Trichilia pleeana* (Meliaceae), *Tetragastris altissima* (Burseraceae), *Catostemma commune* (Bombacaceae), *Virola surinamensis* (Myristicaceae), *Alexa imperatricis, Mora excelsa* (Caesalpiniaceae), *Sterculia pruriens* (Sterculiaceae), *Carapa guianensis* (Meliaceae), and *Peltogyne venosa* (Caesalpiniaceae).

Along water courses and in low valleys, a peculiar evergreen forest type predominates in which *Mora excelsa* (Caesalpiniaceae) is by far the dominant tree. These *Mora* forests, which also extend into the lowlands of adjacent Guyana and are locally predominant there, are medium to tall (to 25 m) in size, with dense crowns, but contain relatively few species (Fanshawe 1952). Other frequent tree species are *Eschweilera decolorans* (Lecythidaceae), *Licania alba* (Chrysobalanaceae), *Clathrotropis brachypetala* (Fabaceae), *Manilkara bidentata* (Sapotaceae), *Terminalia amazonia* (Combretaceae), and *Simarouba amara* (Simaroubaceae).

The inland forests of the Orinoco Delta plains show various degrees of human intervention. Those growing on nonflooded sites in the Upper Delta have been more seriously affected by shifting cultivation and land clearing than the more flood-prone forests of the Middle and Lower deltas. In the latter areas, the main forest use is the harvest of palm products by the indigenous Warao population. By far the most important species are *moriche* (*Mauritia flexuosa*) and *palmito* or palm hearts (*Euterpe oleracea*).

In the southern plains of the Delta Amacuro region, the inland forests seem to have remained under relatively pristine conditions, although military settlement programs near the coast have begun to produce large forest clearings for agricultural land use, especially along the lower courses of the main rivers.

Forests of the Cuyuní-Caroní lowlands. Between the middle Río Cuyuní in the east and the lower Río Caroní in the west, the landscape consists mainly of gently rolling hills; the vegetation is composed of large extensions of tall forests that alternate with shrub savannas. There is a more pronounced dry season in the north, which produces a complex mosaic of evergreen and semideciduous forests. The latter are usually confined to hilltops or other sites with low water-retention capacity.

In the premontane zone that extends over the foothill landscape to the north of Sierra de Lema and La Escalera, tall evergreen forests with three tree layers reaching up to 30 or more meters are common. These forests are characterized by a high diversity of trees, which has been reported to range between 80 and 100 species per hectare (CVG-TECMIN 1987). The most common tree species of this forest type (***unit 6*** on map) are *Catostemma commune* (Bombacaceae), *Alexa imperatricis* (Fabaceae), *Endlicheria* sp. (Lauraceae), *Crudia oblonga* (Caesalpiniaceae), *Eschweilera decolorans* (Lecythidaceae), *Inga alba* (Mimosaceae), *Clathrotropis brachypetala* (Fabaceae), *Brownea coccinea* subsp. *capitella* (Caesalpiniaceae), *Anaxagorea* sp. (Annonaceae), *Sterculia pruriens* (Sterculiaceae), *Aspidosperma marcgravianum* (Apocynaceae), and *Lonchocarpus* sp. (Fabaceae).

In the early 1990s, the evergreen forests of the Río Cuyuní drainage began to be invaded by miners and settlers, especially along the main roads leading south from El Dorado to Las Claritas and east from Tumeremo towards the region of Bochinche on the border with Guyana. The main vegetation uses are shifting cultivation and timber production.

Farther north and west, somewhat drier forest types extend roughly from Tumeremo to Upata and then west to the lower Paragua and Aro rivers. These forests are semideciduous, mostly two-storied, and tall to medium-sized (15–25 m or more tall; see ***unit 7*** on map and Plate 4). They are found on nonflooded ground and usually contain a significant proportion of deciduous trees, making them easy to recognize during the dry season from January to March or April. A detailed reconnaissance of the vegetation around Lago Guri (Rosales and Briceño 1990) provides information on the floristic composition of these forests. In moist sites, the dominant tree species are *Tetragastris panamensis* (Burseraceae), *Pouteria egregia* (Sapotaceae), *Sterculia pruriens* (Sterculiaceae), *Protium* sp. (Burseraceae), *Chaetocarpus schomburgkianus* (Euphorbiaceae), *Aspidosperma marcgravianum* (Apocynaceae), *Licania densiflora* (Chrysobalanaceae), *Manilkara bidentata* (Sapotaceae), and *Erisma uncinatum* (Vochysiaceae). On drier sites, a lower and more open forest type occurs, formed mainly by *Peltogyne floribunda* (Caesalpiniaceae), *Tabebuia capitata* (Bignoniaceae), *Cochlospermum orinocense* (Bixaceae), *Piptadenia* sp. (Mimosaceae), *Cordia alliodora* and *Bourreria cumanensis* (Boraginaceae), *Ceiba pentandra* (Bombacaceae), *Tournefortia punctata* (Boraginaceae), and *Spondias mombin* and *Tapirira guianensis* (Anacardiaceae).

Finally, a low, deciduous forest occurs on dry, sandy or rocky habitats, usually on the tops of the numerous hills to the south of the lower and middle Río Orinoco (***unit 8*** on map). The dominant tree species are *Bursera simaruba* (Burseraceae), *Bourreria cumanensis* (Boraginaceae), *Copaifera pubiflora* (Caesalpiniaceae), *Anadenanthera peregrina* (Mimosaceae), *Tabebuia capitata* (Bignoniaceae), *Erythroxylum* sp. (Erythroxylaceae), and *Pachira quinata* (Bombacaceae) (Rosales and Briceño 1990).

Most of the forests between the Río Cuyuní in the east and the lower Paragua and Aro rivers in the northwest have been heavily logged and farmed during the late 1900s, causing severe and often irreversible reductions and degradations of the original forest cover. Forests continue to be cleared for pasture (around Upata), large-scale agriculture (La Paragua, San Francisco), logging (south of El Manteco), and charcoal production for the steel plants near Puerto Ordaz (especially around Lago Guri). Following construction of the gigantic hydroelectrical complex at Guri on the lower Río Caroní between 1970 and 1990 and the subsequent filling of the artificial lake that ultimately covered 4250 km^2, large extensions of several forest types (ca. 1000 km^2 according to Pernía 1985) were lost to flooding.

Riparian forests of the lower and middle Orinoco. The forests lining both banks of the Río Orinoco differ substantially from those located farther south, evidently because of their different inundation regimes and soil conditions, especially nutrient levels. Because of their many similarities with riparian forests of the Amazon River basin, authors such as Colonnello et al. (1986) have called them *várzea* forests.

In the lower section of the Río Orinoco, these periodically flooded riparian forests (***unit 9*** on map) are generally of medium size (15–20 m) and are semideciduous, dominated by *Piranhea trifoliata* (Euphorbiaceae), *Homalium racemosum* (Flacourtiaceae), and *Sclerolobium guianense* (Caesalpiniaceae).

Along the rocky and sandy shores of the Río Caroní below the dam at Guri, Rosales et al. (1993) found low (3–12 m tall) riparian forests, with species adapted to the river's oligotrophic black-water conditions and therefore resembling Amazon *igapó* forests. The dominant species are *Campsiandra* sp. and *Cynometra parviflora* (Caesalpiniaceae), *Acosmium nitens* and *Andira retusa* (Fabaceae), *Piranhea trifoliata* and *Mabea nitida* (Euphorbiaceae), *Gustavia augusta* (Lecythidaceae), and *Licania apetala* (Chrysobalanaceae).

In the more western section of the lower Río Orinoco, the floristic composition of the riparian forests is more diverse. A detailed study by Rosales (1988) in the partially flooded forests on the northern bank of the Orinoco around Mapire (across from the mouth of the Río Caura) reported 110 tree species, of which the most important are *Combretum frangulifolium* (Combretaceae), *Gustavia augusta* (Lecythidaceae), *Pterocarpus* sp. and *Etaballia dubia* (Fabaceae), *Albizia corymbosa* and *Inga* spp. (Mimosaceae), *Spondias mombin* (Anacardiaceae), *Mabea nitida* (Euphorbiaceae), *Homalium racemosum* (Flacourtiaceae), *Pouteria orinocoensis* (Sapotaceae), *Symmeria paniculata* (Polygonaceae), *Copaifera pubiflora* and *Campsiandra laurifolia* (Caesalpiniaceae), *Eschweilera tenuifolia* (Lecythidaceae), and *Astrocaryum aculeatum* (Arecaceae). These generally low to medium forests (***unit 10*** on map) are evergreen and seasonally flooded up to several meters high. They are not heavily used by the local population, whose main activity is fishing. Small-scale shifting cultivation is carried out, especially on seasonally flooded islands in the Orinoco, where high nutrient levels in the soils allow substantial crop yields.

Forests of the Caura-Paragua peneplains. The gently rolling lowlands and hill-lands of the middle and lower Paragua and Caura river basins are covered by an impressive, unbroken mantle of dense forests, constituting perhaps the most extensive continuous forest region of the country. When flying over this area (which local pilots are reluctant to do because of the lack of emergency landing sites), a remarkable heterogeneity of forest types can be readily seen. The only serious attempt to classify the main forest types of this region was made by the PIRNRG group in 1989 and 1990, but their data are still unpublished.

At least two major forest types can be distinguished in the area. The first type is composed of tall, evergreen lowland forests that cover extensive *tierra firme* or briefly flooded terrains in the lower Caura and middle to upper Paragua river basins, with dense crowns and numerous emergent trees (***unit 11*** on map). According to Williams (1942), one of the few botanists to study these forests along the lower Caura up to Salto Pará in 1939, the tallest trees belong to the Anacardiaceae (*Anacardium*), Clusiaceae (*Calophyllum*), Sapotaceae (*Manilkara*), Burseraceae (*Protium heptaphyllum*), Mimosaceae (*Inga, Parkia*), Caesalpiniaceae (*Copaifera*), and Fabaceae (*Erythrina, Dipteryx punctata*). There is as yet no botanical information published on the forests of the upper Paragua drainage.

The second type of forest is composed of tall, evergreen lowland forests that cover partly flooded terrains and have a predominance of palms (*Oenocarpus, Euterpe* sp.), which are particularly evident in the middle Caura and Erebato river basins (***unit 12*** on map). Since these palm forests are also frequently found on the slopes of lower hills in the upper Río Caura drainage, flooding does not appear to be the main condition for their existence. At this time, however, there is no botanical or ecological information available on these extensive forests.

Although many of the forest types described above are included in the El Caura Forest Reserve (a reserve designated for timber harvest), they have not yet been subject to commercial logging and are therefore in virtually pristine condition. During the first half of the 20th century, however, there was a boom in the exploitation of tonka beans or *sarrapia* (*Dipteryx odorata* and *D. punctata,* Fabaceae) and of *balatá* (*Manilkara bidentata* and *Pradosia surinamensis,* Sapotaceae) in the forests of the lower and middle Caura, where these species are particularly abundant in nonflooded forests (Williams 1942).

Forests of the northwestern piedmont. The lowlands that extend between the middle Río Orinoco and the massive hill-lands and uplands of the Guayana Shield to the south are covered by a mosaic of forests and savannas typical of the transition between the Llanos and the Guayana floristic regions. Some of the forests in the lower Río Cuchivero basin have already been described by Passarge (1903). Subsequently, Brian Boom made a forest inventory along the lower Río Maniapure in 1985 and 1986 (Boom 1990a), and Huber and Guánchez (1988) made another inventory as part of an environmental impact study of Los Pijiguaos bauxite mines on the lower Río Suapure. The predominant forests are of a semideciduous, medium to low premontane type (***unit 13*** on map), in which *Swartzia laevicarpa* (Fabaceae), *Anadenanthera peregrina* (Mimosaceae), and *Cassia moschata* and *Copaifera pubiflora* (Caesalpiniaceae) are the dominant elements together with *Tapirira guianensis* (Anacardiaceae), *Tabebuia ochracea* subsp. *heterotricha* (Bignoniaceae), *Bursera simaruba* (Burseraceae), *Cochlospermum vitifolium* (Bixaceae), and *Attalea maripa* (Arecaceae). In riparian areas towards the Río Orinoco, *Vochysia venezuelana* (Vochysiaceae) is very conspicuous in the dry season with its showy yellow flowers. Further studies of these interesting transition forests will likely show a greater floristic and physiognomic variability along the east-west gradient from the mouth of the Río Caura to the mouth of the Río Sipapo.

Until the late 1980s, this region was only sparsely populated by Panare Amerindians and a few settlers, but population pressure has increased dramatically following the completion of a paved highway from Caicara to Puerto Ayacucho and the opening of one of the world's largest bauxite mines at Los Pijiguaos in 1989 (Plate 79). Many forest areas have begun to be converted into pasture for agriculture, and logging has increased for house construction. Consequently, most of the forests of this region are likely to suffer severely from these activities as the population increases.

Lowland forests of Venezuelan Amazonas. The lowlands that stretch from the Río Ventuari basin in the north to drainages of the upper Orinoco, Atabapo, Casiquiare, and Negro rivers in central and southern Amazonas state consist of gently undulating peneplains and flat plains of alluvial, depositional, or erosional origin (MARNR-ORSTOM 1987). The extensive forests of this region differ both floristically and physiognomically from site to site. This variability is due mainly to the different soil conditions, the complex pattern formed by the underlying parent rock, the length and intensity of seasonal inundation, and the overall climatic gradients with increasing precipitation from north to south and from east to west. The result is an unusually complex mosaic of forest types and associations that cover nearly the entire Amazonas state, interrupted only occasionally by savannas, shrublands, and rivers.

There are few forest inventories reported from this region (see Brünig et al. 1978; Herrera et al. 1978; Chesney 1979; Aymard et al. 1989), so it is too soon to propose a sound classification of the many forest types in Amazonas state. When the results of the large-scale inventories made by PIRNRG (CVG-TECMIN) from 1989 to 1991 are published, they should provide more reliable and detailed information about these Amazonian forests.

All lowland forests of Venezuelan Amazonas have a macrothermic, ombrophilous climate and are nearly all evergreen; they grow under macroclimatic conditions with an annual average temperature of more than 24°C and an average annual rainfall of more than 2000 mm that is more or less evenly distributed throughout the year (see MaO on Figure 1-5, and the discussion of climate in Chapter 1). A major division of forest types distinguishes between those growing on flooded terrain (mainly floodplains) and those growing on nonflooded (*tierra firme*) sites. In flooded forests, the length of seasonal or permanent inundation is the most important criterion for identifying local forest types, whereas in nonflooded forests, particular soil conditions such as nutrient availability and soil texture are the main differentiating factors between forest types.

Forests on nonflooded terrain (***unit 14*** on map; Plate 9) predominate in the lowlands of Amazonas state and are very heterogeneous in their floristic and structural composition. In the northern Manapiare basin, for instance, where a drier and seasonally more pronounced climate prevails, patches of semideciduous, premontane forest types occur. These probably share floristic affinities with forests in the northwestern lowlands (unit 13). On the hilly peneplains of central and southeastern Amazonas state, tall evergreen, ombrophilous forests commonly occur, made up of at least three distinct tree layers up to 40 m tall. Since the crowns of the upper trees are usually large and rather dense, the interior of these forests is relatively dark and lacks a well-developed understory. Among the many tree species growing in these variable forests, the most common belong to the Fabaceae (e.g., *Lecointea amazonica, Clathrotropis glaucophylla*), Caesalpiniaceae (*Peltogyne venosa*), Lauraceae (*Ocotea* spp., *Nectandra* sp.), Chrysobalanaceae (*Licania* spp.), Meliaceae (*Trichilia, Guarea*), Sapindaceae (*Toulicia*), Vochysiaceae (*Erisma uncinatum, Ruizterania* spp.), Burseraceae, Myristicaceae, Sapotaceae, Annonaceae, and Moraceae. Tall palms are also frequent, especially species of *Oenocarpus, Socratea, Leopoldinia,* and *Bactris,* whereas smaller shade-tolerant species of *Geonoma* are common in the understory. Lianas and epiphytes are less frequent, as are ground herbs, which are formed mainly by occasional small colonies of Marantaceae (*Ischnosiphon, Monotagma*), Rapateaceae (*Rapatea, Spathanthus*), Poaceae (*Olyra, Pariana*), and Cyperaceae (*Diplasia karataefolia*).

True flooded forests growing on oxisols or ultisols in various floodplains of the Venezuelan Amazon are often recognizable from the air through their usually high palm richness, especially the genera *Mauritia, Euterpe,* and *Manicaria.* The dominant trees, which reach heights of 30 to 40 m, usually have dense, rounded, or umbrella-shaped crowns and belong mainly to the Apocynaceae (*Aspidosperma*), Fabaceae (*Pterocarpus*), Mimosaceae (*Parkia*), Lecythidaceae, Myrtaceae, and Sapotaceae. Due to the variety of different inundation regimes (from only shortly to permanently flooded) in different floodplains, there are probably many other ecologically and floristically diversified flooded forests. Since they are mostly inaccessible, no single predominant forest type has yet been identified. Therefore, ***unit 15*** on the vegetation map tentatively indicates the main distribution of all flooded evergreen forests in Amazonas state.

In southwestern Amazonas, mainly in the drainage of the Atabapo, Guainía, and Negro rivers, another very peculiar forest type predominates. It is characterized by large stands of low- to medium-sized trees with sclerophyllous leaves and small, open crowns. The local name of this forest type is *caatinga,* first reported by Richard Spruce and derived from the indigenous (Tupi) words *caa* (plants, forest) and *tinga* (white, clear), in allusion to the typically open, clear crowns or perhaps to the white sandy soil usually associated with it (Romero 1993). Since the same term is generally applied also to the arid, thorny forests and shrublands of northeastern Brazil (*caatinga nordestina*), it is necessary to specify "Rio Negro caatinga" to avoid confusion between these two completely distinct vegetation types. In current English, German, and Spanish literature, the term *Amazon caatinga* has been used consistently for this forest complex, but in view of the new phytogeographical evidence of a western Guayana province including part of the Río Negro basin (see Chapter 4), "Rio Negro caatinga" should instead be used. On the other hand, in several publications (Takeuchi 1961, 1962; Goulding et al. 1988), the term *caatinga* is replaced by *campina* or *campinarana.* However, according to Rodrigues (1961a) or Prance and Schubart (1978), *campina* usually refers more to lower shrubby or herbaceous vegetation types, while *campinarana* refers to dense, tall woodlands; in both instances the vegetation grows on white sands in the same region.

The Rio Negro caatinga formation is widespread throughout the Río Negro basin, ranging from the upper Río Guainía in Colombia and south nearly to Manaus (Ducke and Black 1953; Rodrigues 1961a, 1961b; Pires and Rodrigues 1964; Pires and Prance 1985). In Venezuela (***unit 16*** on map), it commonly occurs in an area bounded by the Río Atacavi to the north, the middle Casiquiare and lower Siapa rivers to the east, and the Guainía and Negro rivers to the west. Isolated patches are also found along the lower Río Atabapo and in Yapacana National Park (Huber 1982) and around La Esmeralda (D. Coomes, personal communication).

Rio Negro caatinga forests are typical oligotrophic ecosystems. They are always associated with podsolized white-sand soils of extremely low nutrient content, and they occur in areas with pronounced internal fluctuations of the water table, often resulting in superficial flooding. Although Rio Negro caatinga forests grow in areas with very high annual rainfalls (up to 4000 mm per year) and little seasonality, even short dry periods of only a few days can cause considerable hydrological stress in the vegetation due to the low water-retention capacity of the sandy soils (Medina 1983; Medina et al. 1990). These forests consist of one to two tree layers, usually not exceeding 25–30 m in height, with relatively thin stems and small, open crowns. The foliage of most tree species is coriaceous and grayish green in color, making it is easy to recognize from the air and causing the forest to appear to be deciduous.

Although comparatively poor in species, these forests have a relatively high number of endemic species for lowland areas and contain botanically interesting families, such as the Lissocarpaceae (*Lissocarpa benthamii*) and the Tetrameristaceae (*Pentamerista neotropica;* Plate 69). The dominant Rio Negro caatinga trees belong to the Caesalpiniaceae (*Eperua leucantha*), Euphorbiaceae (*Hevea pauciflora, Micrandra spruceana, M. sprucei*), Sapotaceae

(*Manilkara* spp., *Pradosia schomburgkiana*), Rubiaceae (*Calycophyllum obovatum, Retiniphyllum* sp., *Pagamea coriacea*), Clusiaceae (*Clusia* spp.), Aquifoliaceae (*Ilex* spp.), Annonaceae, and Myristicaceae. Because of the more favorable conditions of illumination near the ground, the understory is relatively well developed, whereas lianas and epiphytes are almost absent.

Depending on the predominance of certain tree species, Klinge (1978) distinguishes two caatinga forest types: the "Yaguácana" type, dominated by *Eperua leucantha,* and the "Cunuri" type, dominated by *Micrandra sprucei.* Both grow typically in a complex mosaic with *tierra firme* forests dominated by *Eperua purpurea* (*yévaro*), and *Sclerolobium dwyeri* (Caesalpiniaceae), and *Monopteryx uacu* (*guaco,* Fabaceae). The spatial relationship of these forest types and of the typically associated scrub types is illustrated in Figure 3-2.

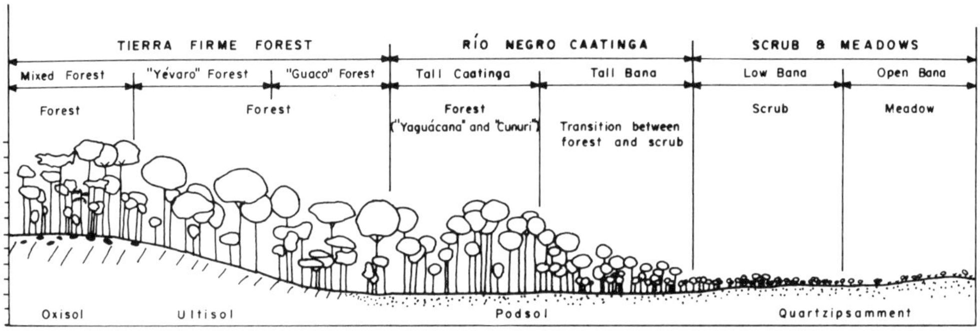

Figure 3-2. A topographical sequence of soil types and vegetation types in the vicinity of San Carlos de Río Negro. Besides the gradation between meadow, scrub, and forest, there is a clear physiognomic and floristic transition of forest types, from tall *bana* to mixed forest on oxisols. The "0" level on the vertical scale represents *tierra firme* forest, with areas below that level subject to periodic flooding or periods of standing water. Adapted from Moyersoen (1993).

From 1972 to 1985, the UNESCO Man and the Biosphere program (MAB) sponsored an international and multidisciplinary long-term study of the ecology of the Amazon caatinga near San Carlos de Río Negro. This project produced many new and important findings on the structure and functioning of these peculiar edaphic plant communities (e.g., Klinge et al. 1977; Brünig et al. 1978; Herrera et al. 1978; Klinge 1978; Klinge and Medina 1979; Jordan 1987, 1989; Medina et al. 1990). An annotated checklist of the plants of the region around San Carlos de Río Negro was prepared by Howard Clark and Ron Liesner, but only an incomplete, preliminary version was published in Jordan (1989).

Another peculiar forest type is restricted to permanently inundated floodplains on oxisols; these are the dense palm swamps mixed with low- to medium-sized treelets. The dominant palms seem to belong either to the genus *Mauritia* or to *Mauritiella aculeata,* whereas species of *Aspidosperma*

(Apocynaceae) and *Ormosia* (Fabaceae) along with *Chaunochiton loranthoides* (Olacaceae) have been reported as some of the more frequent trees and treelets. However, these plant communities are still largely unexplored. In some cases they occupy considerable extensions, as near the Orinoco-Casiquiare divide, in the middle Caño Yagua region, and in the upper Río Mavaca basin (***unit 17*** on map).

Although flooded forests are frequent in Venezuela's Amazonas state, they are much less studied than similar ecosystems in the Brazilian Amazon (Prance 1979; Kubitzki 1989b). With the numerous botanical explorations made since 1986 in the flooded forests of the Casiquiare basin by Basil Stergios, Gerardo Aymard, and other collaborators of UNELLEZ's herbarium (PORT), and visits by Klaus Kubitzki, results should be forthcoming to fill this information gap.

Finally, the riparian forests along large rivers with white or clear waters, such as the Orinoco, Ventuari, and the upper Casiquiare, represent another important forest type of Amazonas state (***unit 18*** on map). These forests usually grow along elevated riverbanks, although in some instances they may become temporarily flooded during high water periods. They are tall, dense, evergreen forests with little understory vegetation. By far the most important arboreal element is the genus *Campsiandra* (Caesalpiniaceae), accompanied by *Pterocarpus amazonica* (Fabaceae), *Tabebuia barbata* (Bignoniaceae), *Ceiba pentandra* (Bombacaceae), *Buchenavia tetraphylla* (Combretaceae), *Myrciaria dubia* and *Calycolpus calophyllus* (Myrtaceae), and *Gustavia hexapetala* (Lecythidaceae).

The lowland forests of Amazonas state are used mainly by Amerindian populations for their slash-and-burn agriculture. Except in the surroundings of Puerto Ayacucho and San Fernando de Atabapo, no large forest clearings are known. Several secondary forest products are exploited on a small scale mainly by Amerindians, such as the oil of the *seje* palm (*Oenocarpus bataua* and *O. bacaba*), *chiqui-chiqui* or *piassaba* fiber (*Leopoldinia piassaba,* Arecaceae; Plate 76), the *sarrapia* fragrance of the tonka bean (*Dipteryx odorata,* Fabaceae), and the aerial roots of *Heteropsis* spp. (Araceae) for wicker furniture, brooms, and other woven articles. In the past, many of these forests were more intensively exploited for wild rubber or *caucho* (*Hevea* spp., Euphorbiaceae), which was intensively tapped during the final part of the 19th century and the first three decades of the 20th century in southern and central Amazonas state. Other natural products such as ornamental flowers of lowland Rapateaceae (Plate 77) have some commercial potential.

Montane Forests (Uplands and highlands, ca. 500–3000 m elevation)

There is a wide variety of mountain systems in the Venezuelan Guayana, from the conspicuous high tepuis to extensive granitic ranges and isolated low massifs. Although forests are the predominant vegetation formation throughout this area, there is even less information available for montane forests than for lowland forests, and much of the following account should be considered provisional. Because of the complexity of the montane forests in the Venezuelan Guayana, they are treated here in geographical sequence, from east to west and then north to south. Upland and highland forests are treated together, allowing for descriptions of the altitudinal zonation in each area, following a general sequence from basimontane to lower montane, montane, upper montane, and high-tepui life zones (see Figure 3-1).

Basimontane and submontane forests of the Serranía de Imataca. The mainly hilly, granitic Serranía de Imataca is located in the northeastern corner of Bolívar state and southwestern Delta Amacuro state. Although the range barely reaches an elevation of 650 m, its dense forest cover is differentiated into a lower basimontane forest type and a more humid submontane forest type at higher elevations. According to Steyermark (1968), the submontane forests (***unit 19*** on map) are macrothermic, ombrophilous, evergreen communities with two tree layers, the upper one reaching 25–30 m tall. The dominant species are *Virola surinamensis* (Myristicaceae), *Licania alba, L. densiflora,* and *Parinari excelsa* (Chrysobalanaceae), *Inga punctata* (Mimosaceae), *Mora gonggrijpii, Hymenaea courbaril*, and *Clathrotropis brachypetala* (Caesalpiniaceae), *Lecointea amazonica* (Fabaceae), *Cedrela odorata* (Meliaceae), *Erisma uncinatum* (Vochysiaceae), *Catostemma commune* (Bombacaceae), *Vochysia tetraphylla* and *Qualea dinizii* (Vochysiaceae), *Aspidosperma* spp. (Apocynaceae), and *Tabebuia stenocalyx* (Bignoniaceae).

In the southeastern part of the Serranía de Imataca, there is a high plain between 400 and 600 m elevation called the Altiplanicie de Nuria; at least six endemic species have been found there in small patches of cloud forest. Large parts of the rich Imataca forests have been logged intensively for almost 30 years, which has led to a considerable decline in the forest cover. Much of the area is now covered by secondary forests in various stages of succession, especially on the more accessible and drier western slopes.

Forests of the Gran Sabana uplands and adjacent tepuis. South-central Bolívar state contains the largest continuous extent of upland and highland landscapes in the Venezuelan Guayana. This area includes the upland plains of the Gran Sabana and is surrounded by the spectacular highlands of the eastern tepui chain (Roraima–Ilú) to the southeast, the mid-altitude Cerro Venamo and Sierra de Lema to the north, the chain of the Aparamán range (Los Testigos) to the north, the Auyán-tepui massif to the northwest, and the huge Chimantá massif to the west. Most of this area is now included in Canaima National Park and is covered by a variety of dense upland and highland forests, except for open meadows in parts of the Gran Sabana and some northern mid-elevation plains.

Although there have been several floristic and ecological studies in the forests of this region since the 1940s, such as Steyermark (1966), Fölster (1986), CVG-TECMIN (1987, 1989), and Hernández (1992), botanical knowledge is still fragmentary. The following account provides a general view of the main groups of forest types found in the different altitudinal life zones of the area. This is illustrated by a schematic transect from the northern Sierra de Lema across the Gran Sabana to the summits of the Chimantá massif (see Figure 3-3).

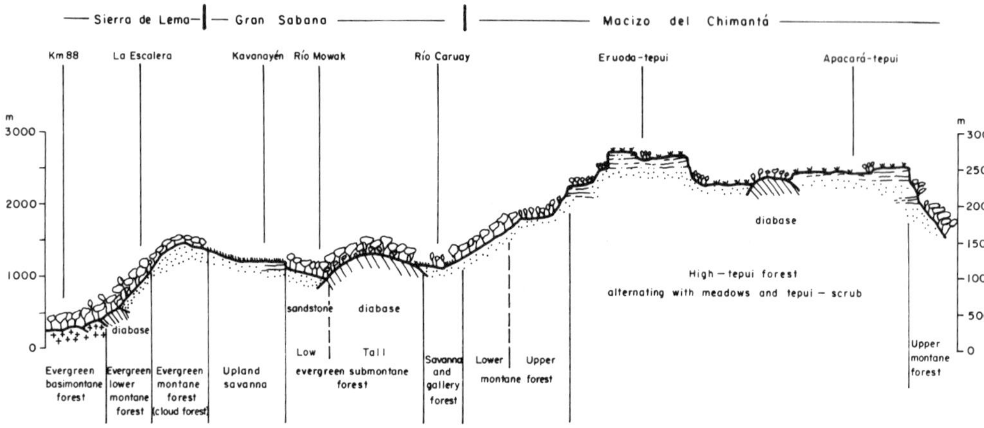

Figure 3-3. A sequence of vegetation types through different life zones and substrates along a roughly north-south schematic transect through the Gran Sabana (vertical scale exaggerated). Elevations above sea level.

On the lower and mid-elevation slopes of the Sierra de Lema along the highway that climbs from Las Claritas (Kilómetro 88) to La Escalera and the Gran Sabana, there are dense, medium-high to tall, evergreen rain forests. Some of these forests also extend in isolated patches into the northern Gran Sabana. Because of their numerous emergent tree crowns, these submontane forests (***unit 20*** on map; Plate 17) usually have an irregular physiognomy, which favors the growth of a dense understory. The dominant trees are *Virola surinamensis* (Myristicaceae), *Protium heptaphyllum* (Burseraceae), *Tabebuia insignis* (Bignoniaceae), *Anaxagorea petiolata* (Annonaceae), *Alexa confusa* (Fabaceae), *Ruizterania ferruginea* (Vochysiaceae), *Licania micrantha* (Chrysobalanaceae), *Simarouba amara* (Simaroubaceae), *Minquartia guianensis* (Olacaceae), *Pourouma* spp. (Cecropiaceae), and *Byrsonima stipulacea* (Malpighiaceae).

Evergreen submontane forests are widespread in the northern sections of the Sierra de Lema, Cordillera Epicara, and the lower talus slopes of the adjacent tepuis, such as Amaruay-tepui. In 1986, Ronald Liesner and Bruce Holst of the Missouri Botanical Garden made the first detailed botanical inventory of the dense slope forests that grow between 500 and 1000 m elevation on this low tepui. The dominant tree species there are *Aspidosperma* spp. (Apocynaceae), *Caraipa densifolia* (Clusiaceae), *Cassipourea guianensis* (Rhizophoraceae), *Clathrotropis macrocarpa* (Fabaceae), *Erisma uncinatum* and

Ruizterania ferruginea (Vochysiaceae), *Lecythis zabucaja* (Lecythidaceae), *Protium cuneatum* (Burseraceae), and *Tapirira guianensis* (Anacardiaceae).

A different forest type, in which the trees are generally lower and their crowns more flattened and more densely arranged into two strata, occurs on the summit of La Escalera (ca. 1400 m elevation), in the northern part of Gran Sabana, and on the slopes of surrounding tepuis as high as 1600 m. These are ombrophilous, evergreen, montane forests (***unit 21*** on map; Plate 22), which have a high abundance of epiphytes due to the frequent occurrence of mist and a sparse undergrowth formed by shrubs and large, terrestrial herbs. Some of the important trees of these forests are *Dimorphandra macrostachya* (Caesalpiniaceae), *Byrsonima stipulacea* (Malpighiaceae), *Sloanea pittieriana* (Elaeocarpaceae), *Platycarpum rugosum* (Rubiaceae), *Endlicheria nilssonii* (Lauraceae), *Sterigmapetalum guianense* (Rhizophoraceae), *Caryocar montanum* (Caryocaraceae), *Moronobea ptaritepuiana* (Clusiaceae), and *Podocarpus magnifolius* (Podocarpaceae). These data are taken from Steyermark, who made extensive collections on the slopes of Ptari-tepui in 1944 (Steyermark 1966) and in the region of Cerro Venamo, at the eastern end of Sierra de Lema, in 1960 (Steyermark and Nilsson 1962).

Between roughly 1600 and 2000 m elevation, there is a belt of low, upper montane forests that usually extends along the upper slopes until it reaches the base of the vertical cliffs of the tepuis (***unit 22*** on map; Plates 22 and 34). In the southeastern Venezuelan Guayana, this is the true cloud forest zone, characterized by a very high frequency of orographic mist during much of the year. This same zone also occurs on the summits of some low tepuis, such as Sororopán-tepui, Carrao-tepui, and Uaipán-tepui, where there are no vertical rock walls separating the summit from the slopes.

A predominant cloud forest component in these areas is the genus *Bonnetia* (Theaceae), which often forms large, dense colonies. *Bonnetia tepuiensis* and *B. roraimae* are particularly abundant on the upper slopes of the eastern tepui chain, Carrao-tepui, and on the Chimantá massif, whereas *B. steyermarkii* is dominant on the upper slopes of Auyán-tepui (Steyermark 1967). Other important trees are *Podocarpus* spp. (Podocarpaceae), *Magnolia ptaritepuiana* (Magnoliaceae), *Schefflera* spp. (Araliaceae), and *Weinmannia* spp. (Cunoniaceae). Most of these cloud forests resemble elfin forests, with the tree trunks and branches covered densely by lichens, mosses, ferns, and other epiphytes. The understory is also very dense, with giant rosette herbs such as *Orectanthe ptaritepuiana* (Xyridaceae), *Brocchinia tatei* (Bromeliaceae), *Didymiandrum stellatum* (Cyperaceae), and bambusoid grasses (*Myriocladus* spp.), as well as numerous low shrubs.

These low, evergreen, high-tepui forests (***unit 23*** on map; Plates 36 and 38) grow mostly on organic soils (peat) overlying sandstone, but in some cases they are also found on mineral soils derived from diabase intrusions, especially in the Chimantá massif (Huber 1992b). Their physiognomy and floristic composition varies accordingly; tepui forests on peat are always dominated by *Bonnetia roraimae, B. tepuiensis,* or *B. wurdackii* (Theaceae), whereas those growing on diabase are often dominated by *Stenopadus chimantensis* (Asteraceae) and *Spathelia ulei* (Rutaceae).

The uppermost forest belt in the eastern Río Caroní drainage is always located above 2000 m and is best developed in low depressions or along creeks and small rivers of the summit plateaus of large tepui massifs such as Auyán-tepui and the Macizo del Chimantá. Smaller tepui summits usually have only patches of forest that are restricted to wind-protected sites or depressions, as on Kukenán-tepui or on Kamarkawarai-tepui (Plate 32). This high montane forest belt is known to occur as high as 2600 m elevation on Murey-tepui in the Chimantá massif.

Compared with the lower elevation cloud forests, high-tepui *Bonnetia* forests have fewer species and are physiognomically more homogeneous (Vareschi 1992b). Their single tree layer is usually 6 to 12 m tall, with dense, microphyllous crowns, twisted stems and branches, few epiphytes, and a relatively open undergrowth dominated by large rosette herbs (especially *Brocchinia* spp.) and other low herbs and shrubs. Tepui forests growing on diabase are much more open and irregular than those on sandstone, but they are too poorly studied to allow a more detailed description.

Montane forests of the Paragua-Caura river basins. There are many large mountain systems in the Caura and Paragua river basins, although none are as high as those of the Caroní basin. Since almost nothing is known about the forests of the slopes and summits of these mountains, only a brief account of the different forest belts that can be recognized from the air is provided here.

In the wide plains between the lower Caura and Aro rivers lies the Serranía del Trueno, a series of steep-sloped, rocky, intrusive mountains that stand out against the surrounding lowlands. Although the highest peaks of this range reach altitudes of approximately 800 m, most of the peaks probably do not exceed 600 m elevation. During the lengthy dry season (up to six or seven months), the slopes and summits of these mountains are completely gray, because of the total deciduousness of their basimontane and lower montane forests, which contrast sharply with the surrounding semideciduous lowland forests. To date, there is no botanical information available on these interesting islands of dry forests (***unit 24*** on map). A similar azonal forest type probably also occurs on the lower granitic inselbergs (*lajas* or rock outcrops) that are scattered to the east as far as the lower Río Caroní.

Farther south towards the center of the Río Paragua basin, there are medium to tall, evergreen, basimontane and lower montane forests along the lower and middle slopes of Cerro Guaiquinima, Cerro Camarón, Cerro Ichún, and Cerro Guanacoco (***unit 25*** on map). These occur mostly between elevations of 300 and 700 m; similar forests are probably also found on the eastern and northern slopes of the Cerro Jaua–Cerro Sarisariñama complex in the Río Caura basin. Although Félix Cardona made some botanical collections in this region during the late 1930s, and Julian Steyermark made intensive collections along the Paragua and the lower Ichún rivers in the 1960s, there are no published data on these forests.

The huge Cerro Guaiquinima plateau is moderately inclined from north (1600 m elevation) to south (750 m) and is largely covered by a dense, medium-sized, evergreen montane forest (***unit 26*** on map). This forest type ranges in elevation between roughly 800 and 1500 m; Steyermark and Dunsterville (1980) indicated its approximate extension in a small vegetation map of the plateau.

The southern mountains of the upper Paragua basin, Cerro Guanacoco, Cerro Ichún, Serranía Marutaní, and Chaco-tepui are all rather low in elevation (ca. 800–1500 m) and have densely forested upland plateaus (***unit 27*** on map). Since this area is one of the wettest in the Venezuelan Guayana, the montane forests are all ombrophilous, evergreen, and medium-sized, with characteristically dense, flattened crowns and coriaceous leaves. There is no botanical information available on the forests of these mountains yet, but on the nearby Sarisariñama plateau, Steyermark found a similar forest type at 1400 m around the famous sinkholes or *simas* (Steyermark and Brewer-Carías 1976). The dominant trees in this medium-sized (20–25 m), evergreen, montane forest are *Pradosia beardii* (Sapotaceae), *Oedematopus duidae* (Clusiaceae), *Virola pavonis* (Myristicaceae), *Clusia* sp. (Clusiaceae), *Myrcia revolutifolia* (Myrtaceae), *Persea grandiflora* (Lauraceae), *Pithecellobium* (=*Abarema*) *longipedunculatum* (Mimosaceae), *Licania discolor* (Chrysobalanaceae), *Couma rigida* (Apocynaceae), *Conceveiba ptariana* (Euphorbiaceae), and *Stenopadus talaumifolius* (Asteraceae). Numerous epiphytic mosses, ferns, and orchids as well as climbers grow on the stems and in the crowns, whereas the ground is covered by large Araceae (*Stenospermation, Philodendron,* and *Anthurium*), Cyclanthaceae (*Sphaeradenia*), and other shade-loving herbs.

Interesting upper montane forests occur on the higher summits of the Jaua-Sarisariñama massif in the southern Caura basin (***unit 28*** on map). According to Steyermark and Brewer-Carías (1976), who collected there in 1974, these are low to medium (15–25 m tall), dense, evergreen, humid forests, in which the dominant trees belong to *Sloanea* (Elaeocarpaceae), *Matayba* (Sapindaceae), *Vochysia* (Vochysiaceae), *Podocarpus* (Podocarpaceae), *Perissocarpa* (Ochnaceae), *Hieronyma* (Euphorbiaceae), *Prunus* (Rosaceae), *Weinmannia* (Cunoniaceae), *Cecropia* (Cecropiaceae), *Hedyosmum* (Chloranthaceae), *Digomphia* (Bignoniaceae), *Vismia* (Clusiaceae), *Schefflera* (Araliaceae), and *Kotchubaea* (Rubiaceae). The understory is well developed with many shrubs and herbs, and with large tufts of the bambusoid grass *Neurolepis glomerata*.

Due to a very low population density of indigenous inhabitants in the Paragua and Caura river basins (mainly Pemón, Sanema, and Yekuana Amerindians), the montane forests of this area have been very little affected by human intervention. In the early 1990s, however, mining activities were reported from the slopes of Cerro Guaiquinima, Serranía Marutaní, and southern Cerro Ichún.

Forests of the Sierra de Maigualida. The huge Sierra de Maigualida mountain system forms the northeastern divide between Amazonas and Bolívar states. It is almost entirely covered by dense forests from its base to approximately 2000 m elevation, at which point the forests are replaced by more open summit vegetation types that reach the highest peaks of nearly 2400 m. Curiously, this enormous mountain region had not been visited by scientists before March 1988, probably because of its very difficult access even by helicopter and its lack of allure relative to the more exotic, isolated tepuis.

In the late 1980s and early 1990s, Otto Huber organized several trips to the summit regions of Sierra de Maigualida. Although there is no published information available yet on the upper portions of this range, it is possible to make a clear, though simplified altitudinal zonation of the main forest types, based on observations from numerous helicopter flights over most parts of the massif.

Ascending the east-facing slopes of the Sierra de Maigualida, there is a variable belt of tall, evergreen, basimontane forests. These are probably very similar to the lowland forests of the adjacent Caura peneplains (unit 11 on map). On steeper slopes, mainly between 600 and 1500 m elevation, another forest type clearly appears, characterized by an unidentified, salmon-red flowering, large tree with rounded crowns. These medium-sized, evergreen montane to lower montane forests (***unit 29*** on map) have an irregular, open canopy structure.

Finally, between 1500 and 2000 m elevation, a still poorly differentiated mosaic of well-developed montane and upper montane cloud forests (***unit 30*** on map; Plate 47) occupies the third and uppermost forest belt in the Sierra de Maigualida. This belt appears to reach its upper limit at an elevation around 2000–2100 m and coincides with a clear, though somewhat fluctuating tree line.

Two of these interesting forests have been visited by Otto Huber and other collaborators, one in the Sierra de Maigualida at 2000 m elevation, the other in the more southern Serranía Uasadi, at about 1900 m. In the first area, a very moist, upper montane forest type occurs on moderately inclined terrain with dark brown topsoil. This forest consists of low trees (4–8 m tall) abundantly covered with mosses and other epiphytes; the canopy is very irregular with many small gaps, allowing the growth of large colonies of terrestrial tank bromeliads (*Brocchinia tatei*). The dominant trees are *Cyrilla racemiflora* (Cyrillaceae), with stem diameters of up to 70 cm, *Clusia* spp. (Clusiaceae), and *Perissocarpa* (Ochnaceae), accompanied by *Ecclinusa ulei* (Sapotaceae), *Gongylolepis* sp. (Asteraceae), *Byrsonima* sp. (Malpighiaceae), and various species of *Schefflera* (Araliaceae). A beautiful white-flowered member of the Magnoliaceae occasionally grows in these low forests as well.

The upper montane forest visited on Serranía Uasadi occurred on a steep slope and was dominated almost exclusively by low to medium-tall (6–12 m) trees of *Perissocarpa* (Ochnaceae) with straight boles, only occasionally mixed with *Clusia* (Clusiaceae), *Schefflera* (Araliaceae), and many palms (*Euterpe*?)

occurring in small colonies. Although the canopy of this forest is quite dense, the large, typically light green leaves of *Perissocarpa* are readily distinguished from a distance and allow high levels of light to penetrate to the forest interior. Still, the ground layer is remarkably scarce, possibly due to allelopathic effects of the decomposing litter of *Perissocarpa* leaves. The trunks and branches of the trees were abundantly covered by dense mats of mosses and filmy ferns, suggesting very moist meso- and microclimatic conditions.

To the west of the main ridges of the Sierra de Maigualida lies the wide valley of Caño Iguana. In this area, clearly drier forest types prevail on parts of the west-facing slopes (***unit 31*** on map). There is evidently a strong rain-shadow effect, because some of the montane forests of this section are semi-deciduous. Some forests are interrupted by open areas with bracken ferns, indicative of sporadic fires. There is so far no floristic information for any of these forests.

Forests of the northwestern uplands. There are extensive uplands in northwestern Bolívar state that stretch west from the lower Río Caura to the Río Parguaza and further to the Río Cuao in Amazonas state. These areas are covered by different forest types, the distribution of which is influenced mainly by the macroclimate, orographic position, and local edaphic conditions. The region has a strong rainfall gradient that increases from north to south and from east to west. There is also an altitudinal gradient that ascends steadily from approximately 100 m in the north near the middle Orinoco lowlands to nearly 1000–1300 m in the south, along the base of Cerro Guanay, Cerro Yaví, and Cerro Yutajé.

Although there have been several forest inventories in the northwestern uplands, especially in the Parguaza and Suapure river basins, little information has been published. This makes it difficult to produce a reliable classification of the predominant forest types in the area, but at least three different upland forest types can be recognized at elevations between 200 and 1300–1500 m: (1) tall to medium-sized, semideciduous to subevergreen, basimontane forests, (2) medium-sized, evergreen, basimontane forests, and (3) medium-sized, evergreen, submontane to montane forests.

The tall to medium-sized, semideciduous to subevergreen, basimontane forests are found from the lower Río Caura in northern Bolívar state to the lower Río Cuao in northwestern Amazonas state, on piedmont hills and low mountains between elevations of approximately 200 and 600 m (***unit 32*** on map). According to Boom (1990a), who made a detailed floristic study of a forest near the Río Maniapure between 100 and 500 m elevation, the following tree species were dominant: *Parinari excelsa, Licania canescens,* and *L. cruegeriana* (Chrysobalanaceae), *Vochysia glaberrima* (Vochysiaceae), *Pouteria* sp. and *Elaeoluma glabrescens* (Sapotaceae), *Cochlospermum orinocense* (Bixaceae), *Terminalia amazonia* (Combretaceae), *Galipea davisii* (Rutaceae), and *Macrolobium* cf. *bifolium* (Caesalpiniaceae). In another forest inventory made by Finol and collaborators in 1970 in the middle Parguaza basin at approximately 400 m elevation, the following dominant tree species were found: *Trichilia* sp. (Meliaceae), *Eschweilera subglandulosa* (Lecythidaceae), *Erisma uncinatum* (Vochysiaceae), *Licania densiflora* (Chrysobalanaceae), as well as several legume trees (Finol 1974).

Medium-sized (12–25 m), evergreen, basimontane forests are found growing on bauxite soils of moderately dissected uplands (400–600 m elevation) in the lower Suapure and Parguaza river basins. This forest type (***unit 33*** on map) was described during inventories made during the development of the bauxite mine at Los Pijiguaos (Huber and Guánchez 1988). Its dominant species, *Caryocar pallidum* (Caryocaraceae) and *Qualea paraensis* and *Vochysia surinamensis* (Vochysiaceae), are indicators of high aluminum contents in the soil. Other important elements of this forest type are the palms, especially *Oenocarpus bacaba, Attalea maripa,* and *Astrocaryum gynacanthum.*

Finally, medium-sized, evergreen, submontane to montane forests extensively cover the dissected uplands and isolated mountains between 600 and 800 m elevation in the north to 1300 to 1500 m in the south, along the north-facing lower slopes of the Guanay-Yutajé-Yaví mountain chain and east to the west-facing slopes of Serranía Nichare (***unit 34*** on map). At the northern base of Cerro Yaví, many areas affected by windfall were observed in this forest type. There is no botanical information available yet on these forests.

Many of the forests of this upland region in northwestern Bolívar have been affected by human impacts. Intensive gold and diamond mining has been carried out for many years in the Guaniamo region, and the large bauxite mine of Los Pijiguaos has attracted numerous workers with their families. Several urban centers are currently under construction for the workers. In the early 1990s, the Venezuelan government opened international licitations for intensive logging of the Parguaza forests. Finally, the unplanned construction of penetration roads that branch off from the newly concluded highway between Caicara and Puerto Ayacucho has led to considerable deforestation for agriculture.

Forests of the Yaví-Yutajé-Guanay mountain complex. There has been very little botanical research in this imposing series of tepuis that rise more than 2000 m above the broad plains of the Manapiare and Parucito valleys in northern Amazonas, but the expected altitudinal zonation of forest types must vary considerably on the different slope exposures. The northern base of the three mountains lies mostly between 600 and 1400 m elevation, whereas the southern base starts at close to 200 m. The southern slopes lie in a rain shadow, resulting in savannas and shrublands in some areas.

Tentatively, three montane forest types can be recognized on this ecologically heterogeneous mountain complex: (1) basimontane, semideciduous forest, (2) dense, medium-sized, submontane to montane evergreen forests, and (3) dense, low-sized, wet, evergreen upper montane forests. The first forest type, a basimontane, semideciduous forest belt, extends along the southern slopes between approximately 200 and 500 m elevation (***unit 35*** on map).

The second type, dense, medium-sized, submontane to montane evergreen forests (***unit 36*** on map), are found between approximately 500 and 1500 m, especially on Cerro Yaví and in the ascending interior uplands of the Yutajé–Coro Coro massif. In 1987, Ronald Liesner and Bruce Holst collected the following common, large trees in the valley between Cerro Yutajé and Cerro Coro Coro at elevations of 500 to 1100 m: *Caraipa* sp. (Clusiaceae), *Caryocar pallidum* (Caryocaraceae), *Chrysophyllum sanguineolentum* subsp. *balata* (Sapotaceae), *Dimorphandra* sp. (Caesalpiniaceae), *Dictyocaryum ptariense* (Arecaceae), *Eschweilera coriacea* (Lecythidaceae), *Ocotea oblonga* and *O. flavantha* (Lauraceae), and *Richeria grandis* (Euphorbiaceae).

Finally, dense, low-sized, wet, evergreen upper montane forests are found ranging from approximately 1500 to 2000 or 2200 m elevation. In some cases, these are true cloud forests, such as on the summits of Cerro Yaví and of Cerro Yutajé (***unit 37*** on map). On Cerro Yaví, the dominant trees are *Schefflera hitchcockii* (Araliaceae), *Clusia* (Clusiaceae), *Ilex retusa* (Aquifoliaceae), *Ternstroemia* (Theaceae), *Cyrilla racemiflora* (Cyrillaceae), and *Weinmannia* (Cunoniaceae), whereas on Cerro Yutajé, at an elevation of approximately 1850 m, there are palm-rich, low forests with *Micropholis* sp. (Sapotaceae), *Clusia pachyphylla* (Clusiaceae), *Ilex* (Aquifoliaceae), *Hedyosmum* (Chloranthaceae), *Schefflera hitchcockii* (Araliaceae), and *Geonoma appuniana* and *Euterpe* (Arecaceae) among the most frequent trees. A similar forest type also occurs in small pockets on the summit of Cerro Guanay.

Forests of the Cuao-Sipapo massif. This large mountain system lies between the Río Ventuari to the east and the Río Orinoco to the west and southwest. It is very diverse orographically and geologically, with typical sandstone tepuis such as Cerro Cuao and a series of other high mountain peaks with an igneous-metamorphic geology. Although these mountains can be seen in the distance from Puerto Ayacucho, very little botanical exploration has been made in this huge area. Bassett Maguire and Louis Politi visited a section of Cerro Sipapo (which they called Cerro Paraque) in 1947, and Julian Steyermark collected on the summit of Cerro Autana in 1971 (Steyermark 1974, 1975). In the late 1980s, the ethnobiologist Stanford Zent made an ethnobotanical survey among the Piaroa Amerindians of the upper Cuao mountain region, and Angel Fernández made helicopter visits to sections of Cerro Cuao around 1500 m elevation. Otto Huber also visited the area by helicopter in 1993. Although a number of new and interesting species have been described from these botanists' collections, there is very little detailed information on any of the dense forests types that cover almost the entire Cuao-Sipapo massif.

Based mainly on physiographic and climatic criteria and notes made during overflights, at least two forest types can be distinguished in this area. The extensive uplands between 400 and 1000 to 1200 m elevation are covered by dense, tall to medium-sized, evergreen submontane and montane forests (***unit 38*** on map). These forests are the main homeland for the Piaroa Amerindians.

At higher elevations, between 1200 and 1600 to 1800 m, there are other, somewhat lower forests with irregular canopies, corresponding to the evergreen upper montane (cloud) forest type (***unit 39*** on map). Near the summit of Cerro Cuao, at approximately 1800 m elevation, patches of high-tepui forests dominated by *Graffenrieda fantastica* (Melastomataceae), *Clusia* sp. (Clusiaceae), *Phyllanthus* sp. (Euphorbiaceae), and *Spathelia ulei* (Rutaceae) were observed, with a dense understory formed by giant bambusoid grasses (*Neurolepis, Myriocladus*) and tank bromeliads (*Brocchinia tatei*).

In his account of the vegetation and flora of the summit of Cerro Autana, Steyermark (1974, 1975) described a small forest community growing along the edge of the plateau; it was composed of low trees to 10 m tall, including two species of *Clusia* (Clusiaceae), *Capirona decorticans* and *Ladenbergia lucens* (Rubiaceae), *Sloanea* sp. (Elaeocarpaceae), *Weinmannia sorbifolia* (Cunoniaceae), and *Miconia roraimensis* (Melastomataceae). This azonal fringe forest, though, is not likely to be representative of the overall forest cover of the Cuao-Sipapo mountain system.

Much of the Cuao-Sipapo massif (1,215,000 hectares) was declared a forest reserve in 1963, but so far there has been no commercial logging, mainly because of the area's difficult access. Except for the traditional land use practices of the forest-dwelling Piaroa Amerindians, no other human-induced impacts have been reported from this mountain area.

Forests of Cerro Yapacana. The isolated, oblong outline of Cerro Yapacana rises from the lowlands close to the confluence of the Orinoco and Ventuari rivers, where it reaches an elevation of close to 1300 m. The mountain is almost entirely forested, except for the southwestern slope and for the cliffs of the northwest- and southeast-exposed walls. Surrounding the mountain is a complex mosaic of lowland forests, shrublands, and broad-leaved meadows.

Despite two botanical explorations made on Cerro Yapacana by Bassett Maguire, Richard Cowan, and John Wurdack in 1951, and by Julian Steyermark and George Bunting in 1970, no information on the vegetation types of this interesting mountain has been published. From the botanical specimens and their label data, it appears that there are two types of montane forests. Tall to medium-sized, dense, evergreen, lower montane forests occur on the slopes between approximately 200 and 800 m (***unit 40*** on map), while low (up to 15 m), dense, evergreen, montane (cloud) forests cover the summit plateau between 1000 and 1300 m (***unit 41*** on map). The latter, a wet forest, has approximately eight endemic species and is dominated by *Tepuianthus yapacanensis* (Tepuianthaceae), together with *Bonnetia tristyla* (Theaceae), *Symplocos yapacanensis* (Symplocaceae), and *Gongylolepis yapacana* (Asteraceae).

Although the lower slopes of Cerro Yapacana are located in a national park, they have been invaded by gold miners since at least the 1980s. Occasional fires have climbed up the drier southwestern slope, where open secondary scrub has partially replaced the original forest. So far, the summit forest appears to remain in pristine condition.

Forests of the Parú massif. The Parú massif is located near the southern headwaters of the Río Ventuari and consists of three mountains, Cerro Parú, Cerro Asisa, and Cerro Euaja. It ranges in elevation from 200 to approximately 2000 m or more. Because of its intermediate location between the Jaua-Sarisariñama massif to the northeast and the Duida-Marahuaka complex to the south, Cerro Parú is a phytogeographically interesting area, and it has a high level of endemism on its summits. The present botanical knowledge of this large mountain system is still scarce, despite several explorations begun in 1951 by Richard Cowan and John Wurdack, and continued by Otto Huber in 1978 and by Paul Berry and Judith Rosales in 1991.

Due to the lack of published data, it is only possible to make a tentative classification of forest types in the Parú massif. Since large parts of the massif lack the sheer sandstone walls characteristic of many tepuis, the altitudinal zonation is more difficult to recognize. In general, a belt of tall to medium-sized, evergreen submontane forests extends along the lower slopes and ascends partially to the several extensive interior plateaus, probably up to 800 to 1000 m elevation (***unit 42*** on map). No botanical collections have been made yet in this forest type.

Above approximately 1000 m elevation, the general climate on Cerro Parú becomes markedly more humid. Lower, evergreen, montane forests occur there, probably up to elevations of 1800 m or more, especially on Cerro Euaja (***unit 43*** on map). A small forest island, situated at approximately 1200 m in the southern interior plateau of Cerro Parú, was inventoried in March 1991 by Paul Berry and Judith Rosales. It was a medium-sized, relatively open forest, growing in a depression with trees up to 15 m tall. The dominant species was *Richeria grandis* (Euphorbiaceae), accompanied by several species of Clusiaceae, *Simarouba amara* (Simaroubaceae), *Podocarpus tepuiensis* (Podocarpaceae), and palms (*Euterpe* sp.). In some areas, the understory was dominated by large clumps of two species of *Saxofridericia* (Rapateaceae; Plate 55).

Forests of the Duida-Marahuaka massif. Cerro Duida was the first tepui to be climbed in the Venezuelan Amazon (by the Tyler-Duida expedition in 1928), and has since been visited numerous times by botanists, the most important of them being Julian Steyermark (1944) and Bassett Maguire (1949, 1950). Maguire, together with Richard Cowan and John Wurdack, was also the first to ascend to the summit plateau of the nearby Cerro Huachamacari in 1950. The summit of the much higher Cerro Marahuaka (2800 m elevation) is accessible only by helicopter; it was first visited in 1974 by Stephen Tillett, but since then it has been the center of more than a dozen expeditions organized first by Charles Brewer-Carías and later by the Fundación Terramar. Only three papers provide useful notes on the vegetation of these mountains, Tate (1931), Tillett and Steyermark (1982), and Delascio and Steyermark (1989). The only paper with information on the forests is Tate's, although it contains very little floristic information.

Both Cerro Marahuaka and Cerro Huachamacari are imposing tepuis with their summits separated from the surrounding lowlands by sheer walls. In contrast, Cerro Duida has a continuously ascending interior plateau from north to south, where its highest point is located at about 2360 m. As a result, Duida is extensively covered by different kinds of forests, whereas on Cerro Marahuaka and Cerro Huachamacari, the forests are limited mainly to the slopes.

Based upon our present state of knowledge, the forests of this mountain complex can be classified into four types: (1) tall, evergreen, lower montane forests, (2) medium, evergreen, montane forests, (3) low, evergreen, upper montane forests, and (4) low, evergreen high-tepui forests.

Tall, evergreen, lower montane forests occur along the talus slopes of all three mountains, extending probably up to 800 m (***unit 44*** on map). There is no botanical information currently available on these forests.

Medium, evergreen, montane forests occur between 800 and 1500 m on the upper talus slopes of Cerro Marahuaka, as well as in the northern section of the Duida plateau (***unit 45*** on map). *Perissocarpa* (Ochnaceae) and *Dimorphandra* (Caesalpiniaceae) appear to be frequent and often locally dominant trees in this forest type on Cerro Duida (N. Dezzeo and O. Huber, personal observation). Typically lowland plants such as the palm *Manicaria saccifera* have been collected here as high as 1100 m elevation.

Low (7–10 m), evergreen, upper montane forests between 1500 and 2200 m extensively cover the central and southern section of Cerro Duida (***unit 46*** on map). The dominant trees here are *Tyleria floribunda* and *T. spathulata* (Ochnaceae), *Neotatea longifolia* (Clusiaceae), *Gongylolepis* sp. (Asteraceae), and *Schefflera* sp. (Araliaceae). Inside these open forests are many large herbs such as *Saxofridericia duidae* and *Stegolepis grandis* (Rapateaceae), *Everardia* (Cyperaceae), and *Brocchinia* (Bromeliaceae). *Gleasonia duidana* (Rubiaceae) and *Archytaea multiflora* (Theaceae) are common along rivers and creeks.

Finally, according to Delascio and Steyermark (1989), a low, evergreen high-tepui forest occurs in occasional depressions on the summit of Cerro Marahuaka above 2600 m (***unit 47*** on map). These authors mentioned *Podocarpus roraimae* (Podocarpaceae), *Schefflera umbellata* (Araliaceae), *Daphnopsis steyermarkii* (Thymelaeaceae), *Psychotria jauaensis* (Rubiaceae), *Befaria sprucei* (Ericaceae), and *Weinmannia velutina* (Cunoniaceae) as the dominant tree species.

Few other tepuis of the Guayana highlands have a greater variety of forest types on their summit than Cerro Duida. Much more study is needed to better understand their ecological and floristic characteristics, which appear to be strongly influenced by particular geologic and edaphic conditions not found on other mountains in the Guayana Shield.

Forests of the Parima uplands. The mainly granitic Sierra Parima, which forms the eastern border of Amazonas state with Brazil, consists of a series of moderately to heavily dissected, low- and mid-altitude uplands, ranging in elevation mainly from 750 to 1300 m. The southwest-facing slopes descend gradually from the uplands to the upper Orinoco lowlands. On either side of the lower Río Matacuni are three small tepuis that reach altitudes between 900 and 1300 m.

This physiographically and geologically complicated mountain range can be considered a southern prolongation of the larger and higher Maigualida-Uasadi range; it is mostly covered by several types of dense forests. Only the upper forests have been explored botanically, by Julian Steyermark in 1972 and 1973. The results of these explorations, together with studies by Ghillean Prance on the Brazilian side of Sierra Parima and by Otto Huber in the savannas of the Parima uplands, have been published in Huber et al. (1984).

Along the western slopes of the Sierra Parima, tall, dense, evergreen submontane forests predominate between approximately 300 and 800 m (***unit 48*** on map). These forests are still completely unexplored and are characterized by an irregular canopy usually 20–30 m tall. The uplands between 800 and 1300 m elevation are covered by a different kind of forest formed essentially by medium-sized (15–25 m tall), semideciduous to sub-evergreen montane forests (***unit 49*** on map; Plate 23). The most important tree species of these forests, which were studied between 800 and 1100 m near Simarawochi in the upper Río Matacuni drainage, are *Eschweilera roraimensis* (Lecythidaceae), *Guarea guidonia* (Meliaceae), *Tapirira guianensis* (Anacardiaceae), *Cordia nodosa* (Boraginaceae), *Miconia* spp. (Melastomataceae), *Vochysia venezuelana* (Vochysiaceae), and *Protium calanense* (Burseraceae). The undergrowth of these generally dense forests is remarkably scarce, as are epiphytes and lianas.

Along the west-central edge of Sierra Parima, two interesting azonal forest types (one edaphically and the other climatically conditioned) have been detected, though not yet visited. The first forest type occurs in variably sized patches on the summits of the three sandstone mesetas located in the lower Matacuni valley; it is a dense, medium-sized (15–20 m), evergreen montane forest (***unit 50*** on map), with the canopy dominated by the broad, flattened crowns of a still unidentified species of *Dimorphandra* (Caesalpiniaceae), together with *Gleasonia* (Rubiaceae) and *Gongylolepis* (Asteraceae).

The second azonal forest type is a low (10–15 m?), deciduous, basimontane forest occurring in a locally restricted sector of the lower Río Ocamo valley, between approximately 300 and 500 m elevation (***unit 51*** on map). This interesting forest is the only entirely deciduous one in Amazonas state; it is probably the result of locally dry climatic conditions caused by a rain shadow. The nearby presence of a dry savanna type and dry open vegetation on rock outcrops appears to confirm this hypothesis.

The forests of the Parima uplands are the homeland of part of the Yanomami Amerindians. In the late 1980s and early 1990s they were invaded by a large number of illegal miners from Brazil, who introduced diseases and overhunted much of the indigenous people's wild game supply. There has also been a worrisome increase in deforestation caused by burning by the increasing indigenous population in the southern Parima forests (Plate 82); this has resulted in a marked increase of open savanna habitats, which may have profound implications for the hydrology of the Orinoco headwaters (see Smole 1976; Huber et al. 1984).

Forests of the southern Amazonas uplands and tepuis. To the south of the Casiquiare, Pasimoni, and upper Orinoco river systems, there is another large and extremely diverse complex of mountain systems that extends southward until forming the border of the Venezuelan Amazonas state with Brazil. The main mountain systems of this complex are the high tepui Sierra de la Neblina and its northern neighbors, Cerro Aracamuni and Cerro Avispa; Sierra Imerí, Sierra Tapirapecó, Sierra Curupira, and Sierra Urucusiro along the border to the east; and Sierra Unturán and Cerro Vinilla to the northeast. In this last area lies the spectacular granitic inselberg of Cerro Aratitiyope, with its widely visible needle-like summit. This region has a striking orographic and geologic diversity, with elevations that vary from 300 m on the northern base of Cerro Vinilla to 3014 m elevation on the (Brazilian) summit of Pico da Neblina in the Neblina massif. Most of the mountains and uplands, however, have an average altitude of between 800 and 2000 m.

Although this whole mountain complex was one of the last to be discovered in Venezuela (Maguire 1955), hundreds of scientists have since visited this remote region, including dozens of botanists of many major national and international botanical institutions. By late 1993, however, there has been no comprehensive publication concerning the vegetation types that occur on this biologically extremely interesting and important area. From available sources, by far the largest part of this mountain region is covered by dense forests up to elevations of 1800 to 2000 m. Because of the general lack of information, only a preliminary classification of these forests is proposed here. The three forest types are (1) tall to medium-sized, evergreen basi- and lower montane forests, (2) tall to medium-sized, evergreen montane forests, and (3) low, evergreen, upper montane forests.

Tall to medium-sized, evergreen basi- and lower montane forests cover the lower uplands and talus slopes of the sandstone tepuis (***unit 52*** on map). These forests reach up to 600 to 800 m and occur mainly in the Cerro Vinilla–Sierra Unturán range, as well as in the Siapa/Matapire uplands and at the base of the Aracamuni-Neblina system.

The second forest type, tall to medium-sized, evergreen montane forests, occur between approximately 800 and 1500 m (***unit 53*** on map). John Wurdack (personal communication) reported very tall trees (to 60 m or more tall) in this forest type on the northern slope of Sierra de la Neblina. Steyermark and Holst (1989) cited the following species for low forest patches growing near the upper limit of this unit on the summit of Cerro Aracamuni at approximately 1500 m elevation: *Neblinanthera cumbrensis* (Melastomataceae), *Aegiphila roraimensis* (Verbenaceae), *Clusia* sp. (Clusiaceae), *Diacidia glaucifolia* (Malpighiaceae), *Phyllanthus vacciniifolius* (Euphorbiaceae), *Psycho-*

tria duricoria and *P. tapajozensis* (Rubiaceae), and *Tyleria silvana* (Ochnaceae). Because of the lack of *Bonnetia* species, this is unlikely to be a true high-tepui forest in the sense of the following unit.

Finally, low (8–15 m tall), evergreen, upper montane forests occur between 1500 and 1800 to 2000 m, especially in the Neblina massif (***unit 54*** on map). According to Maguire (1972a), the dominant trees in low forests ("woodlands") on the summit of Sierra de la Neblina between 1700 and 2000 m include *Bonnetia neblinae* (Theaceae) and *Neotatea neblinae* (Clusiaceae), which reach heights up to 15 m.

Considering the enormous physiographic and geologic diversity of this large southern Amazon mountain region, many more forest types are expected to occur there. The inclusion of the entire area in the new Alto Orinoco–Casiquiare Biosphere Reserve will hopefully stimulate the production of more detailed vegetation maps reflecting more closely the vegetation diversity of the area.

Shrub Formations

Shrublands play a particularly important role in all landscape units of the Venezuelan Guayana. In this area, the shrubby growth form has attained an unparalleled degree of physiognomic and floristic diversity. Surprisingly, shrublands were not even recognized as a discrete category on modern vegetation maps until the 1980s (Huber 1982; MARNR 1982; Huber and Alarcón 1988). It is difficult to find a precise definition of the terms *scrub, shrubland, thicket,* or *woodland,* mainly because of the many intermediate stages that occur in nature. Whereas growth forms such as trees or herbs are usually easily recognized, shrubs are often more problematical to distinguish, especially in tropical vegetation types where a wide variety of growth and life forms occur.

In this volume, scrub (or the equivalent term shrubland) is defined as a natural vegetation type in which a low (usually 0.5–5 m tall), woody compartment is composed of shrubs and shrub-like plants that constitute the main functional unit of the ecosystem. This ecological characterization emphasizes that, although trees and/or patches of herbaceous vegetation may be present, the shrub layer predominates in terms of principal energetic processes and biomass.

A shrub (or bush) is defined here as a plant usually 0.2–5 m tall, with woody (sometimes barely woody) stems and twigs that branch predominantly from the base or else do not branch at all. This definition includes woody and subwoody plants that are monopodial, mostly unbranched, and are often monocarpic; the leaves of these plants usually are distributed along the main stem or else concentrated towards the apex as in stem rosettes with a stem-rosette growth form.

Woodland and thicket are terms not used here, because both are overly general and ambiguous. Woodland refers to any woody vegetation type, including low forests, whereas thicket usually refers to any low-sized, impenetrable vegetation type that is often spiny or secondary.

In Spanish, the word *arbustal* is the equivalent of scrub or shrubland. The commonly used term *matorral,* on the other hand, is used here only for obviously secondary, shrubby vegetation types, such as early successional regrowth of intervened forest ecosystems or degraded low forests in which trees have been removed or destroyed by human intervention (see Huber and Alarcón 1988).

Both the composition and distribution of shrub-dominated plant communities appear to be more closely correlated with particular edaphic conditions than in either forests or grasslands. Most shrublands are restricted to peculiar types of substrate, such as rock outcrops, sandy soils, or peat. For that reason, they rarely occupy large and continuous extensions, but occur often as azonal islands interspersed within the forest or savanna biomes of the Venezuelan Guayana.

Other physical factors that help differentiate shrubland units include climate, hydrology, and altitude. According to their physiognomic, floristic, and ecological characteristics, shrublands can be grouped into macrothermic (lowland), mesothermic (upland), and submicrothermic (tepui summit) scrub. Whereas the limits between macrothermic and mesothermic scrub are often diffuse, the differentiation between mesothermic and submicrothermic scrub is usually very neat, especially at the floristic and ecological level. Because of its strong correlation with tepui summit habitats characteristic of the Pantepui floristic province (Huber 1987), the submicrothermic scrub can also be called Pantepui scrub.

The following overview provides a general classification of the shrublands of the Venezuelan Guayana, with notes on their geographical distribution and principal floristic composition. As with forest types, a geographical sequence is followed from east to west in Bolívar state, and from north to south in Amazonas. The altitudinal sequence proceeds from the (macrothermic) lowlands to the (submicrothermic) highlands. Due to limitations of scale, not all scrub areas can be shown on the accompanying vegetation map. The reader can refer to Huber (1986, 1989) for additional information on the Guayana shrubland formation.

Shrublands of the Caroní-Paragua Drainage

The region along the middle Caroní and Paragua rivers, which extends roughly from the Río Antavari in the north to the town of Urimán in the south and eastward to the western edge of Sierra de Lema, contains one of the largest shrubland areas of the Venezuelan Guayana. Between 400 and 1000–1200 m elevation there is a very interesting mosaic of tall lowland and upland scrub on the numerous and extensive rocky plateaus that surround the base of Auyán-tepui and of the Guaiquinima piedmont (***unit 55*** on map). These usually dense and homogeneous scrub communities, 2–7(–10) m tall, are formed by typically sclerophyllous shrubs and low trees, and grow almost exclusively on open level to inclined sandstone strata. The dominant species are *Platycarpum rhododactylum* (Rubiaceae), *Terminalia quintalata* (Combretaceae), *Clusia* spp. and *Caraipa* spp. (Clusiaceae), *Dacryodes microcarpa* (Burseraceae), *Licania* spp. (Chrysobalanaceae), *Bonnetia sessilis* and *Ternstroemia pungens* (Theaceae), *Humiria balsamifera* (Humiriaceae), *Blepharandra fimbriata* (Malpighiaceae), *Ruizterania ferruginea* (Vochysiaceae), *Himatanthus articulatus* (Apocynaceae), *Ilex retusa* (Aquifoliaceae), *Emmotum* sp. (Icacinaceae), *Taralea crassifolia* (Fabaceae), and *Vellozia tubiflora* (Velloziaceae). Occasionally, the phytogeographically interesting *Pakaraimaea dipterocarpacea* (Monotaceae) also occurs in these shrublands, where it grows as a small, solitary shrub and differs sharply in habit from the large, gregarious trees of the same species that grow in forests of the southern Gran Sabana and adjacent Guyana.

In the Gran Sabana, another type of sclerophyllous upland scrub frequently occurs at elevations between 800 and 1500 m. This shrubland (***unit 56*** on map; Plate 18) usually grows on a rocky, sandstone substrate, but in some instances also occurs on deep white sands of alluvial origin. Its density and height vary greatly according to the substrate, but scrub growing on sandstone is generally more evenly spaced, whereas scrub on sandy soils tends to be clustered into many small, dense shrub islands that are separated by bare white sand areas. The following species dominate in these shrublands: *Euphronia guianensis* (Euphroniaceae), *Bonyunia minor* (Loganiaceae), *Bonnetia sessilis, Ternstroemia pungens,* and *T. crassifolia* (Theaceae), *Clusia* spp. (Clusiaceae), *Gongylolepis benthamiana* (Asteraceae), *Macairea parvifolia* (Melastomataceae), *Humiria balsamifera* and *Vantanea minor* (Humiriaceae), *Ochthocosmus roraimae, O. attenuatus,* and *Cyrillopsis micrantha* (Ixonanthaceae), *Thibaudia nutans, Notopora schomburgkii,* and *Befaria sprucei* (Ericaceae), *Spathelia ulei* (Rutaceae), and *Byrsonima concinna* (Malpighiaceae).

On the summit of Cerro Guaiquinima, at elevations between 800 and 1200 m, there are floristically very diverse, tall upland scrub communities that occur especially on the rocky, southern section of the mountain (***unit 57*** on map). Some of the most interesting elements are various species of *Terminalia* (Combretaceae), *Bonnetia lanceifolia* (Theaceae), *Blepharandra fimbriata* (Malpighiaceae), *Stomatochaeta cylindrica* and *Stenopadus colveei* (Asteraceae), and *Marlierea pudica* (Myrtaceae).

Another upland region in the Paragua drainage with extensive shrublands occurs on the summits of the Serranía Marutaní along the Brazilian border, at elevations between 1000 and 1400 m (***unit 58*** on map). This 1.5- to 4-m tall scrub was discovered and described by Steyermark and Maguire (1984b); it has a continuous, rather homogeneous cover dominated by *Tyleria floribunda* (Ochnaceae), together with species of Humiriaceae, Theaceae, Cyrillaceae, Malpighiaceae, and Asteraceae.

At higher elevations in the meso- and submicrothermic life zones of the Río Caroní basin, the most important shrublands occur on the summit plateaus of Auyán-tepui and especially on the Chimantá massif (***unit 59*** on map; Plate 35). On Auyán-tepui, a relatively homogeneous 1- to 3-m tall scrub occurs between 1600 and 2400 m elevation. It usually grows on deep organic soils (histosols, peat). The predominant shrubs are low, with rounded or stunted crowns and belong to the Theaceae (*Bonnetia*), Rubiaceae (*Maguireothamnus, Pagameopsis*), Tepuianthaceae (*Tepuianthus auyantepuiensis*), Ericaceae (*Notopora, Thibaudia*), Malpighiaceae (*Blepharandra hypoleuca*), and Melastomataceae (*Macairea*). Plants with stem rosettes, such as *Achnopogon steyermarkii* (Asteraceae), are infrequent.

The huge, fragmented Chimantá massif probably contains the widest variety of shrubby life forms and vegetation types in Pantepui Province. At least three physiognomically distinct scrub types, with numerous local floristic variants on each of the different summit sections, occur between 1900 and 2600 m elevation (Huber 1992b): (1) low, dense shrublands on rocky slopes and outcrops, (2) tall, paramoid scrub on peat, and (3) low, dense scrub on peat.

The first scrub type, low, dense shrublands on rocky slopes and outcrops, is formed by a species-rich assemblage of bushes and subshrubs belonging mainly to the Theaceae, Asteraceae, Ochnaceae, Ericaceae, Myrsinaceae, and Rubiaceae. The second type, tall, paramoid scrub on peat, is dominated by various endemic species of the genus *Chimantaea* (Asteraceae), especially *C. mirabilis* (Plates 51 and 52). This spectacular scrub type (Plate 37) presents many striking analogies with certain upper Andean shrublands (páramos), because of the predominance of the caulirosulate growth form in both areas (*Espeletia* sensu lato in the Andes, *Chimantaea* on Chimantá). This is the only case where such a conspicuous phenomenon of convergent evolution of high mountain ecosystems has been observed in the Guayana highlands. The third scrub type, low, dense scrub on peat, is dominated by Theaceae (especially *Bonnetia multinervia*), Melastomataceae (*Mallophyton chimantense*), Rubiaceae (*Aphanocarpus*), *Adenanthe bicarpellata* (Ochnaceae), and Ericaceae (*Ledothamnus*).

Several other scrub types occur on many of the remaining tepui summits of the Caroní drainage, but these are always restricted to small patches of peat accumulated in depressions on the rock surface. The dominant elements of these low bush islands are *Bonnetia* (Theaceae), *Comolia* (Melastomataceae), *Ledothamnus* (Ericaceae), *Stomatochaeta* (Asteraceae), and *Maguireothamnus* (Rubiaceae).

Shrublands of the Caura Basin

In the mainly forest-covered basin of the Río Caura, well-developed shrublands occur only on the summit plateaus of the Jaua-Sarisariñama massif, between 1300 and 2400 m elevation. Steyermark and Brewer-Carías (1976) described extensive areas of dense upland scrub on Cerro Sarisariñama around the mountain's large sinkholes (*simas*), at an elevation of 1300 to 1400 m. Although Steyermark originally called this characteristic vegetation type *selva achaparrada o enana* (stunted or dwarf forest), it is a true shrubland, apparently similar in physiognomy to that found on Serranía Marutaní. These generally very dense and species-rich shrublands (***unit 60*** on map) usually grow on rocky soils; they are mostly between 2 and 8 m tall, but often are much smaller and less dense. The most frequent species of sclerophyllous shrubs or low trees recorded by Steyermark and Brewer-Carías (1976) are *Terminalia quintalata* (Combretaceae), *Gongylolepis benthamiana* (Asteraceae), *Celianella montana* and *Phyllanthus vacciniifolius* (Euphorbiaceae), *Clusia pusilla* and *Oedematopus duidae* (Clusiaceae), *Blepharandra hypoleuca* (Malpighiaceae), *Bonnetia tristyla* and *B. jauaensis* (Theaceae), *Ilex retusa* (Aquifoliaceae), and *Tepuianthus sarisarinamensis* (Tepuianthaceae).

The shrublands on the southern summit of Cerro Jaua grow at elevations of 1800 to 2400 m (***unit 61*** on map). They are usually less than 3 m tall, much lower than those on Sarisariñama. Although the Jaua shrublands share many species with Cerro Sarisariñama, they are physiognomically quite distinct, due to the striking predominance of *Bonnetia jauaensis* (Theaceae). They are here considered as true Pantepui (high-tepui) scrub. Other codominant species on rocky exposures are *Gongylolepis pedunculata, G. jauaensis* (Plate 56), and *Stenopadus jauaensis* (Asteraceae), *Maguireothamnus jauaensis* (Rubiaceae), *Tyleria breweri* (Ochnaceae), and *Blepharandra hypoleuca* (Malpighiaceae). On peat deposits, almost monospecific, impenetrable communities of *Archytaea multiflora* (Theaceae) occur in isolated patches.

In the Sierra de Maigualida, there are occasional, small scrub islands in the summit areas, mainly in rocky depressions. These communities are dominated by *Gongylolepis jauaensis* (Asteraceae; Plate 56) and species of *Schefflera* (Araliaceae).

Shrublands of the Northern Amazonas Tepuis

Continuing to the west along the mountains that form the border between northern Amazonas and Bolívar states, there are several important and very characteristic scrub types that grow exclusively on rock surfaces at intermediate and upper elevations of the Yutajé–Coro Coro massif and the adjacent Cerro Guanay (***unit 62*** on map).

The heavily broken and irregular summit of Cerro Guanay ranges in altitude from 1600 to 2000 m and is covered by an almost continuous, dense, sclerophyllous scrub 0.5–4(–6) m tall. It is dominated mostly by *Bonnetia crassa* (Theaceae; Plate 58), together with many other low, sclerophyllous bushes in the Melastomataceae (especially *Graffenrieda*), Aquifoliaceae (*Ilex*), Araliaceae (*Schefflera*), and Asteraceae (*Stenopadus, Gongylolepis*). Similar shrublands also occur on the ascending interior slopes of the Yutajé–Coro Coro massif (Plate 25). Towards the summits (at approximately 2000 m elevation), the stem-rosette shrub *Gongylolepis jauaensis* (Asteraceae) forms small colonies with its large, coriaceous leaves and enormous whitish flower heads (Plate 56), whereas other species of *Bonnetia* (*B. kathleenae, B. celiae,* Theaceae) replace the more widely distributed *B. crassa.*

Several types of shrublands occur on the summits of the Cuao-Sipapo massif. They are dominated by various species of *Graffenrieda* and *Acanthella sprucei* (Melastomataceae), *Bonnetia* spp. (Theaceae), and *Phyllanthus* spp. (Euphorbiaceae). On granite rock surfaces in the Sipapo drainage a striking red-flowered species of *Decagonocarpus* (Rutaceae) is one of the most common shrubs.

Shrublands of the Northwestern Piedmont

The most important shrublands of the somewhat drier piedmont region of northwestern Bolívar are located in the hilly uplands between the lower and middle Suapure and Parguaza rivers (***unit 63*** on map). These shrublands grow on plinthic and bauxite substrates derived from the underlying granitic rocks of the Guayana Shield and are restricted to the drier hilltops of the piedmont region between 400 and 800 m elevation. They are always surrounded by dense sclerophyllous or mesophyllous evergreen forests (unit 33 on map). Physiognomically, these are low, stunted woodlands 3–7 m tall, with relatively open, small crowns, a few lianas, and a sparse understory. According to the inventories made by Huber and Guánchez (1988 and unpublished), the dominant families are Humiriaceae (*Sacoglottis, Humiria*), Olacaceae (*Chaunochiton*), Apocynaceae (*Lacmellea*), Chrysobalanaceae (*Licania, Hirtella*), Annonaceae (*Xylopia*), Flacourtiaceae (*Casearia*), Fabaceae (*Ormosia*), and Caesalpiniaceae (*Macrolobium*).

Shrublands of the Sipapo, Atabapo, and Guainía Lowlands

Numerous peculiar scrub islands of varying density and extension are interspersed in the lowland (50–200 m elevation) drainage area of these three river systems (***unit 64*** on map; Plate 14). This scrub type is restricted to dune-like elevations of white sand that rise up to 2 m above the surrounding, often flooded, forested terrain. It typically consists of dense, sclerophyllous shrubs 1–5(–8) m tall that form small bush islands separated by patches of bare white sand. The most frequent species of these interesting shrublands are *Humiria balsamifera* (Humiriaceae), *Ilex divaricata* (Aquifoliaceae), *Heteropterys oblongifolia* (Malpighiaceae), *Emmotum glabrum* (Icacinaceae), *Pradosia schomburgkiana* (Sapotaceae), *Ormosia macrophylla* (Fabaceae), *Calliandra tsugoides* (Mimosaceae), *Pagamea guianensis* (Rubiaceae), *Tepuianthus savannensis* (Tepuianthaceae; Plate 68), *Biophytum* sp. (Oxalidaceae), and *Simaba* sp. (Simaroubaceae). Sometimes there are surprisingly high concentrations of orchid species, especially in the transition area between shrublands and adjacent forest, as well as other epiphytes such as *Aechmea brevicollis* (Bromeliaceae; Plate 70).

Farther south, especially in the drainages of the upper Negro and Guainía rivers, there is another peculiar scrub type, locally called *bana* (***unit 65*** on map). It occurs in small patches on seasonally flooded, deep sandy podsols surrounded by Rio Negro caatinga forests (unit 16 on map; see Plate 13 and Figure 3-1). The *bana* could be considered the ultimate seral stage in reduction caused by extreme levels of nutrient deficiency and unfavorable soil-water conditions. The floristic and ecological aspects of these shrublands were intensively studied during the 1970s and 1980s (Klinge et al. 1977; Klinge and Medina 1979; Bongers et al. 1985). They are typically formed by 0.5- to 3-m tall dense bush islands growing on low mounds and ridges, separated from each other by shallow, temporarily inundated depressions with ephemeral herbaceous vegetation. Dominant families in the Río Negro *bana* are Bombacaceae (*Pachira*), Rubiaceae (*Pagamea, Retiniphyllum, Remijia*), Melastomataceae (*Macairea, Tibouchina*), Humiriaceae (*Humiria, Sacoglottis*), Nyctaginaceae (*Neea*), Fabaceae (*Aldina, Ormosia*), and Mimosaceae (*Pithecellobium*).

Along the many black-water creeks and rivulets of southwestern Amazonas, there is a characteristic local riparian scrub (***unit 66*** on map; Plate 12) locally called *boyal*. These riparian communities may also be regarded as low, open forests and are formed by large colonies of *Molongum laxum, Parahancornia negroensis,* and several species of *Malouetia,* especially *M. glandulifera* (Apocynaceae), and also *Heteropetalum brasiliense* (Annonaceae), all characterized by an extremely lightweight wood collectively called *palo de boya*. The same kinds of species in adjacent Brazil and in southernmost Venezuela are also known as *molongó*. These areas are seasonally flooded and look very peculiar at low water, when they show the up to 10-m tall stems to be thickened at their base and with only a few, small coriaceous leaves at their poorly branched apex. Other interesting companion species are *Lissocarpa benthamii* (Lissocarpaceae), *Terminalia ramatuella* (Combretaceae), *Eschweilera tenuifolia* (Lecythidaceae), *Leopoldinia pulchra* (Arecaceae), and the herbaceous *Thurnia polycephala* (Thurniaceae; Plate 71), which often forms large, submersed colonies.

Upland and Highland Scrub of the Central and Southern Amazonas Tepuis

Much of the large and dissected internal plateaus of Cerro Parú is covered by a variety of scrub types, usually associated with very broken, rocky terrain. These almost unexplored, tall, dense shrublands (***unit 67*** on map) occupy an altitudinal belt between 800 and 1500 m elevation, and they appear quite heterogeneous in their physiognomy and floristic composition. Dominant families are Malpighiaceae (*Diacidia ferruginea*), Tepuianthaceae (*Tepuianthus sarisarinamensis*), Clusiaceae (*Neotatea longifolia, Bonnetia crassa, Ternstroemia* sp.), Ochnaceae (*Tyleria spathulata*), and Aquifoliaceae (*Ilex* spp.). According to John Wurdack (personal communication), who explored the higher summits of Cerro Parú (Cerro Asisa) around 2000 m elevation, low scrub grows there on a peaty substrate, probably corresponding to the dwarf high-tepui scrub type.

The summits of Cerro Duida and Cerro Huachamacari are extensively covered by tall, dense shrublands (***unit 68*** on map; Plates 42 and 43). Especially the higher, southern section of Cerro Duida has a well-developed mosaic of scrub communities that gradually change from tall upland (at 1500 m elevation) to low-sized, high-tepui scrub types (at 2200 m elevation). The upland scrub is 3–6 m tall, dense, and dominated by *Tyleria grandiflora* (Ochnaceae) and *Neotatea longifolia* and *Clusia* spp. (Clusiaceae). The high-tepui scrub is less than 3 m tall, but is extremely dense and dominated by *Bonnetia crassa* and *B. tristyla* (Theaceae), *Bonnetia duidae* (to be transferred to *Neotatea* in the Clusiaceae), *Tyleria linearis* (Ochnaceae), *Blepharandra hypoleuca* (Malpighiaceae), *Duidania montana* (Rubiaceae), *Gongylolepis* spp., *Duidaea* sp. (Plate 59), and *Stenopadus* sp. (Asteraceae). The dense shrublands on the western side of the summit of Cerro Huachamacari at approximately 1800 m elevation are very similar to those on Cerro Duida, with *Bonnetia crassa* (Theaceae), *Tyleria spathulata* (Ochnaceae), and several species of *Gongylolepis* (Asteraceae) as a prominent feature of the tepui scrub.

Finally, on the summits of the large Sierra de la Neblina massif, including Cerro Avispa and Cerro Aracamuni to the north, there are several different scrub types at altitudes between 1600 and 2500 m. These grow mainly on organic peat soils and on rocky substrates. Although very little is known about the vegetation of these mountains, the "Neblinaria scrub" has been widely publicized through a number of popular articles that followed a series of explorations (see Brewer-Carías 1988). This particular scrub type is restricted to the Sierra de la Neblina, where it grows on shallow organic soils at elevations between 1600 and 2200 m (***unit 69*** on map). It is a dense plant community approximately 2–3 m tall dominated by *Bonnetia maguireorum* (Theaceae), a species previously treated as *Neblinaria celiae.* These beautiful plants have delicate, lavender flowers and a striking stem-rosette habit, with thick, coriaceous leaves densely clustered at the apex of a corky, brittle stem (Plate 64).

According to a hypothesis by Givnish et al. (1986), a series of peculiar anatomical structures and survival strategies in *Bonnetia maguireorum,* such as the thick, corky bark and a rapid resprouting activity after burning, indicate a specific adaptation to fire-prone tepui habitats. However, more convincing studies carried out on a number of different tepuis, such as comparative ecophysiological measurements between similar growth forms in a variety of high-tepui taxa, will be needed to further test the validity of this hypothesis (see also "Burning" in Chapter 5).

Herbaceous Formations

Herbaceous formations occur in almost all geographical regions and physiographical landscapes of the American tropics, where they are known by many local names, the most common ones being *sabana* (savanna), *campo* (field), *pajonal* or *pastizal* (grassland), and *puna* (high-altitude grasslands). The great majority of these widespread and often very diverse plant communities is characterized by a more or less continuous grass layer, which may contain isolated woody elements, such as shrubs, trees, or small forest islands.

In the Guayana Shield, however, other types of herbaceous communities are frequently found. In contrast to the more common grasslands, these are dominated not by grasses or sedges, but rather by herbaceous plants of entirely distinct floristic affiliation and physiognomic appearance. Because of the high physiognomic, ecological, and floristic diversity of the herbaceous ecosystems of the Venezuelan Guayana, they require a more elaborate and specifically adapted scheme of classification than other areas. As a result, some of the terms mentioned below may sound unfamiliar to the reader, since the current terminology does not adequately characterize herbaceous communities.

The term *meadow* here collectively designates any herbaceous vegetation type, although it does not correspond precisely to the more appropriate and concise Spanish term *herbazal*. Meadows can be divided into two broad, floristically distinct categories: (1) gramineous meadows or grasslands (*herbazales graminosos*), dominated by grasses and/or sedges, and (2) nongramineous meadows (*herbazales no-graminosos*), consisting mainly of other herbs without a gramineous morphology.

Gramineous meadows or grasslands are further divided into two categories. The first includes savannas (*sabanas*), which are grasslands dominated principally by grasses with a C_4 photosynthetic pathway, restricted to tropical (macrothermic) lowland, or lower upland (submesothermic) environments. Many savannas also contain more or less isolated woody elements, such as shrubs, trees, or small forest or palm islands. The second kind of grasslands are the montane grasslands (*praderas*), which are dominated essentially by grasses with a C_3 photosynthetic pathway, occurring sporadically in upper montane (mesothermic and submicrothermic) localities of the Guayana highlands (but extensively in the meso- and, especially, microthermic high-Andean life zones); woody elements are either scarce or absent.

Nongramineous meadows are very characteristic of the vegetation of the Venezuelan Guayana. Four types of such meadows, differing in their floristic composition and physiognomic structure, can be recognized: (1) broad-leaved meadows, (2) tubiform meadows, (3) rosette meadows, and (4) fruticose meadows.

Broad-leaved meadows (*herbazales de hoja ancha*) are typically dominated by various species of Rapateaceae, which are widespread in the Guayana highlands but also in some upland and lowland localities. Tubiform meadows (*herbazales tubiformes*) are dominated by peculiar tubiform herbs of the Bromeliaceae or Sarraceniaceae and are frequent in uplands and highlands of the Guayana. Rosette meadows (*herbazales arrosetados*), in which dense rosette herbs of Xyridaceae and Eriocaulaceae dominate, are locally frequent on certain tepui summits. Finally, fruticose meadows (*herbazales fruticosos*) have the herbaceous layer mixed with numerous short subshrubs.

As with forests and shrublands, the following account of the herbaceous ecosystems in the Venezuelan Guayana is based on the major altitudinal subdivisions of the region: lowlands (macrothermic), uplands (submesothermic), and highlands (mesothermic to submicrothermic).

Lowland and Upland Grasslands (Savannas)

Lowland savannas are most widespread in the northeastern parts of the Venezuelan Guayana, but they also occur discontinuously in the northwestern sector of Bolívar state, as well as in northern and central Amazonas. All Guayana savannas are floristically and physiognomically very similar to those from the adjacent Llanos to the north and northwest. In most savannas, there are two distinct compartments, a more or less continuous herbaceous layer and a shrub or tree layer that varies greatly in size and density. Floristic richness is usually higher in the herbaceous stratum, especially among the grasses and leguminous herbs, which often represent the principal families in these savannas.

Upland savannas are restricted mainly to the plateaus of the Gran Sabana (southeastern Bolívar) and the Sierra Parima (eastern Amazonas). Their physiognomy is strikingly uniform, and they generally lack any emergent trees or shrubs.

Lowland savannas of the Orinoco Delta. In some parts of the upper and middle Orinoco Delta, large areas with seasonally flooded shrub and/or palm savannas occur (***unit 70*** on map). Their herbaceous layer is dominated by flood-resistant grasses such as *Leersia hexandra* and *Imperata brasiliensis.* Shrubs are scarce, whereas *Mauritia* palms are quite abundant and often form extensive colonies.

Lowland savannas of northeastern Bolívar state. This large area, ranging roughly from Tumeremo in the east to the lower Río Caura in the west, has a great variety of savanna types that extensively cover the gently rolling hills and peneplains. The most widespread savanna type is a low to medium (2–4 m tall) shrub savanna (***unit 71*** on map; Plate 4) that usually grows on oxisols with superficial ferruginous concretions (*ripio*). The main herbaceous species are *Trachypogon plumosus, Axonopus canescens, Andropogon* spp., *Aristida tincta, A. setifolia,* and *Thrasya petrosa* (Poaceae), *Bulbostylis* spp., *Rhynchospora* spp., and *Scleria cyperina* (Cyperaceae), *Polygala* spp. (Polygalaceae), and *Buchnera* spp. (Scrophulariaceae). Among the subshrubs, there are many leguminous species in genera such as *Galactia, Chamaecrista, Eriosema,* and *Stylosanthes,* as well as several Euphorbiaceae, Malvaceae, and *Krameria ixine* (Krameriaceae). *Curatella americana* (Dilleniaceae), *Byrsonima crassifolia* and *B. coccolobifolia* (Malpighiaceae), *Palicourea rigida*

(Rubiaceae), and *Roupala montana* (Proteaceae) predominate among the taller shrubs. A spiny palm, *Acrocomia aculeata,* also occurs frequently in these relatively dry savannas. This savanna type has been studied and mapped in the region around Guri by Ramia (1961), CVG-TECMIN (1989, 1991d), and Rosales and Briceño (1990).

In shallow depressions with poor drainage, this shrub savanna often intergrades with a peculiar scrub type dominated by *Curatella americana* (Dilleniaceae), locally called *chaparral.* The dense stands of *Curatella* are up to 7 m tall and almost monospecific, only occasionally accompanied by treelets of *Guettarda divaricata* (Rubiaceae) or *Calliandra* sp. (Mimosaceae). Because of the relatively open crowns of the shrub layer, however, enough light reaches the ground to allow the growth of a dense, savanna-like herbaceous stratum. *Chaparrales,* therefore, represent a transitional category between dense shrub savanna and a true shrubland. Similar *chaparrales* are also found on stony soils of slightly inclined slopes in the lower Caura region and extend farther west to the Río Maniapure region (Hueck 1961).

West of El Dorado to the lower Río Paragua region, there is an open, treeless savanna type that grows discontinuously, usually on acidic, sandy, and seasonally water-logged soils (***unit 72*** on map). The thin herbaceous layer is mainly formed by *Panicum caricoides, Paspalum plicatulum, Mesosetum* spp., *Sorghastrum setosum,* and *Trachypogon plumosus* (Poaceae), *Eleocharis* sp. and *Rhynchospora* spp. (Cyperaceae), *Xyris jupicai* (Xyridaceae), *Caperonia paludosa* (Euphorbiaceae), and *Polygala* spp. (Polygalaceae).

Another treeless, open savanna type occurs between 300 and 600 m elevation along the middle Río Caroní, extending around the northern and southern bases of Auyán-tepui (***unit 73*** on map). These savannas have a dense, 10- to 60- (to 80-) cm tall herbaceous layer formed by *Trachypogon plumosus, Axonopus pulcher, Panicum olyroides, Thrasya* sp., *Paspalum lanciflorum, Echinolaena inflexa* (Poaceae), *Scleria cyperina* and *Bulbostylis paradoxa* (Cyperaceae), and *Perama hirsuta* (Rubiaceae). Of particular interest in these savannas is the isolated occurrence of small populations of *Monotrema bracteatum* and *Cephalostemon squarrosus* (Rapateaceae), which are otherwise known only from the lowlands of Venezuelan Amazonas or from savannas in Amazonas state of northern Brazil. Some shrubs or low trees such as *Byrsonima crassifolia* (Malpighiaceae), *Palicourea rigida* (Rubiaceae), and *Curatella americana* (Dilleniaceae) occur very sporadically in this open savanna type. *Curatella* reaches its southern limit in Bolívar in this area. More or less extensive stands of *moriche* palms (*Mauritia flexuosa*) typically occur along river courses and in small depressions.

Another physiognomically very peculiar savanna type has been found growing in small patches interspersed among the forests of the eastern piedmont region around Cerro Guaiquinima (***unit 74*** on map). This shrub or tree savanna is restricted to acidic sandy soils which are periodically water-logged. The usually sparse and low herbaceous layer is dominated by *Bulbostylis lanata* and *Lagenocarpus* sp. (Cyperaceae), as well as by *Mesosetum* spp. and

Echinolaena inflexa (Poaceae). The shrub layer is formed by relatively dense, regularly spaced individuals of *Euphronia guianensis* (Euphroniaceae) up to 6–8 m tall, accompanied by *Pera schomburgkiana* (Euphorbiaceae), *Meriania urceolata* and *Tococa nitens* (Melastomataceae), and *Hirtella bullata* (Chrysobalanaceae).

Lowland savannas of northwestern Bolívar state. In the area between the lower courses of the Caura and Maniapure rivers, there is a noticeable transition zone between eastern and western savanna elements in northern Bolívar. Besides modern climatic, geologic, and edaphic factors, other historic and as yet unidentified evolutionary factors must have contributed to this transition zone.

The main savanna type south of the riparian forests of the Río Orinoco and north of the extensive pre- and basimontane forests to the south is a dense shrub and palm savanna that alternates with more or less extensive stands of *chaparrales* (***unit 75*** on map). This interesting savanna mosaic stretches in a belt of variable width from the lower Río Caura in northern Bolívar westward to the northwestern tip of Amazonas. It generally occupies nonflooded oxisols and ultisols on low hills and premontane hill-lands, but some of the *chaparrales* also occur in seasonally flooded depressions. The usually dense and tall herb layer is almost always dominated by *Trachypogon plumosus,* which is the most important and widespread grass species in northern South American savannas. The most characteristic elements of this savanna type, however, are the woody species, particularly the abundant *Attalea* palms and the numerous stands of *Caraipa llanorum* (Clusiaceae). Both these species are absent from the savannas east of the Río Caura. *Bowdichia virgilioides* (Fabaceae), a common treelet in the shrub savannas of the Llanos, also occurs more frequently here than in the northeastern Bolívar savannas.

The alluvial plains that border the southern bank of the Río Orinoco are periodically flooded and have another savanna type (***unit 76*** on map) in which the sparse herbaceous layer is formed by small bunch grasses such as *Panicum orinocanum, P. micranthum, Mesosetum rottboellioides,* and sedges (*Rhynchospora filiformis, R. globosa, Bulbostylis lanata*), together with annual herbs such as *Abolboda* sp. and *Xyris* sp. (Xyridaceae), *Utricularia* spp. (Lentibulariaceae), and *Perama galioides* (Rubiaceae). Taller, woody species are completely absent, but some low subshrubs such as *Comolia leptophylla* (Melastomataceae) frequently occur within the herbaceous layer. This savanna type also occurs sporadically on similar sites in northwestern and central Amazonas.

Lowland savannas of Amazonas state. Grass savannas only occupy small areas in Venezuelan Amazonas. They are clearly more common in the northern half of the state, where the climate is much more seasonal, and thus correspond well to the category of seasonal savannas as defined by Sarmiento

(1983). In the eastern Río Manapiare valley, there is a large area of periodically flooded savannas, which corresponds to the hyperseasonal savannas of Sarmiento (1983).

By far the predominant type of lowland savanna in Amazonas is the nonflooded shrub and tree savanna (***unit 77*** on map). Shrub and tree savannas are well developed in the region both north and south of Puerto Ayacucho, in the upper Río Ventuari region, and around La Esmeralda. They also occur in more isolated parts of the state, where they usually grow on hilly terrain with oxisols and ferruginous concretions or on level sandy soils. The shrub and tree savannas around Puerto Ayacucho have a dense, 0.3- to 1-m tall herbaceous stratum dominated by *Trachypogon plumosus, Axonopus canescens, Panicum cervicatum, Mesosetum rottboellioides,* and *Echinolaena inflexa* (Poaceae), many species of Cyperaceae (*Bulbostylis, Rhynchospora, Scleria*), and other herbs and forbs belonging to the Fabaceae, Euphorbiaceae, Polygalaceae, Solanaceae, Melastomataceae, Caryophyllaceae, Malvaceae, and Sterculiaceae. The most important shrubs between 1 and 4 m tall are *Byrsonima crassifolia* (Malpighiaceae), *Bowdichia virgilioides* (Fabaceae), *Casearia javitensis* (Flacourtiaceae), *Psidium salutare* (Mytaceae), and *Protium heptaphyllum* (Burseraceae). Among the trees, *Platycarpum orinocense* (Rubiaceae) is both the most common and the most characteristic species, growing up to 12 m tall. It often occurs in nearly pure, regularly spaced stands; it is occasionally accompanied by *Chaunochiton angustifolium* (Olacaceae), another typical woody element of this savanna type, which seems to be restricted to northern Amazonas and adjacent Colombia (Huber 1990).

Interestingly, *Curatella americana* (Dilleniaceae), which is so common in savannas of northwestern Bolívar state, is almost absent from the shrub and tree savannas of northwestern Amazonas. It reappears in large numbers, however, in the hill savannas of the northern Manapiare and the upper Ventuari regions (unit 82). It also occurs in a completely isolated, small savanna along the lower Río Ocamo in the upper Orinoco area. There is so far no satisfactory explanation for such an irregular distribution of this common neotropical savanna treelet in the Venezuelan Amazonas. According to Eden (1974), some of the isolated savannas in the mainly forested Amazonas are relict areas of a more widely distributed savanna complex that covered this region during drier climatic periods of the Pleistocene or early Holocene. Although this hypothesis has not yet been confirmed by paleoecological or palynological records from this area, it does help explain certain distribution patterns of these savannas and their component species.

The flooded savannas of the central and eastern Río Manapiare basin cover wide extensions of the alluvial plains and are completely treeless (***unit 78*** on map; Plate 15). Their 1–1.5 m tall herbaceous layer is formed by grasses such as *Sorghastrum setosum, Panicum tricholaenoides, Echinolaena inflexa, Hypogynium virgatum,* and many other herbs and subshrubs. Occasionally, small stands of the palm *Attalea* grow near the river courses. This savanna type is unique to Amazonas but is still largely unexplored botanically. It may be expected to contain numerous additional and new species, which would notably increase the savanna flora of this region.

Upland savannas of southeastern Bolívar state. The large upland plateau covering most of southeastern Bolívar state is called the Gran Sabana, due to the wide extensions of grasslands that cover the undulating landscape of this very picturesque region. These savannas were first mentioned by M. R. Schomburgk (1847–1848) in a volume devoted to the fauna and flora of Guayana and were studied in detail by Francisco Tamayo in 1946, who published an extensive account on their floristics and physiognomy (Tamayo 1961). Since 1982, several large research projects have focused on the Gran Sabana grasslands, with the aim of establishing their causal and dynamic relationships with the adjacent forest and scrub vegetation (see Fölster 1986; Huber 1986; CVG-TECMIN 1987, 1989; Hernández 1987). Three main savanna types occur in this area at elevations of 750 to 1300 m (Huber 1986; Schubert and Huber 1990): (1) open, treeless savannas, (2) palm savannas, and (3) shrub savannas.

Open, treeless savannas have an herbaceous stratum of variable density up to 1 m tall (***unit 79*** on map; Plate 20). The dominant grasses are *Axonopus pruinosus, A. kaietukensis, Trachypogon plumosus, Echinolaena inflexa,* and *Leptocoryphium lanatum,* and the most common sedges are *Bulbostylis paradoxa, Rhynchospora globosa, Hypolytrum pulchrum,* and *Scleria cyperina.* Other occasional herbs in these rather species-poor communities are *Buchnera weberbaueri* (Scrophulariaceae), *Polygala* spp. (Polygalaceae), and *Sisyrinchium vaginatum* (Iridaceae). Woody elements are rare and usually do not emerge above the herb layer; they are mostly low, stunted individuals of *Palicourea rigida* (Rubiaceae), almost stemless leaf rosettes of *Byrsonima verbascifolia* (Malpighiaceae), or small plants of *Eugenia punicifolia* (Myrtaceae) or *Casearia sylvestris* var. *lingua* (Flacourtiaceae). This savanna type is widespread throughout the Gran Sabana, where it usually grows on highly weathered, poor oxisols that are often covered by dense ferruginous concretions (*ripio*).

Palm savannas grow in the wide alluvial plains of the lower Kukenán, Yuruaní, Aponguao, and Caruay valleys, between 750 and 1000 m elevation (***unit 80*** on map). These seasonally flooded savannas were studied by Terán and Duno de Stefano (1988); they are formed by a dense, up to 1.5–2 m tall herb layer, with large stands of the *moriche* palm (*Mauritia flexuosa*) frequently interspersed. The species-rich herbaceous stratum is dominated by tall grasses (*Hypogynium virgatum, Andropogon* spp., *Panicum* spp.), sedges (mainly various species of *Rhynchospora* and *Bulbostylis*), and many herbs and forbs, such as *Byttneria genistella* and *Waltheria* sp. (Sterculiaceae), *Eriocaulon* sp. (Eriocaulaceae), *Xyris* spp. (Xyridaceae), or *Phyllanthus* spp. (Euphorbiaceae). Low shrubs such as *Miconia stephananthera* (Melastomataceae), *Mahurea exstipulata* (Clusiaceae), or the endemic *Piper sabanaense* and *P. tamayoanum* (Piperaceae) also occur, although they are restricted mainly to the interior of the extensive *moriche* palm colonies. *Mauritia flexuosa* seems to reach its upper altitudinal limit in the Gran Sabana at about 1100 m elevation.

The third type, shrub savannas, grow on extremely nutrient-poor, acidic soils derived from the decomposition of quartzites and sandstones belonging to the Roraima Group. They occur primarily on rocky hills and slopes in the southern Gran Sabana, at elevations between 800 and 1100 m (***unit 81*** on map). The herbaceous layer, which appears very sparse and irregular, is dominated by sedges (*Lagenocarpus* sp., *Bulbostylis paradoxa* and *B. capillaris*), together with small bunches of *Axonopus kaietukensis, A. pruinosus, Trachypogon plumosus,* and *Paspalum lanciflorum* (Poaceae). Other herbs or low subshrubs, such as *Buchnera* sp. (Scrophulariaceae) or *Macairea lasiophylla* (Melastomataceae), are rare. The taller shrubs (1–3 m) are usually widely spaced on the open slopes, but they are more frequent near the intermittent water courses. The most common shrubs are *Euphronia guianensis* (Euphroniaceae), *Bonyunia minor* (Loganiaceae), *Miconia* spp. (Melastomataceae), *Cybianthus fulvopulverulentus* (Myrsinaceae), *Trattinnickia burserifolia* (Burseraceae), *Palicourea rigida* (Rubiaceae), *Bonnetia sessilis* and *Ternstroemia pungens* (Theaceae), and *Calea divaricata* (Asteraceae).

Opinions vary considerably concerning the origin of the grasslands in the Gran Sabana plateau. From a general climatic point of view, one would expect to find the entire upland covered by forests or azonal shrublands on rock outcrops; although a slight rain-shadow effect intensifying the seasonality of the local climate in the southern Gran Sabana may be caused by the high mountains of the eastern tepui chain (Roraima–Ilú), this phenomenon would not explain the even broader savanna areas in the northern Gran Sabana. Some authors, such as Volkmar Vareschi (personal communication), believed that these savannas were caused mainly by the numerous fires that are lit annually by the indigenous population in the forests and grasslands. Considering that human occupation of this area is assumed to have occurred only in relatively recent times (approximately 300 years ago, according to Butt-Colson 1985), this hypothesis appears to be too generalized, especially if one takes into account the great present-day extensions of the grasslands. Based on the results of field studies, it seems more likely that restricted savanna areas existed there in equilibrium with the surrounding forests prior to the human invasion phases, and that human activity has vigorously induced the expansion of the grasslands through the massive use of fire (Fölster 1986; Huber 1986). The presence of several savanna types with specialized or endemic plants also suggests an older, nonanthropogenic origin of the Gran Sabana grasslands.

Upland savannas of the Guanay-Cuao massifs. Some small savanna areas occur at elevations of 300 to 800 (to 1000) m between the southern piedmont area of Cerro Guanay and the northeastern piedmont hills of the Cuao massif in northwestern Amazonas. These peculiar shrub or tree savannas (***unit 82*** on map) are not found elsewhere in the Venezuelan Guayana. They usually grow on rocky substrates derived from quartzite, and both their herbaceous and woody layers are relatively open, averaging 1–1.5 m and 4–6 m tall, respectively. The dominant herbs are *Trachypogon plumosus, Axonopus canescens,* and *Paspalum lanciflorum* (Poaceae), *Bulbostylis lanata* and *Rhynchospora barbata* (Cyperaceae), accompanied by colonies of *Vellozia tubiflora* (Velloziaceae), and by the large rosette plants of *Navia hohenbergioides* (Bromeliaceae). The shrub or low tree layer is dominated by *Caraipa llanorum* (Clusiaceae), *Byrsonima amoena* and *Mezia huberi* (Malpighiaceae), *Dioclea guianensis* (Fabaceae), and *Arrabidaea nigrescens* (Bignoniaceae). Towards the northern limit of this savanna type, the shrub layer becomes increasingly mixed with *Curatella americana* (Dilleniaceae).

Savannas of the Sierra Parima uplands. The large and mainly forested plateau of the Sierra Parima extends along the eastern border of Venezuelan Amazonas state with the Brazilian state of Roraima. It has two types of savannas, one natural and one clearly anthropogenic (Huber et al. 1984). Natural savannas occur in small patches in the northern half of the Parima uplands (***unit 83*** on map) and consist mainly of a dense, 0.5–1.5 m tall herbaceous stratum and a conspicuous shrub layer up to 5 m tall. The dominant grasses are *Axonopus chrysites, Echinolaena inflexa, Panicum rudgei, Ichnanthus breviscrobs, Aristida recurvata,* and *Andropogon leucostachyus,* and the main sedges are *Bulbostylis capillaris, Scleria hirtella,* and *S. bracteata.* The most important shrubs are *Byrsonima chrysophylla* (Malpighiaceae), *Trattinnickia burserifolia* (Burseraceae), *Clusia* spp. (Clusiaceae) and several species of *Miconia* (Melastomataceae). A typical example of this savanna type is found at Simarawochi (Shimada-wochi), a small Sanema Amerindian settlement in the headwaters of the Río Matacuni, at approximately 750 m elevation.

Human-induced savannas are found in the southern section of the Sierra Parima (***unit 84*** on map; Plate 23), mainly in the headwaters of the Río Putaco, where a steadily increasing population of Yanomami Amerindians lives. Due to externally induced changes in their lifestyles, these traditionally forest-dwelling tribes are clearing extensive forest areas and converting them, by repeated burning, first into dense secondary meadows dominated by bracken ferns (*Pteridium caudatum*) and finally into species-poor, treeless grasslands (Huber et al. 1984). These savannas are extensively dominated by *Bulbostylis paradoxa* (Cyperaceae), accompanied by *Axonopus anceps, Leptocoryphium lanatum,* and *Andropogon bicornis* (Poaceae), and some subshrubs such as *Eriosema crinitum* (Fabaceae), *Achyrocline satureioides,* and *Calea abelioides* (Asteraceae).

The limited presence of two distinct savanna types in the Parima uplands may have broader implications for other parts of the Venezuelan Guayana. The semideciduous montane forests (unit 49 on map; Plate 23) are apparently highly vulnerable to fire, which leads to a characteristic sequence of progressively degraded ecosystems; from disturbed forest they pass to secondary scrub (*matorrales*), then to bracken fields (*helechales*), and finally to low, species-poor secondary grasslands. The savanna-producing process that is taking place today in the Sierra Parima may be similar to what might have occurred in the Gran Sabana uplands several centuries ago as a consequence of large-scale human settlement. In both cases, it is important to recognize that natural savannas exist in both regions and may have acted as genetic sources for some of the more aggressive colonizer species of the newly created open spaces that were formerly covered by forests.

High-tepui Grasslands

Montane grasslands are rare in the Venezuelan Guayana, but in some cases they form important, though geographically restricted ecosystems within the diversified high-tepui landscape. Because of their small extension and fragmented occurrence, these montane grasslands do not appear on the accompanying vegetation map. Although grasses generally play a clearly subordinate role in most of the herbaceous plant communities found on the tepui summits, there are some particular site conditions where the gramineous life form is successful and even dominant. This occurs with some periodically waterlogged areas of peat that are generally located in wide valley bottoms on several of the largest tepui massifs.

Some of the best examples of high-tepui grasslands are found on Auyán-tepui, the Chimantá massif, Sierra de Maigualida, and on Cerro Marahuaka. In most cases, these grasslands are dominated by *Cortaderia roraimensis,* which forms dense and almost pure colonies in water-logged or frequently flooded sites near shallow water courses. This is a medium-sized bunch grass of Andean affinity, and it is often accompanied by small colonies of *Aulonemia* sp. (Poaceae, Bambusoideae) or of *Cladium costatum* and *Rhynchocladium steyermarkii* (Cyperaceae). Examples of these *Cortaderia* grasslands are found on the northwestern summit of Auyán-tepui, between 1600 and 1800 m elevation, in the central summit plains of Chimantá at 2200 m, and in small patches on the summit of Cerro Marahuaka, at 2500 to 2650 m.

Completely different high-montane grasslands occur in valley bottoms of the northern summits of Sierra de Maigualida, at elevations above 2100 m. There, a densely pubescent, still unidentified species of *Axonopus* dominates on water-saturated peats, together with colonies of *Orectanthe sceptrum* (Xyridaceae) and occasional low shrubs. Curiously, this grass species uses the C_4 photosynthetic pathway, which is rather atypical for such a high mountain grass, but is consistent with all other members of the genus (E. Medina, personal communication).

Lowland Meadows

Nongramineous meadows in the lowlands of the Venezuelan Guayana are restricted mainly to the Orinoco Delta and central Amazonas. In both areas, these peculiar ecosystems represent highly specialized plant communities that are adapted to severely limiting environmental conditions. They are also very interesting for their floristic endemism as well as their ecological and evolutionary complexity.

Broad-leaved meadows of the Orinoco Delta. According to Canales (1985), there are two distinct types of meadows in the interior of the Orinoco Delta. The first type occurs in the Middle Delta, where dense, herbaceous communities 1–2 m tall grow on permanently water-logged, organic soils (histosols, peat). These broad-leaved marsh meadows (***unit 85*** on map) are composed of *Blechnum serrulatum* (Blechnaceae), *Acrostichum aureum* (Pteridaceae), and *Cyperus articulatus* and *Scleria* spp. (Cyperaceae). The second type of meadow grows in permanently flooded openings of the Lower Delta, sometimes forming extensive floating mats in abandoned river channels (***unit 86*** on map). These broad-leaved swampy meadows have an odd appearance due to the dominance of *Montrichardia arborescens* (Araceae), a giant herb with erect, cordate leaves that often grows in nearly pure stands. These areas are locally called *rabanales* because *M. arborescens* resembles cultivated radish plants. Other frequent herbs are *Thalia geniculata,* (Marantaceae), *Typha domingensis* (Typhaceae), *Cyperus giganteus* (Cyperaceae), *Heliconia psittacorum* (Heliconiaceae), and *Costus arabicus* and *Renealmia alpinia* (Zingiberaceae). In more open places, *Eichhornia crassipes* (Pontederiaceae) and *Paspalum fasciculatum* (Poaceae) form floating communities. *Moriche* palms (*Mauritia flexuosa*) are also often mixed in these meadows, forming the transition to palm swamps.

Lowland meadows of the upper Orinoco basin. In the upper Orinoco and Negro river basins, there are some peculiar types of meadows that are exclusively associated with deep, sandy, quartzitic soils. These are possibly the ecosystems with the highest number of endemic species per area in the Guayanan lowlands. They are often called white-sand savannas, white-sand *campinas,* or *campinarana* in Brazil. Although similar plant communities are known from other parts of the Guayana Shield, such as Guyana, Brazil, and Colombia, nowhere else have they reached such a pronounced degree of local differentiation as in southern Venezuela. Especially along the middle and lower Río Ventuari, the lower Casiquiare, and the Atabapo and Guainía rivers, there is an astonishing variety of broad-leaved meadows that were studied in the late 1970s by Otto Huber (see Huber 1985a).

The upper Orinoco lowland meadows (***unit 87*** on map; Plates 13 and 14) always grow on level, highly acidic, and extremely nutrient-poor soils (quartzipsamments) that are often poorly drained, despite their deep sandy texture.

The plant cover is extremely variable, ranging from a few isolated tufts of low, stunted herbs to dense meadows to 1 m tall. In most cases, the genus *Schoenocephalium* (Rapateaceae) forms the main herbaceous component; *S. cucullatum* (Plate 77) is more widespread in central Amazonas state, whereas the slender *S. teretifolium* predominates in the Guainía and lower Casiquiare meadows. A third species of *Schoenocephalium* (*S. martianum*) occurs in similar vegetation types in Colombia (near Araracuara). Other members of Rapateaceae frequently found in these meadows include *Monotrema, Cephalostemon,* as well as the beautiful endemic *Guacamaya superba,* which grow together with many other species of Xyridaceae (*Xyris, Abolboda*), Eriocaulaceae (*Syngonanthus, Paepalanthus, Eriocaulon*), and Cyperaceae (*Bulbostylis, Rhynchospora, Lagenocarpus*). Grasses are remarkably scarce, represented mainly by some species of *Panicum* and *Axonopus* and the bambusoid *Steyermarkochloa angustifolia.*

These meadows usually also have a rich and highly specialized woody flora, which consists mainly of low shrubs and subshrubs. There are numerous endemics found here, belonging to groups such as *Ouratea* (Ochnaceae), *Ochthocosmus* (Ixonanthaceae), *Lasiadenia* (Thymelaeaceae), *Simaba* (Simaroubaceae), *Ecclinusa* (Sapotaceae), *Macrolobium* (Caesalpiniaceae), *Tepuianthus* (Tepuianthaceae), *Archytaea* (Theaceae), and many other genera. Several of these low ligneous plants have developed xylopodia or taproots, as in certain species of *Mabea* (Euphorbiaceae), *Tetrapterys* (Malpighiaceae), *Dulacia* (Olacaceae), and *Ouratea* (Ochnaceae). Generally, these shrubs have very few, loosely arranged sclerophyllous leaves; otherwise the thick, coriaceous leaves are densely clustered at the top of the branches, as in *Terminalia yapacana* (Combretaceae) and *Pachira sordida* (Bombacaceae). Most of the shrubs of the upper Orinoco lowland meadows have showy yellow, white, or reddish flowers. When flowering with herbs of the Rapateaceae and Xyridaceae, the result is an extremely colorful meadow.

Upland and Highland Meadows

Almost all herbaceous plant communities that grow in the uplands and highlands of the Venezuelan Guayana have one or more species of Rapateaceae predominating in the herb layer. In contrast to the lowland meadows of southern Venezuela, where just two species of *Schoenocephalium* represent the most important floristic elements, the rapateaceous taxa in the montane meadows have reached a much higher degree of differentiation, as evidenced by the genus *Stegolepis,* which has to be considered as the most characteristic floristic element of this biome in the Guayana region. Of the 31 species of *Stegolepis* to be recognized in this flora (P. Berry, personal communication), at least 17 species play a dominant role in the various upland and high-tepui meadows. Four additional genera of Rapateaceae (*Amphiphyllum, Kunhardtia, Marahuacaea,* and *Phelpsiella*) are endemic to the flora area and are local dominants in meadows of several tepui summits of Amazonas. Because of their clear altitudinal differentiation, species of Rapateaceae provide one of the main floristic criteria for distinguishing lowland meadows from upland and highland meadows in Venezuelan Guayana (Huber 1988a).

Upland meadows of the Gran Sabana. Between 950 and 1400 m elevation, there are more or less extensive areas of broad-leaved meadows on the plateau of the Gran Sabana (***unit 88*** on map; Plate 19) in southeastern Bolívar. These meadows are interspersed in the wide grasslands of the Gran Sabana; they always grow on dark organic soils (peat) and are most widespread in the upper basins of the Aponguao and Kamoyrán rivers. The dominant herbs are *Stegolepis ptaritepuiensis* and *S. angustata* (Rapateaceae), which grow sympatrically and form dense colonies to 1.5 m tall. They are widely visible when their bright yellow flowers begin to bloom after the start of the rainy season in late April or early May. Other common herbs are *Xyris* spp. and *Abolboda macrostachya* (Xyridaceae), *Mapania tepuiana* and *Lagenocarpus guianensis* (Cyperaceae), *Brocchinia reducta* (Plate 21) and *B. steyermarkii* (Bromeliaceae), *Nietneria paniculata* (Liliaceae; Plate 50), and *Trimezia fosteriana* (Iridaceae). Low shrubs and subshrubs commonly grow in these meadows, with *Chalepophyllum guianense* (Rubiaceae), *Poecilandra pumila* (Ochnaceae), and *Clusia pusilla* (Clusiaceae) among the most interesting endemic taxa.

Ecologically, *Stegolepis* seems to reach its lower distribution level as a community-forming taxon in the broad-leaved meadows of the Gran Sabana. A dense meadow of *Stegolepis angustata* occurs in the Antavari plateau as low as 850 m. Since the altitudinal optimum of this genus lies clearly in a belt between 1500 and 2500 m elevation, its widespread, though presently somewhat fragmented occurrence in the significantly lower Gran Sabana plateau is not easily understood. A possible explanation of this phenomenon could be that these Gran Sabana meadows represent relictual communities resulting from late Quaternary climatic oscillation phases, when the high-tepui meadows were forced to migrate towards lower elevations. Although there are no convincing facts that corroborate this assumption yet, results from paleoecological research in the Gran Sabana documented the presence of alternating climatic phases during the Holocene (Rull 1991).

Highland meadows of the southeastern Bolívar tepuis. There are more or less extensive herbaceous ecosystems on all tepui summits in southeastern Bolívar. Their flora, physiognomy, and extension vary considerably from one massif to another. Due to the usually limited extensions of these high-tepui meadows, their separate cartographic representation on the accompanying vegetation map at a scale of 1:2,000,000 is not always possible. For Auyán-tepui and the Chimantá massif, however, the herbaceous vegetation is included in unit 59 (low tepui-summit scrub and meadows on peat and rock; Plates 35 and 39). Overall, four kinds of highland meadows can be recognized in the southeastern tepuis of Bolívar state. The first group of highland meadows in this region belong to the eastern tepui chain. Here, on the rocky, open, and windswept plateaus of Roraima-, Kukenán-, Yuruaní-, and Ilú-Tramen-tepuis, there are small patches of mixed herbaceous and subshrubby vegetation between 2400 and 2750 m elevation. Many of these communities are peculiar rosette meadows dominated by the large, sharp-tipped rosettes of *Orectanthe sceptrum* (Xyridaceae), as well as the smaller rosettes of *Connellia augustae, C. caricifolia,* and *C. quelchii* (Bromeliaceae) and the densely clustered, fine rosettes of many species of Eriocaulaceae (*Rondonanthus, Syngonanthus*). On the lower, more irregular summits of Karaurín-tepui and Wadakapiapué-tepui, rosette meadows are formed mainly by large colonies of *Brocchinia tatei* (Bromeliaceae), accompanied by *Neurolepis angusta* (Poaceae, Bambusoideae), *Heliamphora heterodoxa* (Sarraceniaceae), *Orectanthe sceptrum, Xyris* spp. (Xyridaceae), many species of Eriocaulaceae, and dense tufts of *Everardia* spp. (Cyperaceae).

The herbaceous vegetation on the spectacular summit of Uei-tepui differs notably from that of all other eastern tepui summits. There, at approximately 2000 m elevation, a dense broad-leaved meadow formed mostly by *Stegolepis guianensis* alternates locally with rich montane grasslands made up of *Panicum chnoodes* and *Axonopus caulescens* (Poaceae), together with species of Cyperaceae and Eriocaulaceae. *Orectanthe sceptrum* (Xyridaceae) is completely absent here. Possibly, occasional fires ascending from the adjacent Arabopó uplands have modified the floristic assemblage of these high-tepui meadows, but the summit plateau of Uei-tepui is covered by a relatively thick brown soil that is not found on the summits of the other tepuis of this region. In the uppermost section of the conical peak of the summit plateau, at approximately 2150 m elevation, there are dense rosette meadows of *Connellia augustae* (Bromeliaceae) on the steep rock outcrops.

The second category of highland meadows in the high Bolívar tepuis occurs on Ptari-tepui and Los Testigos massif. In general, the herbaceous vegetation of these tepui summits are rosette meadows similar to those of the eastern tepuis, but here low shrubs and subshrubs are more frequent. The most notable shrubs are several species of *Bonnetia* (Theaceae) on Ptari-tepui, as well as a small form of *Chimantaea lanocaulis* (Asteraceae) on the summit of Murisipán-tepui. On the higher summits of the Los Testigos massif (Kamarkawarai-tepui and Murisipán-tepui), there is yet another species of *Stegolepis, S. humilis,* which dominates in small pockets of meadows on the rock surface (Plate 39). Larger colonies of *Brocchinia tatei* (Bromeliaceae) also grow in areas with poor drainage.

A third type of dense broad-leaved meadow occurs on the northwestern branches of Auyán-tepui, where *Stegolepis humilis* (Rapateaceae) forms large colonies with *Cladium costatum* (Cyperaceae) and many other herbaceous species of Cyperaceae, Xyridaceae, and Eriocaulaceae. There are also low shrubs belonging to the Rubiaceae, Ericaceae, and Melastomataceae. These meadows grow on deep peat that occurs mainly between 1600 and 1900 m elevation. Numerous small islands of mixed herbaceous and shrubby vegetation are also scattered over the wide, rocky plateaus of the southern and eastern branches of the mountain, between 1800 and 2400 m elevation. In these cases, rosette meadows dominated by *Orectanthe sceptrum* (Xyridaceae; Plate 50), with species of Eriocaulaceae and Liliaceae, are more abundant than the broad-leaved meadows.

Finally, the summits of the many interior plateaus of the Chimantá massif have a great variety of herbaceous vegetation types (Huber 1992b). In this area, the endemic, narrow-leaved *Stegolepis ligulata* (Rapateaceae) is by far the dominant element. It is found in a wide range of plant communities, from almost monospecific, relatively open meadows on shallow sands overlying bare sandstone rocks to extremely dense, species-rich, shrubby meadows on deep peats. The most important accompanying herbs belong to the Cyperaceae, Xyridaceae, Eriocaulaceae, Bromeliaceae, Sarraceniaceae, Orchidaceae, Iridaceae, Liliaceae, and Lentibulariaceae, whereas the most common and significant shrubs and subshrubs are in the Theaceae, Ericaceae, Melastomataceae, Asteraceae, Ochnaceae, Rubiaceae, and Santalaceae. Huber (1992b) gave detailed descriptions of the flora and vegetation of these meadows, and Cuevas (1987, 1992) made detailed analyses of the nutrient contents in plants and soils of these specialized tepui ecosystems.

Upland meadows of Cerro Guaiquinima. On the western and northern sections of the summit plateau of Cerro Guaiquinima, there are large areas of herbaceous vegetation between 1200 and 1600 m elevation. The dominant herb of these meadows is the endemic *Stegolepis squarrosa* (Rapateaceae; ***unit 89*** on map). These broad-leaved meadows grow on peat accumulations, are species-rich, and usually are surrounded by montane forests and shrublands. Other common ground herbs are *Navia ovoidea* and *Brocchinia acuminata* (Bromeliaceae), *Xyris* spp. and *Abolboda* sp. (Xyridaceae), *Everardia* sp. and *Lagenocarpus* sp. (Cyperaceae), and *Panicum chnoodes* (Poaceae). Low shrubs are also present, such as *Bonnetia lanceifolia* (Theaceae), *Blepharandra fimbriata* (Malpighiaceae), *Terminalia quintalata* (Combretaceae), and *Stomatochaeta condensata* (Asteraceae).

Although the broad-leaved meadows of Cerro Guaiquinima are physiognomically similar to the upland meadows of the Gran Sabana, they are floristically quite distinct, which justifies their designation as a separate vegetation unit. Steyermark and Dunsterville (1980) placed great importance on the presence of several lowland elements in these meadows, but in reality, this merely confirms the intrinsic transitional character of the upland life zone of the Guayana mountains.

Highland meadows of the Jaua-Sarisariñama massif. The meadows that occur on the summit plateaus of the Jaua-Sarisariñama massif at 1800–2250 m elevation are dominated by large tubular herbs of *Brocchinia hechtioides* (Bromeliaceae). They are probably the best example of the peculiar tubiform meadows that occur only in the Guayana highlands. These meadows are widely visible because of the characteristic greenish-yellowish color of the dense stands of *Brocchinia,* and they grow mainly on level or slightly inclined rocky soils (***unit 90*** on map). Other common herbs in this unit are *Everardia montana* subsp. *glaucifolia* (Cyperaceae), *Orectanthe sceptrum, Abolboda* spp., and *Xyris* spp. (Xyridaceae), *Nietneria corymbosa* (Liliaceae), *Stegolepis grandis* (Rapateaceae), several terrestrial species of Orchidaceae, and some ferns. Low shrubs usually also occur, which include many endemics such as *Maguireothamnus jauaensis* (Rubiaceae), *Bonnetia jauaensis* (Theaceae), *Stenopadus jauaensis* (Asteraceae), and *Ledothamnus jauaensis* (Ericaceae). Occasionally, these tubiform meadows grow on wet peaty soils. Besides the invariable predominance of *Brocchinia hechtioides,* four species of *Stegolepis* grow sympatrically there (*S. grandis, S. jauaensis, S. microcephala,* and *S. albiflora*); all but *S. grandis* are endemic to the Jaua-Sarisariñama massif. So far, the interesting herbaceous vegetation of the Jaua-Sarisariñama massif has been described only by Steyermark and Brewer-Carías (1976), but a more thorough exploration is needed, especially on the summit of Cerro Jaua.

Upland and highland meadows of the tepuis of northern Amazonas. There is a wealth of different herbaceous vegetation types in the wide arc of mountains that forms the northern and northeastern border of Amazonas state (Sierra de Maigualida, Cerro Yaví, Cerro Yutajé–Coro Coro, and Cerro Guanay), as well as the still largely unexplored summits of the Cuao-Sipapo massif in northwestern Amazonas. Except for Cerro Yaví, all these mountains feature a predominance of the genus *Kunhardtia* (Rapateaceae), which has large, bright red flower heads instead of the typically yellow flowers of genera such as *Stegolepis.*

The broad-leaved meadows that occur above 2000 m elevation on the upper slopes of the granitic Sierra de Maigualida are relatively small in extension and are usually restricted to open rock surfaces (Plate 48). The dominant herbs are *Kunhardtia rhodantha* (Rapateaceae), *Brocchinia melanacra* (Bromeliaceae), *Anthurium* sp. (Araceae), and many species of Eriocaulaceae, Cyperaceae, and Xyridaceae. Low subshrubs such as *Tibouchina huberi* and *Leandra gorzulae* (Melastomataceae) appear occasionally as well.

The relatively small summit plateau of Cerro Yaví is very peculiar, since it is the only mountain of the entire Pantepui Province that completely lacks species of Rapateaceae. There are no high-tepui meadows on this mountain, although locally dense mats of low sedges (*Cephalocarpus, Everardia,* and *Rhynchospora*) occur on moist sites, and large colonies of *Pleurostima celiae* (Velloziaceae; Plate 57) occur on open, flat rock surfaces.

On the summits of the large sandstone massif of Yutajé–Coro Coro, broad-leaved meadows dominated by *Kunhardtia rhodantha* are restricted mainly to small extensions of floodplains in valley bottoms. One meadow on Cerro Yutajé at 1850 m elevation was composed mainly of *Kunhardtia rhodantha* (Rapateaceae), *Myriocladus* sp. (Poaceae, Bambusoideae), *Everardia* sp. and *Lagenocarpus* sp. (Cyperaceae), *Brocchinia melanacra* (Bromeliaceae),

Panicum chnoodes (Poaceae), and other species of Eriocaulaceae, Xyridaceae, Droseraceae, and ferns.

In the Cuao-Sipapo massif, large extensions of broad-leaved meadows dominated by *Kunhardtia rhodantha* (Rapateaceae) occur on rocks in both the uplands and the highlands, from approximately 1300 to 2000 m elevation. The upland meadows on the summit of the nearby Cerro Autana, at approximately 1300 m elevation, were intensively collected by Julian A. Steyermark in 1971. According to his published reports (Steyermark 1974, 1975), these meadows are very dense, up to 1.5 m tall, and consist of a mixture of mainly broad-leaved herbs (*Kunhardtia rhodantha*) and tubular herbs (*Brocchinia hechtioides*).

Due to limitations of scale on the accompanying vegetation map, all upland and highland meadows of the northern Amazon tepuis that are dominated by *Kunhardtia rhodantha* are collectively grouped under ***unit 91*** (Plates 16, 46, 47, 48, and 54).

Upland meadows on Cerro Parú. The wide, almost level, upland plateaus of the interior sections of Cerro Parú in northeastern Amazonas are covered by extensive shrubby meadows (***unit 92*** on map). They are only occasionally interrupted by narrow gallery forests or small forest islands. The predominant herb is the large, broad-leaved *Saxofridericia grandis* (Rapateaceae). This species has broad, distichous leaf bases that form a caudex up to 0.5 m tall; from these arise the massive leaf blades and the slender scapes with showy yellow inflorescences (Plate 55). Other herbs frequently found in these meadows are *Brocchinia hechtioides* and *Lindmania* spp. (Bromeliaceae), *Xyris* spp. and *Abolboda* sp. (Xyridaceae), several species of the Eriocaulaceae, and a common species of *Panicum* (Poaceae). Among the shrubs and subshrubs, which are sometimes prominent, the most important are *Bonnetia crassa* (Theaceae), *Celianella montana* and several endemic species of *Mabea* (Euphorbiaceae), *Tibouchina fraterna,* species of *Clidemia,* and of *Tococa* (Melastomataceae), and an interesting dwarf *Caraipa* (Clusiaceae) that creeps among the herbs.

The upland meadows of Cerro Parú range between 1000 and 1300 m elevation and grow on organic soils. According to John Wurdack (personal communication), at higher elevations (1300–2000 m) on the southwestern rim of the massif, there are more open meadows that grow on rocky substrates; these are dominated by the endemic and monotypic *Phelpsiella ptericaulis* (Rapateaceae).

Highland meadows of the Duida-Marahuaka massif. Although the summit plateau of Cerro Duida is mostly covered by forests and shrublands, there are several small extensions of herbaceous vegetation that grow mostly on peat. Some of these areas were named "Savanna Hills" by George Tate during the first exploration of the mountain in 1928 and 1929 (Gleason 1931). These meadows (***unit 93*** on map) occur irregularly between 1500 and 2400 m elevation and are usually dominated by *Brocchinia hechtioides* (Bromeliaceae) and *Amphiphyllum rigidum* (Rapateaceae), together with other, often endemic, herbaceous and shrubby species. The pitcher plant *Heliamphora tatei* var. *tatei* (Sarraceniaceae) of Cerro Duida is particularly large (to 1.5 m tall) and has a pseudo-branched stem (see Steyermark 1984; Plate 60). More detailed accounts of these meadows were published by Tate (in Gleason 1931) and Steyermark (1966).

The herbaceous vegetation of the summit of Cerro Marahuaka (***unit 94*** on map) differs notably in its northern and southeastern sections, although both are located between approximately 2500 and 2750 m elevation and are separated from each other only by a narrow, forested valley. The northern plateau (Marahuaka Fufha) is extensively covered by organic soils (peat), which are covered by dense meadows dominated by *Stegolepis terramarensis* (Rapateaceae), along with the terrestrial bromeliads *Steyerbromelia discolor* and *Brewcaria marahuacae.* On the mostly rocky southeastern summit of Cerro Marahuaka (Marahuaka Fhuif and Atahua-Shiho), another, more open type of meadow predominates. Here, the endemic, broad-leaved *Marahuacaea schomburgkii* (Rapateaceae) is the most important herbaceous element (Plate 44). It grows together with numerous species of Bromeliaceae, Cyperaceae, Xyridaceae, Eriocaulaceae, Poaceae, Gentianaceae, and Campanulaceae, as well as low shrubs and subshrubs of Ericaceae, Cyrillaceae, Ochnaceae, and Asteraceae. Both vegetation types are described in greater detail by Steyermark and Maguire (1984a) and Colonnello (1984).

Finally, the summit of the smaller Cerro Huachamacari, to the northwest of Cerro Duida, is partly covered by a very dense and species-rich shrubby meadow that grows exclusively on deep peat (***unit 95*** on map; Plate 42). This vegetation type occurs mainly on the southeastern section of the summit, between 1600 and 1800 m elevation. Its most important herbs are *Stegolepis grandis* (Rapateaceae), *Racinaea spiculosa* var. *stenoglossa, Brocchinia tatei,* and *B. acuminata* (Bromeliaceae), *Heliamphora tatei* var. *tatei* (Sarraceniaceae; Plate 60), and *Orectanthe sceptrum* (Plate 50) and *Xyris* spp. (Xyridaceae). Among the shrubs and subshrubs, the most common genera are *Tibouchina* and *Clidemia* (Melastomataceae), *Bonnetia* (Theaceae), *Duidania, Maguireothamnus,* and *Pagameopsis* (Rubiaceae), *Stenopadus* (Asteraceae), *Cyrilla* (Cyrillaceae), and *Ilex* (Aquifoliaceae).

Upland and highland meadows of the tepuis of southern Amazonas. There are several mountain systems south of the upper Orinoco and Casiquiare rivers that have extensive herbaceous vegetation types both in upland and highland life zones. On the southern summit of Cerro Vinilla at approximately 750 m elevation, there are a few small meadows, which are very interesting because they are the only known areas where the large, broad-leaved *Saxofridericia spongiosa* (Rapateaceae) dominates the open herb layer (***unit 96*** on map). These meadows occur on sandstone outcrops together with *Abolboda macrostachya* (Xyridaceae), *Philodendron englerianum* (Araceae), and *Navia culcitaria* (Bromeliaceae) among the herbs, and *Tepuianthus savannensis* (Tepuianthaceae) and *Bonnetia tristyla* (Theaceae) among the small shrubs.

The summit of Cerro Aracamuni lies southwest of the Vinilla-Unturán ranges. This area is largely covered by extensive, broad-leaved meadows mixed with low shrubs (***unit 97*** on map). According to Steyermark and Holst (1989), these meadows occur at approximately 1500 m elevation on peaty soils. The herbaceous layer is dominated by three species of *Stegolepis* (Rapateaceae), Eriocaulaceae, Bromeliaceae, Xyridaceae, and Liliaceae; dominant shrubs include *Bonnetia tristyla* (Theaceae), *Poecilandra retusa* (Ochnaceae), *Pagameopsis maguirei* subsp. *neblinensis* (Rubiaceae), and *Gongylolepis huachamacari* (Asteraceae). Steyermark and Holst (1989) found the flora of the Aracamuni summit to be similar to that found on the adjacent Neblina massif to the south.

There are several herbaceous vegetation types on the large and ecologically very diverse Neblina massif, but despite the many botanical specimens collected there on several expeditions, the lack of specific vegetation studies precludes a more precise classification. Broad-leaved meadows with shrubs and subshrubs occur in the Titiricó plateau on the southeastern (Brazilian) side of the massif around 2200 m elevation (Dunsterville in Brewer-Carías 1978). The dominant herbs are *Stegolepis neblinensis* (Rapateaceae), *Navia aloifolia* (Bromeliaceae; Plate 61), and *Heliamphora tatei* var. *neblinae* (Sarraceniaceae), which occur together with many species of Eriocaulaceae, Xyridaceae, and Cyperaceae. These meadows are indicated collectively under ***unit 98*** on the accompanying vegetation map.

Pioneer Formations

The pioneer formation is widely distributed in the Venezuelan Guayana and includes species that are early colonizers of essentially bare or open habitats. Most of the groups belonging to this category are not vascular plants, but rather more simply structured organisms such as algae, fungi, and lichens. Some isolated communities of higher plants, however, are also included here in the pioneer formation, when they occupy advanced successional niches in the early pioneering phases of open areas and have already displaced the thallose or cryptogamic life forms. Such communities consist of highly specialized herbs or subshrubs that usually cannot be classified into meadows or shrublands.

The most important pioneer communities occur on rocky substrates, which are abundant in both the lowlands and the highlands of the Venezuelan Guayana. Since the geology of the substrates in these distinct altitudinal belts is usually quite different, they are treated individually below. Other pioneer communities occur on sandy soils of riverbanks and temporary islands that emerge during the dry season in large rivers. These consist mainly of ephemeral or annual species of sedges, grasses, Xyridaceae, and Eriocaulaceae.

Lowland Saxicolous Ecosystems (*Lajas*)

Scattered mainly along the western and northwestern edge of the Guayana Shield in Amazonas and Bolívar states are numerous rock outcrops that are interspersed through both forest and savanna formations. These very characteristic black, rounded hills and boulders are locally called *lajas*. Geomorphologically, they are true inselbergs, or island mountains, because of their mostly isolated occurrence in the surrounding lowland plains and peneplains. All *lajas* of the Venezuelan Guayana belong to the Precambrian basement of the Guayana Shield. They consist of igneous-metamorphic rock types, mainly granites and granitoid gneisses that are often grouped collectively under Rapakivi granite (Blancaneaux and Pouyllau 1977).

The bare rock surface of the *lajas* is rough and usually quite irregular, forming many variably sized depressions. The slopes of these hills are variously inclined, and it is usually possible to walk to the summits, which often reach elevations of 100 to 400 (to 500) m. The black coloration of the open rock surface is due to the extensive cover of several genera of crustaceous blue-green algae, or cyanobacteria, such as *Stigonema, Gloeocapsa,* and *Scytonema* (U. Lüttge, personal communication 1992). Depending on the depth of the depressions, there are other, more advanced pioneer communities. These include bright yellow-flowering, ephemeral colonies of *Utricularia* (Plate 67) and *Genlisea* (Lentibulariaceae) that predominate in shallow holes. In larger depressions with dissolved organic materials, they are replaced by low herbs or sedges (*Bulbostylis*), grasses (*Thrasya, Axonopus*), species of Xyridaceae (*Xyris*) and Rubiaceae (*Perama, Borreria*). Along fissures on somewhat level sites of bare rock, larger colonies of several endemic species of *Pitcairnia* (Bromeliaceae) predominate, together with *Vellozia tubiflora* (Velloziaceae), Cactaceae (*Melocactus* spp., *Cereus*), and an interesting endemic dendroid Cyperaceae (*Bulbostylis leucostachya*).

As soon as the organic and mineral detritus accumulated in the depressions allows the formation of a shallow soil profile, small shrub islands with a number of endemic species occupy these habitats. The shrubs belong mainly to the Melastomataceae (*Acanthella, Comolia, Graffenrieda*), Bignoniaceae (*Tabebuia, Jacaranda*), Bombacaceae (*Pseudobombax*), Burseraceae (*Bursera*), Erythroxylaceae (*Erythroxylum*), Violaceae (*Rinorea*), Euphorbiaceae (*Alchornea, Manihot*), Clusiaceae (*Oedematopus, Clusia*), Ochnaceae (*Ouratea*), and Fabaceae (*Clitoria, Swartzia*). During the pronounced dry season, many of these shrubs are deciduous and have massive flowering displays (Huber 1980).

On large, irregularly shaped *lajas* where deeper soils have accumulated in the depressions, there are true forest islands that alternate with low scrub and savannas. The forests are generally just 4–8 m tall and relatively dense; they are formed mostly by deciduous treelets of Bombacaceae (*Pachira*), Rubiaceae (*Simira*), Bixaceae (*Cochlospermum*), and Bignoniaceae (*Tabebuia*), together with the gregarious palm *Syagrus orinocensis*. The understory is often formed by dense colonies of wild pineapple (*Ananas parguazensis*) or other spiny bromeliads, and small populations of *Zamia lecointei* (Zamiaceae) have occasionally been found growing in shady sites.

The flora of these lowland granitic outcrops or *lajas* is most diverse and has the highest levels of endemism in the northwestern corner of Amazonas state and adjacent Bolívar (***unit 99*** on map; Plates 5 and 6). Farther east, there are a number of similar outcrops that reach as far as the lower Río Caroní, but the floristic richness is notably lower, and endemic species are practically absent. This is probably related to the lower levels of precipitation in northern Bolívar, which may also be responsible for the increased presence of species of Cactaceae and other spiny shrubs on these eastern *lajas* (***unit 100*** on map). Other common taxa on these dry *lajas* are *Curatella americana* (Dilleniaceae), several species of legumes, Rubiaceae, Erythroxylaceae, and *Krameria* (Krameriaceae).

A third, large complex of lowland granitic outcrops occurs in the western Casiquiare and lower Siapa and Pasimoni river basins (***unit 101*** on map). These impressive, often massive rocky hills rise abruptly from the surrounding lowland forests to elevations of up to 800 m. Some are widely visible with their steep, black slopes, such as Piedra Cocuy near the border of Venezuela, Colombia, and Brazil. The shrub islands growing on these inselbergs are evergreen and usually taller than those on the northern *lajas*. Their floristic composition is also different, with mostly species of Clusiaceae, Ochnaceae, Melastomataceae, Rubiaceae, and Theaceae. Among the most important colonizing herbs are several species of Bromeliaceae, Rapateaceae, and terrestrial Orchidaceae. Although some of these *lajas* were studied botanically as early as 1854 by Richard Spruce, then a century later by Richard Schultes, Bassett Maguire, and John Wurdack, little information has been published so far on this habitat (see Steyermark and Holst 1989).

Highland Saxicolous Ecosystems

One of the most common substrates on the summits of the sandstone table mountains of the Venezuelan Guayana is open rock surface. This habitat predominates especially on the high plateaus of the eastern tepuis (Roraima, Kukenán, Yuruaní, Ilú, and Tramen) and on Los Testigos range, Ptari-tepui, the eastern branch of Auyán-tepui, Aprada-tepui, and some of the high summits of the Chimantá massif (***unit 102*** on map; Plate 40). Open rock surfaces also occur frequently on the numerous summits of the granitic Sierra de Maigualida, but they are much less abundant on the tepui summits of Amazonas and southern and southwestern Bolívar states.

The first colonizers of highland rock outcrops are cyanobacteria such as *Stigonema,* which is responsible for the characteristic black color of the rock surface, as well as a variety of crustaceous and frutescent lichens such as species of *Siphula, Caloplaca, Xanthoparmelia,* and *Usnea* which are particularly abundant on the Chimantá massif (Ahti 1992). Among the vascular plants, species of Bromeliaceae are the most important rock colonizers, with many endemic species of *Lindmania, Navia,* and *Brocchinia.* On the Chimantá massif, species of *Lindmania* such as *L. subsimplex* form extensive, dense, and almost monospecific carpets on flat rock surfaces (Huber 1992b; Plate 40). On the hump-like, open, rocky boulders of the summits of the granitic Sierra de Maigualida, in contrast, only scattered clumps of the widespread *Racinaea spiculosa* var. *stenoglossa* are found as colonizers, together with cyanobacteria, lichens, and small cushions of moss. So far, very little is known about the colonizing plant communities of the vertical upper walls of the tepuis, but rosette-forming monocots such as Bromeliaceae, Liliaceae, Eriocaulaceae, Xyridaceae, and Rapateaceae appear to dominate among the flowering plants.

Aquatic Vegetation

The extensive river systems of the Venezuelan Guayana comprise the region's main bodies of water. There are no natural lakes, and small lagoons are found only sporadically on a few mountains, such as Laguna Autana or Laguna Leopoldo near the base of Cerro Autana, Laguna Asisa on the Parú massif, and some water-filled sinkholes on the summit of Cerro Guaiquinima. Around 1973, a huge artificial lake was created along the lower Río Caroní to produce hydroelectric power. Now known as Lago Guri, it is the only large body of standing water in the Venezuelan Guayana.

Most of the aquatic flowering plants of the Venezuelan Guayana belong to typical hydrophytic families with broad geographical distributions, such as Alismataceae, Pontederiaceae, Cabombaceae, and Nymphaeaceae. There are also several widespread aquatic ferns such as *Azolla* (Azollaceae), *Ceratopteris* (Parkeriaceae), and *Salvinia* (Salviniaceae). The fern ally *Isoëtes* (Isoëtaceae) has three aquatic species in the flora area, one in the lowlands and the other two on high-tepui summits.

A number of common hydrophytes in the Venezuelan Guayana grow in running waters of black- and clear-water rivers. These belong mainly to Podostemaceae, Eriocaulaceae, and Mayacaceae. Several small, highly specialized species of *Rhyncholacis* (Plate 66), *Mourera,* and *Apinagia* (Podostemaceae) form large colonies that grow firmly attached to rocks in the middle of rapids or swift water currents. A detailed overview of this interesting family in the Orinoco basin of Venezuela was made by Grubert (1974). The tiny, white flowering herbs of *Mayaca* (Mayacaceae) often grow on rocks in small rivulets in both lowland and upland regions. *Tonina fluviatilis* and *Eriocaulon melanocephalum* are aquatic species of Eriocaulaceae. This *Eriocaulon* species has hair-like leaves and forms large, dense colonies in stony riverbeds of both lowland and highland sites (as on Cerro Duida).

In the small, mostly seasonal ponds of the southern Gran Sabana, several species of the Alismataceae and Nymphaeaceae often occur. In the lowlands,

Nymphaea (Nymphaeaceae), *Sagittaria* and *Echinodorus* (Alismataceae), *Eichhornia* (Pontederiaceae), *Pistia stratiotes* (Araceae), and *Ludwigia sedoides* (Onagraceae) are the most common aquatic macrophytes found in small lagoons and quiet river margins with white or clear waters.

There is apparently only one genus of hydrophytes that is endemic to the Guayana Shield. *Jasarum steyermarkii* (Araceae) lives submerged in black-water rivers and small ponds between 1200 and 1300 m elevation in the northern Gran Sabana and down to 500 m elevation along the upper Mazaruni River in Guyana. The monocot family Thurniaceae, with one genus and three species, is nearly restricted to the Guayana Shield, but extends farther south in the Amazon basin. In the black-water rivers of the upper Orinoco and Negro river basins, large colonies of *Thurnia polycephala* frequently grow partially submerged along the streambanks. At low water levels during the dry season, these sedge-like plants are emergent and produce numerous globose inflorescences with tiny white flowers that attract many small black bees and flies (Plate 71).

CHAPTER 4

Floristic Analysis and Phytogeography

Paul E. Berry, Otto Huber, and Bruce K. Holst

Botanical exploration of the Guayana Shield since the 1890s has demonstrated that this area contains a rich and largely unique flora. The actual number of taxa within the area and the level of endemism, however, have mostly been a matter of speculation. One of the botanists most familiar with the region, Bassett Maguire, first estimated that there were close to 8000 species of vascular plants on the Guayana Shield (although he referred to this area of roughly 1,000,000 km^2 as the "Guayana Highland"), with more than 75 percent of the species endemic (Maguire 1970). Since then his figures have been cited repeatedly in botanical literature. This chapter reexamines the number of taxa in the Venezuelan Guayana and reevaluates the levels of endemism.

Species Numbers and Composition

By extracting data from family treatments submitted for the *Flora of the Venezuelan Guayana* and assembling checklists of the remaining families from literature and herbarium searches, we now calculate that there are close to 9400 described and currently recognized species of vascular plants in the Venezuelan Guayana. Since this area covers about half the extension of the Guayana Shield, it is clear that Maguire's estimate for the entire shield area was too low. Table 4-1 provides a summary of the figures from the Venezuelan Guayana divided into three major groupings of vascular plants, and Appendix A presents a list of the number of genera and species for each family in the flora.

With its nearly 9400 species, the *Flora of the Venezuelan Guayana* includes about two-thirds of the total number of vascular plant species in Venezuela. This information is based on an unpublished checklist of 13,400 species of vascular plants for the whole country, assembled by Robert Wingfield (personal communication) during the 1980s. We estimate from our data and Wingfield's that about 4500 species of vascular plants in Venezuela occur both north and south of the Río Orinoco and that the number of species in the

Table 4-1. Number of taxa in the Venezuelan Guayana by major plant groups.

	Families	Genera	Species
Pteridophytes	29	92	671
Gymnosperms	3	3	18
Angiosperms	198	1691	8722
	230	1786	9411

northern half of Venezuela (between 8500 and 9000) is comparable to the number of species in the Venezuelan Guayana. Even allowing for several thousand plant species that may not yet have been recorded in Venezuela, there are clearly far fewer species of vascular plants in the country than Steyermark (1977, p. 130) estimated at "between 20,000 to as many as 35,000."

Strictly cultivated species are excluded from the *Flora of the Venezuelan Guayana,* but non-native species that have either escaped from cultivation or are naturalized in the area are included. There are only 123 non-native species, or 1.3 percent of the flora, 25 of which are introduced grasses that inhabit naturally open or cleared areas. Just 3 of the 230 families in the *Flora of the Venezuelan Guayana* are not native: Brassicaceae (*Cardamine*), Crassulaceae (*Kalanchoe*), and Plantaginaceae (*Plantago*). Unlike modern floras of Hawaii, where 47 percent of the species are introduced (Wagner et al. 1990), and California, where 17.4 percent of the species are non-native (Hickman 1993), the flora of the Venezuelan Guayana has a very low level of alien species, indicating relatively little disturbance and a highly intact natural flora. A checklist of Guyana, Suriname, and French Guiana (Boggan et al. 1992) also cited a low level of 3 percent introduced and naturalized species. For Peru, non-native plants (including cultivated species) comprise 3.5 percent of the flora (Brako and Zarucchi 1993; J. Zarucchi, personal communication).

With better data becoming available for different countries in South America, it is possible to begin to evaluate diversity patterns in the continent. For Peru, with an area of nearly 1,300,000 km^2, Brako and Zarucchi (1993) recorded 17,143 species of seed plants. This number is supplemented by 1060 species of pteridophytes known from Peru (R. Stolze, personal communication). Venezuela, which covers about 950,000 km^2, has close to 13,400 recorded species of vascular plants, about 1100 of them pteridophytes (see above). The *Checklist of the Plants of the Guianas* (Boggan et al. 1992) covers just under 500,000 km^2, an area adjacent to the Venezuelan Guayana and nearly equivalent in size. Since the Guianas checklist includes nonvascular plants and also lists plants expected to occur there but not yet found, only the vascular plants known to occur in the area were extracted from the checklist. This gave a total of 7088 species, which compares with the 9411 vascular plant species known from the Venezuelan Guayana. The 32 percent higher number of species in the Venezuelan Guayana can be attributed in large part to the greater expanse of montane habitats compared to the Guianas.

Table 4-2 lists the 23 vascular plant families with more than 100 species in the Venezuelan Guayana. These are all important neotropical families, but certain families are over- or under-represented with respect to other areas of the continent. In the *Catalogue of the Flowering Plants and Gymnosperms of Peru* (Brako and Zarucchi 1993), families such as Solanaceae, Acanthaceae, Malvaceae, Cactaceae, Scrophulariaceae, Verbenaceae, and Lamiaceae appear among the 23 largest families, but they are much more poorly represented in the Venezuelan Guayana. Conversely, Ochnaceae, Chrysobalanaceae, Clusiaceae, Malpighiaceae, and Myrtaceae are among the largest families in the Venezuelan Guayana, but are less important numerically in the Peruvian flora. Some of these differences may be due primarily to climatic differences in the two areas, for instance the wider array of arid habitats may explain the higher representation of Cactaceae in Peru. Other discrepancies appear to be related to differences in substrates, such as the poor representation in the Venezuelan Guayana of Solanaceae, which rarely occur on the oligotrophic soils that are widespread in the Guayana (W. D'Arcy, personal communication). In both floras, Orchidaceae is the single largest family, and three or four of the ten largest families are monocots. There are only three gymnosperm families native to the Venezuelan Guayana, Gnetaceae, Podocarpaceae, and Zamiaceae, and together they account for just 18 species. Peru and the Guianas are similarly poor in gymnosperm species, with only four species in two families native to the Guianas (Boggan et al. 1992) and 17 species in four families native to Peru (Brako and Zarucchi 1993).

From the *Checklist of the Plants of the Guianas* (Boggan et al. 1992), there are 1726 known genera of vascular plants in the Guianas, of which 85 are pteridophytes. The Venezuelan Guayana is slightly more diverse, with 1786 genera, of which 92 are pteridophytes (Table 4-1). The much larger area

Table 4-2. Families with more than 100 species in the Venezuelan Guayana.

Family	Number of species	Family	Number of species
Orchidaceae	698	Mimosaceae	163
Rubiaceae	530	Apocynaceae	162
Poaceae	420	Lauraceae	143
Melastomataceae	397	Bignoniaceae	132
Fabaceae[1]	319	Clusiaceae	128
Bromeliaceae	273	Piperaceae	126
Asteraceae	257	Malpighiaceae	123
Cyperaceae	243	Dryopteridaceae	115
Euphorbiaceae	237	Chrysobalanaceae	111
Caesalpiniaceae	203	Ochnaceae	107
Myrtaceae	178	Annonaceae	102
Araceae	164		

[1]In the strict sense; when combined with Caesalpiniaceae and Mimosaceae, the legumes total 685 species.

covered by Peru has 2576 genera of vascular plants, of which 118 are pteridophytes (Brako and Zarucchi 1993; R. Stolze, personal communication).

Genera that are much better represented in the Venezuelan Guayana compared to Peru are *Navia, Xyris, Rhynchospora, Ilex,* and *Ouratea* (Table 4-3). In some cases, different genera in the same family displace each other in importance between the two areas, such as in the Araceae, where *Philodendron* is very speciose in the Venezuelan Guayana and *Anthurium* is very rich in Peru. In the Bromeliaceae, *Navia* in the Venezuelan Guayana is substituted by *Tillandsia* and *Pitcairnia* as very species-rich genera in Peru. *Solanum* and *Senecio* are two very speciose genera in Peru (292 and 177 species, respectively); in the Venezuelan Guayana there are only 39 species of *Solanum* and no species of *Senecio*.

There are five large genera that are much more speciose in the Venezuelan Guayana than in the Guianas. These are *Ilex* (69 versus 19 species), *Navia* (70 versus 7 species), *Schefflera* (60 versus 12 species), *Selaginella* (59 versus 32 species), and *Xyris* (74 versus 26 species). Most of these genera, as well as smaller ones like *Bonnetia* (Theaceae), *Lindmania* (Bromeliaceae), and *Stegolepis* (Rapateaceae), are well represented in the Guayana highlands, which are much more extensive in southern Venezuela than in the Guianas. Large genera that have more species in the Guianas than in the Venezuelan Guayana, such as *Eugenia, Licania,* and *Swartzia,* are mainly lowland, forest-dwelling groups.

Levels of Endemism

Several levels of endemism were calculated for the flora, starting with the number of taxa known only from the Venezuelan Guayana. Then it was determined how many of the taxa in the Venezuelan Guayana are endemic to the Guayana Shield, which is a larger but a more coherent physiographic and phytogeographic region. Third, the number of taxa occurring in Pantepui (or

Table 4-3. Genera with more than 50 species in the Venezuelan Guayana.

Genus (Family)	Number of species	*Genus* (Family)	Number of species
Psychotria (Rubiaceae)	128	*Schefflera* (Araliaceae)	60
Miconia (Melastomataceae)	106	*Licania* (Chrysobalanaceae)	59
Xyris (Xyridaceae)	74	*Selaginella* (Selaginellaceae)	59
Piper (Piperaceae)	73	*Ocotea* (Lauraceae)	58
Navia (Bromeliaceae)	70	*Pleurothallis* (Orchidaceae)	58
Rhynchospora (Cyperaceae)	± 70	*Elaphoglossum* (Dryopteridaceae)	57
Ilex (Aquifoliaceae)	69	*Ouratea* (Ochnaceae)	57
Philodendron (Araceae)	68	*Passiflora* (Passifloraceae)	57
Epidendrum (Orchidaceae)	63	*Maxillaria* (Orchidaceae)	55
Panicum (Poaceae)	62	*Peperomia* (Piperaceae)	53

the Guayana Highland), which is here defined as all areas in the Guayana Shield over 1500 m elevation, were counted. Floristically, this is one of the zones of highest endemicity in northern South America. Finally, data on the number and distribution of very localized endemic species were assembled.

Overall levels of endemism in the Venezuelan Guayana are summarized in Table 4-4. More than 20 percent of the species are restricted to the flora area, and 40 percent of the species are endemic to the Guayana Shield. From these figures, it is now clear that Maguire's (1970) estimate of 75 percent species endemism in the Guayana Shield was too high. Because most of the non-Venezuelan part of the Guayana Shield (northern Brazil, southeastern Colombia, and most of the Guianas) is lowland, where endemism is generally less than in the highland parts of the shield, it is likely that the species endemism for the entire shield area is somewhat less than the 40 percent level recorded for the species from the Venezuelan Guayana.

Table 4-4. Endemism levels in the Venezuelan Guayana.

	Total number of taxa	Taxa endemic to Venezuelan Guayana	Taxa endemic to Guayana Shield
Families	230	0	4
Genera	1786	34 (1.9%)	118 (6.6%)
Species	9411	2136 (22.7%)	3763 (40.0%)

There are 34 genera that are endemic to the Venezuelan Guayana and a total of 118 genera from the flora area that are endemic to the Guayana Shield. Outside of Venezuela, an additional 20 genera are known to be endemic to the Shield area. Table 4-8 (see end of chapter) lists all 138 genera now considered to be endemic to the Guayana Shield. For each of these genera, the elevational range, number of species, and distribution by country are included. Many of the larger genera are concentrated in the Venezuelan Guayana, with outliers in adjacent areas of Brazil and Guyana. Table 4-9 (see end of chapter) lists 25 additional genera that are nearly endemic to the Venezuelan Guayana or the Guayana Shield, but with one or more outliers extending somewhat beyond the limits of the Shield. The genera listed in Tables 4-8 and 4-9 are important in defining the phytogeographic character of the region and are further discussed in the following section.

Although no vascular plant family is entirely restricted to the Venezuelan Guayana, four families are endemic to the Guayana Shield, and all occur mainly in the flora area. The monotypic Saccifoliaceae, related to Gentianaceae, occurs just along the border of Brazil and Venezuela on the Sierra de la Neblina. Euphroniaceae, with a single genus of three species, was previously included in either the Trigoniaceae or Vochysiaceae and is centered in the Venezuelan Guayana, but extends short distances into neighboring countries.

Tepuianthaceae, placed in the Sapindales, consists of a single genus with six species (Plate 68), four in the Venezuelan Guayana and one each in Colombia and northern Brazil. The largest of the endemic Guayana Shield families is the Hymenophyllopsidaceae, consisting of a single genus of eight species of high-elevation ferns all found in the flora area with two of them barely extending into neighboring Guyana and Brazil (Plate 62).

Maguire (1970) estimated that the summit area of the tepuis, which is roughly equivalent to the concept of Pantepui used in this flora (see Chapters 1 and 3), would include close to 2000 species, with 90 to 95 percent of the species endemic. The Pantepui flora of the Venezuelan Guayana is now comparatively well known, with 2322 species and 630 genera (Table 4-5). Since there is only a small portion of Pantepui that occurs outside of Venezuela, these figures are very close to the expected totals for the entire area of Pantepui.

Although Maguire was close in his estimate of the total number of species in Pantepui, his estimate of species endemism was too high. In contrast to Maguire's 90 to 95 percent figure, our data indicate that 33 percent of the Venezuelan Pantepui species are endemic to the Pantepui area (Table 4-5). Even at a broader geographical scale, the level of endemism does not approach Maguire's estimate; 47 percent of the Venezuelan Pantepui species are restricted to the flora area, and about 65 percent are restricted to the Guayana Shield. The remaining 35 percent of the Venezuelan Pantepui species that extend beyond the Guayana Shield are restricted to the New World, except for 26 species (just over 1 percent of the Pantepui flora) that also range into the Old World.

Mayr and Phelps (1967) estimated the total area of Pantepui to be close to 5000 km^2, which is still an accurate figure (see Table 1-3). Although this is just over 1 percent of the flora area, it includes 766 endemic species, or 36 percent of the total number of species endemic to the Venezuelan Guayana. Similarly, 23 of the 34 genera that are endemic to the flora area occur only in Pantepui. Given the small size of the Pantepui area, this is still a remarkable concentration of endemic taxa.

Local Endemics

A final stage in the analysis of endemic taxa in the Venezuelan Guayana deals with local endemic species, that is, those occurring either at a single known locality, a single mountain, a particular river basin, or in a localized habitat such as the low inselbergs (*lajas*) along the Río Orinoco. This kind of analysis gives insights into regional patterns of endemism within the flora area.

There are close to 1270 species of local endemics in the Venezuelan Guayana. Amazonas state has 815 local endemic species, nearly twice as many as Bolívar state, with 440. There are only two species endemic to Delta Amacuro state, and these occur along the Bolívar border in the Serranía de Imataca. About 12 more species occur along the border between Amazonas and Bolívar states, mostly in the Sierra de Maigualida. By far the greatest number of local endemics are known from the slopes and summits of individual tepuis. The

Table 4-5. Number of Pantepui taxa (occurring above 1500 m) in the Venezuelan Guayana.

	Taxa present in Pantepui	Taxa endemic to Pantepui	Taxa endemic to Venezuelan Guayana	Taxa endemic to Guayana Shield
Families	158	1	0	4
Genera	630	23 (3.7%)	29 (4.6%)	85 (13.5%)
Species	2322	766 (33.0%)	1088 (46.9%)	1517 (65.3%)

mountains with the highest number of local endemics are Sierra de la Neblina (192 species), Macizo del Chimantá (99), Cerro Duida (76) and Cerro Sipapo (61). These are large, high-elevation tepuis with a wide array of habitat types, and all but the Macizo del Chimantá are in Amazonas state. There, tepuis arise from a base in lowland forests between 100 and 200 m elevation, which provides a much greater gradient for additional species to occur compared to the high tepuis in Bolívar, which arise from a large plateau at 500 to 1200 m elevation (the Gran Sabana). The generally higher rainfall in Amazonas state than in Bolívar state (see Chapter 1) probably contributes to the higher levels of endemism there.

Certain lowland areas in Amazonas state are also rich in local endemics. These include the Río Pasimoni basin (including the Río Yatúa and the Río Baría or Mawarinuma, with 39 endemic species), the lowland savannas and forests surrounding Cerro Yapacana (24 species), the Río Atabapo basin (22 species), and the Río Guainía basin in Venezuela (22 species). These areas all lie within black-water river basins and have a predominant cover of white-sand, oligotrophic substrates. The low inselbergs or *lajas* that dot the landscape along the Río Orinoco in northwestern Amazonas and northwestern Bolívar states support about 50 known endemic species. The families with the highest number of local endemics in the Venezuelan Guayana are the Rubiaceae (124 species), Bromeliaceae (114 species), Melastomataceae (78 species), Asteraceae (56 species), and Araceae (48 species).

Floristic Relationships

During much of the early exploration of the Pantepui zone, great emphasis was placed on the high endemicity, bizarre life forms, and purported isolation of the tepui summit flora (e.g. Brown 1901; Maguire 1970; Brewer-Carías 1978). The flora was characterized as "ancient," "relictual," and one "required to adjust to a constantly and continuously restrictive, isolating and narrowing environmental amplitude of genetic exchange and of habitat" (Maguire 1970, p. 97). Further exploration, however, showed that the isolation of the tepui summits was far less complete than originally depicted in the "islands in time" (George 1988) or Lost World scenario (Doyle 1912). Steyermark and Dunsterville (1980) and Huber (1988a) documented the wide altitudinal range of many highland taxa and the mixture of lowland and upland taxa on tepuis of intermediate elevations. The close phytogeographical connection of lowland Amazonian white-sand savannas to high-tepui meadows and shrublands, although originally noted by Maguire (1970), was emphasized by Huber (1988a) and Kubitzki (1989a, 1990). Paleoecological studies have shown dynamic changes in the spatial and altitudinal zonation of vegetation types on the summits of tepuis as well as in the Gran Sabana during the Holocene and probably through much of the Pleistocene (Rull 1991). Consequently, the focus on the Guayanan flora has changed from an emphasis on evolution in isolation to an understanding of the ecological determinants of the endemism, and the relationships and dynamics of the present-day flora.

Students of the Guayanan flora concur that most endemism is associated with nutrient-poor substrates that are derived from erosion of ancient crystalline and sandstone rocks (Maguire 1970; Steyermark 1986b; Huber 1988a; Kubitzki 1989a, 1990). These white-sand soils have poor water retention and are patchily distributed in the lowlands, where they have been overlain in some areas by more recent lateritic soils that are widespread in the Amazon basin. Kubitzki (1990) referred to the edaphic Guayanan element as the "psammophilous flora of northern South America" and stressed its altitudinal amplitude from lowlands to highlands. Examples of typical Guayanan genera (sometimes including the same species) that occur both on highland summits as well as in lowland areas are *Gongylolepis* and *Stenopadus* (Asteraceae), *Brocchinia* and *Navia* (Bromeliaceae; Plate 61), *Macairea* (Melastomataceae), *Kunhardtia* and *Saxofridericia* (Rapateaceae), *Retiniphyllum* (Rubiaceae), *Tepuianthus* (Tepuianthaceae), and *Archytaea* and *Bonnetia* (Theaceae).

Kubitzki (1989a) hypothesized that the core phytogeographical elements of the Guayana Shield developed and differentiated in lowland areas on substrates derived from Roraima sediments. He suggested that genera now endemic to the highlands are not ancient lineages, but for the most part recent immigrants from the Andes. This theory, though, is based on several very speculative suppositions. One is that the uplift of the Roraima sediments is relatively recent (Neogene and upper Pleistocene), an idea based on geochemical studies by Grabert (1976a, 1976b). Grabert's conclusions, however, are contradicted by an abundance of other kinds of geological evidence (see Chapter 1). Second, Kubitzki assumes that because groups such as Asteraceae (particularly the subtribe Mutiseae sensu lato), Bromeliaceae, and Ericaceae are an important part of the Andean flora, the Guayanan genera in these families are derived from them. Although certain genera present in both areas, such as *Disterigma* and *Gaylussacia* (Ericaceae), are almost certainly recent immigrants from the Andes, there is to date no evidence to support the supposition that the endemic genera of these families are derived from Andean relatives. Phylogenetic studies in these groups should be pursued to test hypotheses of the relationships and place of origin of the endemic genera.

Besides the Guayanan genera with broad elevational ranges, there are large suites of endemic or nearly endemic genera that are constrained, perhaps climatically, to either lowland or highland habitats (see Tables 4-8 and 4-9). In Rapateaceae, for example, the speciose tepui genus *Stegolepis* (Plate 50) and the monotypic genera *Amphiphyllum, Marahuacaea,* and *Phelpsiella* occur exclusively in upland and highland habitats. Genera such as *Guacamaya, Monotrema,* and *Schoenocephalium,* on the other hand, occur only in lowland savannas. With such well-defined floristic assemblages at both elevational levels, the preponderance of endemic taxa in the highlands, and the lack of phylogenetic evidence to determine if lowland or highland groups are generally more recently derived, there is very little basis to support Kubitzki's (1989a) assertion of a youthful Guayana highland flora.

The issue of the age and origin of the Guayana highland flora has partly been confused by the presence of a significant Andean or even wider austral or Holarctic element in the flora. This element helps explain the lower values of endemism obtained for the Pantepui flora compared to those originally predicted by Maguire (1970). The presence of a single species of a typically Andean genus in the Guayana highlands, such as *Oritrophium marahuacense* (Asteraceae) on the summit of Cerro Marahuaka, is a fairly obvious sign of a relatively recent immigration from the Andes. Other genera that are likely recent arrivals from the Andes include *Viburnum* (Caprifoliaceae), *Myrteola* (Myrtaceae), *Monnina* (Polygalaceae), and some of those listed by Steyermark (1979a) and Huber (1988b). In the case of genera like *Schefflera,* however, which is very well-represented in the Andes (and in the Old World) but has 60 species in the Venezuelan Guayana, such groups may have a much longer evolutionary history in the Guayana and are not necessarily derived from Andean ancestors. Demonstrating that interchanges have taken place in both directions, there are several cases of Guayanan-centered genera with single representatives that have emigrated to the Andes, such as *Everardia* (Cyperaceae), *Bonnetia* (Theaceae), *Phainantha* (Melastomataceae), *Pterozonium* (Pteridaceae; Plate 53), and in the Asteraceae, *Gongylolepis* and *Stenopadus* (this last genus with an undescribed species that was discovered in Ecuador in the early 1990s).

Much of the Guayanan flora is composed of genera that are broadly distributed in the Neotropics or even more widely. Particular affinities with other phytogeographical areas will require further analysis, but Maguire (1970), Steyermark (1986b), and Huber (1988b) have discussed some of the patterns. For example, the Brazilian Shield or planalto has a number of its characteristic genera present in the Guayana, such as *Calea* (Asteraceae), *Paepalanthus* and *Syngonanthus* (Eriocaulaceae), *Bertolonia, Marcetia,* and *Microlicia* (Melastomataceae), *Euplassa* (Proteaceae), *Declieuxia* (Rubiaceae), *Thesium* (Santalaceae) and *Vellozia* (Velloziaceae). There are also at least two Old World families each with a single New World genus restricted to the Guayana Shield: *Pentamerista* (Tetrameristaceae; Plate 69) and *Pakaraimaea* (Monotaceae, although a second genus from the Araracuara area of Guayanan Colombia will soon be described).

The following section employs some of the data presented here to develop a hierarchical scheme of formal phytogeographic units in the Guayana Shield.

Phytogeography of the Guayana Region

As early as the 1840s the naturalists Richard and Robert Schomburgk recognized that a distinctive biological region lay within the Guayana Shield, yet until now there has been no systematic attempt to phytogeographically define and characterize this region. Twentieth-century biogeographers have dealt with the area in a variety of ways, but usually quite superficially. Maguire (1956, 1970, 1972b, 1979) devoted the most attention to Guayana, although he was inconsistent in his terminology—using "Guayana Highland" and "Guayana floristic province" sometimes interchangeably, at times referring to Pantepui and other times to the entire Guayana Shield. This same ambiguity between a strictly highland or mixed lowland and highland flora can be seen in other systems such as the Venezuela-Guayana Region of Mattick (1964), the Region of Venezuela and Guiana (Good 1969), the Guayanan Domain of Cabrera and Willink (1973), and the Region of the Guayana Highlands (Takhtajan 1986). Parts of Guayana were also proposed as areas of forest refugia during the drier and cooler climatic phases of the Quaternary, including the Venezuelan Guayana (Pantepui) refuge of Steyermark (1982); the Imataca, West Guiana, East Guiana, and Imerí centers of Prance (1982); and the Roraima, Pantepui, and Guyana dispersal centers of Müller (1973).

The only previous attempt to phytogeographically subdivide the region was made by Maguire (1979). He defined Guayana as the area overlain by the Roraima sandstone formation and its outwash sediments, but limited his phytogeographical circumscription of the Guayana floristic province to those parts of the Shield that contained tepuis (Maguire 1979, p. 230, fig. 1). His system comprised three subprovinces, which were further divided into 11 mountain complexes that he delimited by watershed basins. The Pakaraima–Gran Sabana subprovince contained the eastern tepuis such as Auyán-tepui, Roraima-tepui, and Mount Ayanganna. The Río Caroní–Río Negro subprovince included all tepuis in Venezuelan Amazonas as well as Cerro Guaiquinima and the Jaua-Sarisariñama massif in Bolívar state. The trans-Río Negro–Colombian Guayana subprovince included the Colombian sandstone mountains of Cerro Chiribiquete and Serranía de la Macarena. Maguire intended to elaborate upon this outline in a final part of *The Botany of the Guayana Highland,* but he was unable to accomplish this before his death.

In the following section, a new hierarchical phytogeographical classification for the area is proposed, using a combination of floristic and vegetational criteria supplemented by geological and climatic data. Since considerable parts of the Guayana Shield are still poorly known floristically and vegetationally, especially the forested areas, some of the units proposed here are preliminary and will probably be revised with further study.

Definition and Delimitation of the Guayana Region

In phytogeographical terms, the Guayana Region encompasses a highly characteristic complex of neotropical flora and vegetation growing on the land surface occupied roughly by the Guayana Shield and on some peripheral areas directly influenced by it. It consists of an ancient, geologically stable speciation center that has produced a well-defined autochthonous floristic and ecologic assemblage, but which has also received significant inputs from adjacent, historically more recent speciation centers, such as the Amazon and the Andes.

In the phytogeographical hierarchy, the Guayana Region forms part of the Neotropical Realm (Good 1969). It lies between the Caribbean Region to the north, the Atlantic Ocean to the east, and the Amazon Region (Hylaea) to the south and west (Figure 4-1). Maguire (1979) and some other authors have included in the Guayana Region the Serranía de la Macarena, which is located at the eastern base of the Colombian Andes, but the geological and phytogeographical placement of this mountain range is still in dispute and is not included in the present classification.

One of the most significant traits of the Guayana Region is its peculiar

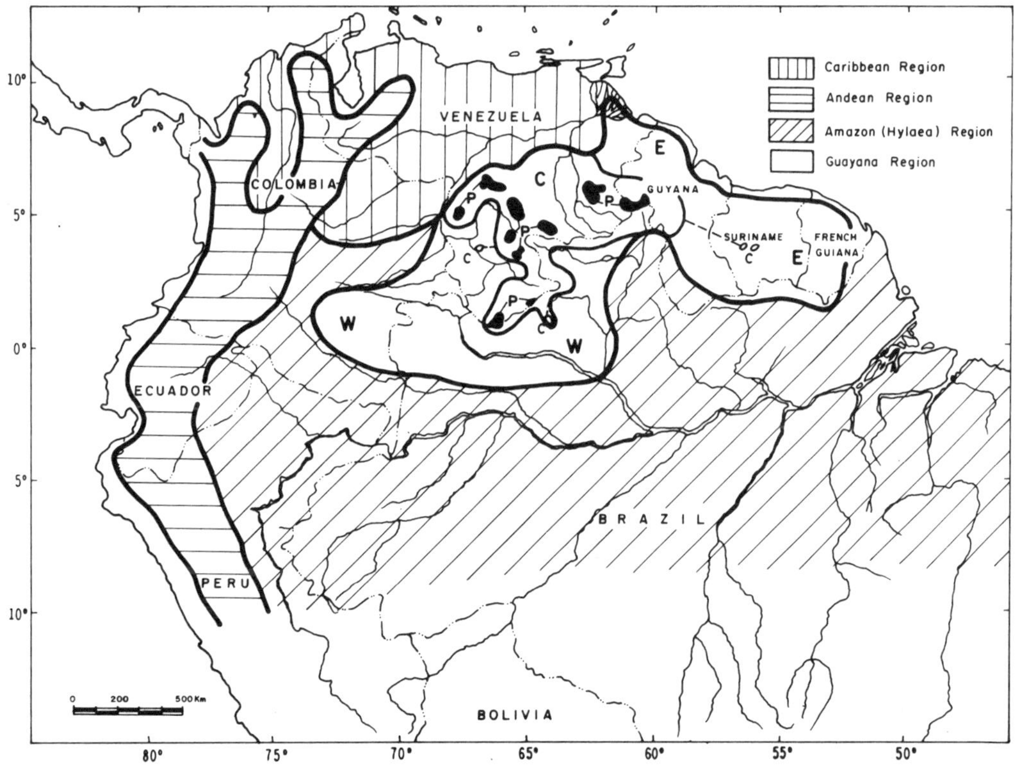

Figure 4-1. Spatial relationship of the four floristic regions of northern South America. The Guayana Region is divided into four provinces: E = Eastern Guayana Province, C = Central Guayana Province, W = Western Guayana Province, P = Pantepui Province.

montane physiography, which strongly characterizes many of its floristic and ecological features. The Guayanan mountain system consists of a loosely arranged series of ancient uplands and highlands between 500 and 3000 m elevation, surrounded by extensive lowland areas. The persistent continentality of most of the Guayana Shield's surface contrasts markedly with the surrounding lowlands of the Orinoco and Amazon plains, which emerged relatively recently from marine or lacustrine environments. There is a predominance of highly weathered parent rocks in the region, both igneous-metamorphic (granites and gneisses) and sedimentary (quartzites and sandstones). Edaphically, most substrates are nutrient-poor and derived from the highly weathered, ancient parent rock.

Floristically, the region has a rich and highly diversified vascular plant flora, which has been characterized earlier in this chapter and partly in Chapter 3. In each of the main physiographical levels there is a series of unique, indigenous vegetation types, such as the Rio Negro caatinga in the lowlands, the Guayana shrublands in the uplands, and the tepui meadows in the highlands. There are also many highly adapted growth forms such as thick-stemmed shrubs (pachycauls), rosette-forming herbs, and tube-leaved herbs.

In addition to the ecologically rich montane core area, the Guayana Region includes several lowland sections both in the middle of the region and along its periphery. Many of the physical conditions of the lowland areas are directly influenced by the adjacent or surrounding mountains, and they often share floristic and vegetational relationships with ecosystems of the upland and highland biota.

The above criteria provide strong support for recognizing the Guayana Region as a major phytogeographical entity (see Table 4-6). As occurs in most large continents, however, the precise limits between adjacent phytogeographical regions are problematical, especially between the Guayana Region and the Amazon Region.

Table 4-6. Criteria used to define the phytogeographical Guayana Region.

Geography	Floristics	Ecology
Location: Guayana Shield of northern South America	*Flora:* Very rich (ca. 15,000 species of vascular plants)	*Characteristic habitats:* Nutrient-poor substrates
Geology: Igneous-metamorphic (granite, gneiss) and sedimentary (quartzite, sandstone)	*Endemism:* Very high (four endemic families, ca. 140 endemic genera of vascular plants)	*Plant characteristics:* High diversity of growth forms in numerous plant families
Physiography: Montane uplands and highlands, peripheral lowlands	*Phytosociology:* Many highly evolved plant communities	*Vegetation:* High diversity of vegetation types (forests, shrublands, meadows)

Subdivisions of the Guayana Region

The Guayana Region can be divided into four phytogeographical provinces based on their geographic location, predominance of key vegetation types, and the presence of characteristic floristic assemblages (Figure 4-1). The main characteristics of each province are summarized below, and an outline of the phytogeographic subdivisions is given in Table 4-7.

Eastern Guayana Province. This province covers the macrothermic lowlands (from sea level to 500 m elevation) of northeastern Venezuela, Guyana, Suriname, and part of French Guiana (Figure 4-1). Its western bor-

Table 4-7. Phytogeographical subdivisions of the Guayana Region.

1. Eastern Guayana Province (lowlands)
2. Central Guayana Province (lowlands, uplands, slopes)
   a. Pakaraima-Gran Sabana District
      Eastern Pakaraima Subdistrict (including Tafelberg outlier)
      Supamo-Lema Subdistrict
      Gran Sabana Subdistrict
      Western Pakaraima Subdistrict (including Tepequém and Uafaranda outliers)
   b. Guaiquinima District
   c. Caura-Paragua District
   d. Nichare-Parima District
      Nichare-Uasadi Subdistrict
      Parima Subdistrict
   e. Guaniamo-Guayapo District
      Guaniamo-Parguaza Subdistrict
      Cuao-Guayapo Subdistrict
   f. Parú-Duida District
      Parú Subdistrict
      Duida Subdistrict
      Yapacana-Cerro Tigre Subdistrict
   g. Unturán District
      Vinilla-Tapirapecó Subdistrict (including Serra Aracá outlier)
      Aracamuni-Imeri Subdistrict
3. Pantepui Province (central highlands)
   a. Eastern Pantepui District
      Roraima Subdistrict
      Chimantá Subdistrict
   b. Jaua-Duida District
      Jaua-Sarisariñama Subdistrict
      Asisa Subdistrict
      Duida-Marahuaka Subdistrict
   c. Western Pantepui District
      Maigualida Subdistrict
      Yutajé Subdistrict
      Cuao-Sipapo Subdistrict
   d. Southern Pantepui District
4. Western Guayana Province (southwestern and western lowlands, uplands)
   a. Upper Río Negro District
   b. Atabapo-Ventuari District
   c. Araracuara District

der is formed by the eastern Llanos of Monagas, the Serranía de Imataca, and the middle Río Cuyuní basin in Venezuela. To the south it is bordered by the piedmont region of the Pakaraima mountains in Guyana, the southern mountain ranges in Suriname, and the hill-lands of French Guiana up to the Oyapoque River, including the Tumuc Humac range.

The province mostly covers alluvial coastal plains, alternating with hill-lands and scattered low mountain ranges. It does not include the littoral plant communities along the seashore such as mangroves and restingas, which may belong in a separate phytogeographic region extending along the entire Atlantic coast of tropical South America that still needs to be properly defined. In the northern part of this province in Guyana and Venezuela (including most of Delta Amacuro state), there are extensive floodplains and swamps interspersed with large areas of white-sand soils, whereas relatively loamy, nonflooded soils predominate in Suriname and French Guiana. The entire area receives abundant rainfall (average >2000 mm per year) and experiences a mild dry season from December through March.

The Eastern Guayana Province is characterized by several evergreen lowland forest types, especially the inundated mora forests dominated by *Mora gonggrijpii* and *M. excelsa* (Caesalpiniaceae), *Catostemma commune* (Bombacaceae), *Triplaris surinamensis* (Polygonaceae), and *Symphonia globulifera* (Clusiaceae), and the greenheart forests dominated by *Chlorocardium rodiei* (Lauraceae) and species of *Eschweilera* (Lecythidaceae). Other characteristic vegetation types of this province are the forests and shrublands on white sand, such as the wallaba forests dominated by *Eperua* and *Dicymbe* species (Caesalpiniaceae), and the muri scrub dominated by *Humiria balsamifera* (Humiriaceae) (see Granville 1988).

The herbaceous vegetation types of this province include the moist coastal savannas found intermittently from northeastern Guyana to Kourou in French Guiana (Fanshawe 1952, Donselaar 1968, Hoock 1971) and the small inland savannas of central Suriname that grow on soils derived from Roraima sandstone (Donselaar 1968). A specialized vegetation has developed on the numerous granitic outcrops (inselbergs) that are scattered throughout the lowlands of Suriname and French Guiana (Granville and Sastre 1973). These communities are ecologically convergent with similar habitats along the northwestern border of the Guayana Shield in eastern Colombia and western Venezuela.

Floristically, the province has a rich forest flora and a specialized inselberg flora with numerous endemic species (Granville 1988). The lowland forests are particularly diverse in families such as Chrysobalanaceae (Prance 1982) and Lecythidaceae (Schulz 1960; Mori and Prance 1987; Mori 1991). Genera endemic to the province are *Polylychnis* (Acanthaceae), *Lembocarpus* (Gesneriaceae), *Maburea* (Olacaceae), *Potarophytum* and *Windsorina* (Rapateaceae), and *Neobertiera* (Rubiaceae).

Mori (1991) recognized a broad phytogeographical unit he called the Guayana Lowland Floristic Province. This coincides in broad terms with the entire

Guayana Region as defined in this volume, except that Mori included all Amazonian Brazil north of the Amazon River and east of the Río Negro (but excluded areas west of the Río Negro). Although many plant taxa in Amapá, northern Pará, and parts of northern Amazonas states in Brazil are shared with the Eastern Guayana Province, much more work is needed to be able to define where the limits of the Guayana Region lie in northern Brazil. Whereas Mori's delimitation of the area is very broad, ours is more conservative, and a better approximation should be sought once more information becomes available from these transitional areas.

Central Guayana Province. This province extends irregularly over the highly fragmented submontane and montane landscapes of the central area of the Guayana Shield. Because of its complicated topography, however, it is difficult to delimit precisely. The province mainly comprises the submesothermic and mesothermic uplands and mountain slopes found roughly between 300 and 1500 m elevation in central-western Guyana and the Venezuelan Guayana (Figure 4-2), with isolated outliers in Suriname (Tafelberg) and northern Brazil (Serra Tepequém and Serra Aracá). It stretches continuously from the Pakaraima mountains in the east to the Cuao-Sipapo massif in the west, and from the Sierra de Maigualida in the north through the Parima uplands to the Neblina massif in the south. This province does not include any mountain summits above 1500 m elevation, since these belong to the phytogeographical province of Pantepui.

Most of the mountain systems included in the Central Guayana Province consist of quartzitic or sandstone rocks of the Roraima Formation, but there are also large areas where granitic rock types predominate, such as in the Maigualida, Sipapo, Parima, and the Unturán mountains. The distribution patterns of some of the plant species in the province coincide with the different geologies. Except for northern Bolívar state, where there is a pronounced seasonal climate, the rest of the Central Guayana Province is humid to perhumid (2000–4000 mm per year or even higher), and submesothermic to mesothermic (average annual temperature between 15 and 22°C).

The two most characteristic vegetation types of this province are montane forests and shrublands. In the montane forest biome, many local forest types occur in the piedmont and along the slopes of the numerous mountain systems; these range from tall, semideciduous premontane or basimontane forests to low, perhumid and epiphyte-laden cloud forests at the base of the sandstone cliffs of the tepuis. This province also includes forest and scrub types that grow on the summits of low-elevation tepuis such as Cerro Guaiquinima, Cerro Ichún, Cerro Morrocoy, Cerro Yapacana, Tafelberg, Serra Aracá, and Serra Tepequém.

The shrublands of this province are very characteristic and are best developed between 500 and 1300 m elevation. Since they are essentially edaphic plant communities, their distribution pattern is irregular and discontinuous throughout the Guayana Region (see Huber 1989). The herbaceous vegeta-

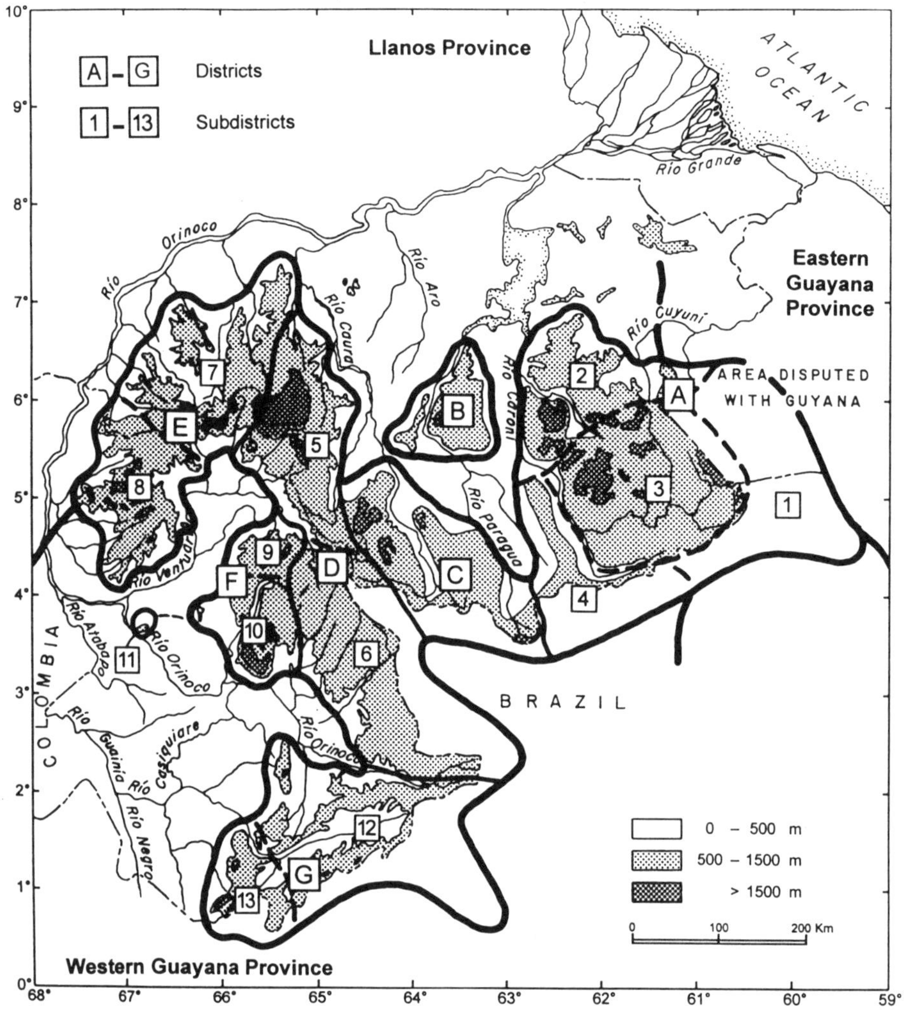

Figure 4-2. The Central Guayana Province (outlined by the heavy line) of the Guayana Region showing its districts and subdistricts. A = Pakaraima-Gran Sabana District; 1 = Eastern Pakaraima Subdistrict (including Tafelberg outlier), 2 = Supamo-Lema Subdistrict, 3 = Gran Sabana Subdistrict, 4 = Western Pakaraima Subdistrict (including Tepequém and Uafaranda outliers). B = Guaiquinima District. C = Caura-Paragua District. D = Nichare-Parima District; 5 = Nichare-Uasadi Subdistrict, 6 = Parima Subdistrict. E = Guaniamo-Guayapo District; 7 = Guaniamo-Parguaza Subdistrict, 8 = Cuao-Guayapo Subdistrict. F = Parú-Duida District; 9 = Parú Subdistrict, 10 = Duida Subdistrict, 11 = Yapacana-Cerro Tigre Subdistrict. G = Unturán District; 12 = Vinilla-Tapirapecó Subdistrict (including Serra Aracá outlier), 13 = Aracamuni-Imeri Subdistrict. Also shown are the names of the three adjacent floristic provinces: the Eastern Guayana Province and the Western Guayana Province are part of the Guayana Region, while the Llanos Province is part of the Caribbean Region.

tion types of this province are not as differentiated or extensive as in the other provinces of the Guayana Region. The main area of grasslands is located in the Gran Sabana uplands in southeastern Bolívar state and in the Parima uplands near the border of Brazil with Venezuela. Most of these grasslands are floristically closely related to similar lowland savanna types such as the Llanos of northern South America.

Floristically, the Central Guayana Province is characterized by a rich montane forest and shrubland flora that comprises probably 3000–4000 species. The dominant tree families are Chrysobalanaceae, Clusiaceae, Fabaceae, Lauraceae, Lecythidaceae, Rubiaceae, Sapotaceae, Sapindaceae, and Vochysiaceae. Some distinctive genera of the montane forests in this province are *Pakaraimaea* (Monotaceae), *Perissocarpa* (Ochnaceae), and *Tepuianthus* (Tepuianthaceae). Other phytogeographically and floristically important families and genera of the Central Guayanan shrublands are Clusiaceae (*Clusia, Neotatea*), Combretaceae (*Terminalia*), Euphroniaceae (*Euphronia*), Humiriaceae (*Humiria, Vantanea*), Melastomataceae (*Graffenrieda, Meriania*), Ochnaceae (*Ouratea, Tyleria*), Rubiaceae (*Chalepophyllum, Platycarpum*), Rutaceae (*Spathelia*), and Theaceae (*Bonnetia, Ternstroemia*). Genera that are endemic to this province include *Salpinctes* (Apocynaceae), *Jasarum* (Araceae), *Thysanostemon* (Clusiaceae), *Rhoogeton* (Gesneriaceae), *Spirotropis* (Fabaceae), *Boyania, Maguireanthus,* and *Tryssophyton* (Melastomataceae), *Pakaraimaea* (Monotaceae), *Holstianthus, Maguireocharis,* and *Yutajea* (Rubiaceae), *Amphiphyllum* (Rapateaceae), *Apocaulon* (Rutaceae), and *Aratitiyopea* (Xyridaceae).

Granville (1991) considered that the montane forest flora of the Guianas shows stronger relationships with eastern Andean forests, and secondarily with forests of the West Indies, than with forests of the Amazon basin. He attributed this pattern to the existence of Pleistocene forest refuges along the slopes of the mountains of the Guianas. However, this proposition is based mainly on the floristic analysis of the undergrowth of certain montane forests, and it is not clear if the tree flora of the Central Guayana Province follows this phytogeographical pattern.

Since knowledge of the forest flora of this province is very fragmentary, the recognition of districts and subdistricts (Table 4-7 and Figure 4-2) is still speculative. These subdivisions are based both on geographic discontinuities and on observed differences in physiognomy from aerial reconnaissances. They are intended here mainly as a framework for future field studies to critically examine and revise.

Pantepui Province. The third province of the Guayana Region includes high mountain ecosystems of the Guayana highland mainly above 1500 m elevation (Figure 4-3, see also Chapter 1). The Pantepui Province is restricted to western Guyana, southern Venezuela, and northernmost Brazil; it extends from the Ayanganna Mountains in the east to the Cuao-Sipapo massif in the west, and from Sierra de Maigualida in the north to Serranía de la Neblina in the south. The province has a discontinuous distribution of often isolated summits that rise out of the underlying Central Guayana Province.

The Pantepui Province is characterized mostly by montane shrublands, meadows, and open rock communities. Forests play a subordinate role in this province, and they are always low, relatively species-poor, and restricted to large depressions or streamsides. There are a great variety of shrublands on almost all tepui summits, ranging from dense high-tepui scrub to paramoid scrub or other types of caulirosulate shrublands, as well as open, sclerophyllous scrub and wiry, ericoid scrub (Huber 1988a, 1992b, 1992c). The high-

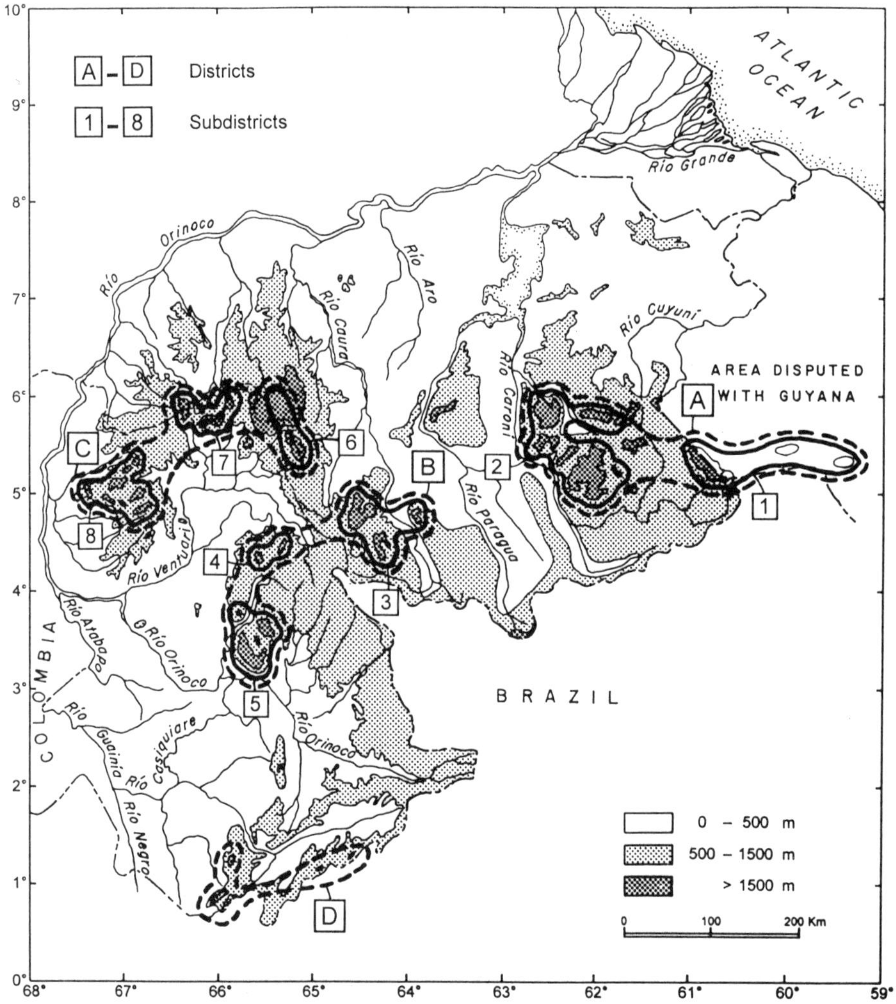

Figure 4-3. Main areas of the Pantepui Province in the Venezuelan Guayana and adjacent Guyana. A = Eastern Pantepui District; 1 = Roraima Subdistrict, 2 = Chimantá Subdistrict. B = Jaua-Duida District; 3 = Jaua-Sarisariñama Subdistrict, 4 = Asisa Subdistrict, 5 = Duida-Marahuaka Subdistrict. C = Western Pantepui District; 6 = Maigualida Subdistrict, 7 = Yutajé Subdistrict, 8 = Cuao-Sipapo Subdistrict. D = Southern Pantepui District.

tepui meadows are well differentiated into broad-leaved, tubiform, and rosette meadows, with a few areas that resemble high-Andean grasslands.

The floristic richness and high degree of endemism of the Pantepui Province are striking (Maguire 1956, 1970; Steyermark 1979a, 1986b). Besides the endemic Pantepui genera listed in Table 4-10 (see end of chapter), the notable absence of legumes and Solanaceae, both widespread and speciose in the Andes, also highlights the unique floristic composition of this province.

Because the Pantepui Province occurs in rather discrete mountain blocks, it can be clearly divided into a series of four phytogeographical districts (each with its own subdistricts), based on floristic and geographical criteria: (1) the Eastern Pantepui District, (2) the Jaua-Duida District, (3) the Western Pantepui District, and (4) the Southern Pantepui District (Figure 4-3).

The Eastern Pantepui District includes the high tepuis of western Guyana and of southeastern Bolívar state in Venezuela, between 1600 and 2750 m elevation. This is the classical section of Pantepui, including Roraima-tepui, Auyán-tepui, and the Chimantá massif. It is readily delimited by the distribution of *Bonnetia roraimae* (Theaceae), a characteristic, often dominant endemic treelet. Relatively speciose genera that are endemic to this district are *Quelchia* (Asteraceae), *Connellia* (Bromeliaceae), and *Tepuia* (Ericaceae). There are two subdistricts, with the Roraima Subdistrict including the high tepuis of Mount Ayanganna and Mount Wokomong in Guyana, and the eastern tepui chain in the southeastern Gran Sabana of Venezuela. The predominant vegetation types on these tepui summits consist of small islands of meadows and scrub interspersed with bare quartzitic rocks. The rest of the Eastern Pantepui District belongs to the Chimantá Subdistrict, located along the western border of the Gran Sabana in southeastern Venezuela and including the huge Auyán-tepui and Chimantá massifs. The Aparamán or Los Testigos range is transitional between this and the Roraima Subdistrict. The Eastern Pantepui District is one of the most ecologically and floristically diverse sections of Pantepui Province. Genera endemic to the Chimantá Subdistrict are *Achnopogon* and *Chimantaea* (Asteraceae), *Mallophyton* (Melastomataceae), and *Coryphothamnus* (Rubiaceae).

The Jaua-Duida District comprises the summits of a series of large sandstone massifs in southwestern Bolívar and east-central Amazonas states, mainly between 1500 and 2800 m elevation. Despite distances of up to several hundred kilometers between massifs such as Cerro Duida and Cerro Jaua, there are clear affinities between them, particularly in their vegetation types. Each massif has its own set of endemic species, yet they also share ecologically important genera such as *Tyleria* (Ochnaceae), *Neotatea* (Clusiaceae), and *Tepuianthus* (Tepuianthaceae), as well as dominant tepui meadow species like *Stegolepis grandis* (Rapateaceae). Three subdistricts can be recognized within the Jaua-Duida District.

The Jaua-Sarisariñama Subdistrict comprises the summits of Cerro Guanacoco, Cerro Sarisariñama, and Cerro Jaua, all in the upper Río Caura drainage of southeastern Bolívar state. The highest parts of the Serranía Marutaní in the upper Paragua drainage may also belong here, but this area is still too poorly explored to allow a precise classification. The characteristic vegetation types are dense to open shrublands on rocky substrates, mixed with tubiform meadows, especially on Cerro Jaua. *Tyleria* (Ochnaceae) is prominent in this subdistrict, and there are a number of endemic *Stegolepis* species (Rapateaceae) growing mostly in peat, such as *S. breweri, S. jauaensis,* and *S. microcephala.*

The Asisa Subdistrict includes the summits of Cerro Asisa (the highest part of Cerro Parú) and nearby Cerro Euaja in the upper Ventuari drainage of east-central Amazonas state. *Phelpsiella* (Rapateaceae) is endemic to Cerro Asisa, and common genera there are *Coccochondra* (Rubiaceae), *Tepuianthus* (Tepuianthaceae), *Celianella,* and *Mabea* (both Euphorbiaceae).

The Duida-Marahuaka Subdistrict consists of Cerro Duida, Cerro Marahuaka, and the nearby Cerro Huachamacari. Although Cerro Duida and Cerro Marahuaka belong to the same mountain system, the latter is much higher (2800 m versus 2350 m for Duida) and has its own set of endemic taxa and vegetation types. Endemic Pantepui genera in this subdistrict include *Salpinctes* (Apocynaceae), *Duidaea* (Plate 59) and *Tyleropappus* (Asteraceae), *Duidania* (Rubiaceae), and *Marahuacaea* (Rapateaceae).

The Western Pantepui District is the most extensive and comprises the series of sandstone and granitic mountains that form the northeastern and northern border of Amazonas state, as well as the westernmost high mountains of the Guayana Shield in the Parguaza and Sipapo river drainages. It is a very irregular unit with elevations ranging from 1300 m to 2350 m. Phytogeographically, the most characteristic genus of this district is the red-flowered *Kunhardtia rhodantha* (Rapateaceae). Shrubby Melastomataceae (especially of the genera *Graffenrieda* and *Meriania*) are also characteristic of this district, as are an unusual number of endemic species of *Phyllanthus* (Euphorbiaceae).

Three subdistricts are distinguishable in the Western Pantepui District, presented below from east to west. The Maigualida Subdistrict, with the summits of the mostly forested, granitic Sierra Maigualida, includes an assemblage of high-tepui ecosystems that occur between 1800 and 2350 m elevation. The easternmost limit of *Kunhardtia* occurs in this subdistrict, together with endemic taxa such as *Huberopappus* (Asteraceae). *Bonnetia* is noticeably absent from this granitic mountain range.

The Yutajé Subdistrict includes Cerro Yaví, Serranía Yutajé–Coro Coro, and Cerro Guanay, located along the northern border of Amazonas state and ranging in altitude between 1800 and 2300 m. These mountains are characterized by a high diversity of high-tepui shrublands and by a relatively high level of endemism. *Pleurostima celiae* (Velloziaceae; Plate 57), a rare species known elsewhere only on Serra Tepequém in northern Roraima state of Brazil, occurs on Cerro Yaví. This mountain and perhaps also Cerro Yutajé are dry tepuis with little or no peat formation on their summits.

The Cuao-Sipapo Subdistrict includes the relatively small sandstone tepui of Cerro Cuao, reaching nearly 2000 m elevation, and a complex of granitic mountain tops of the Sipapo massif that reach 1400–1800 m elevation. Characteristic vegetation types are high-tepui scrub, meadows, and open rock communities. Well diversified genera in this subdistrict include *Graffenrieda* (Melastomataceae), *Phyllanthus* (Euphorbiaceae), *Schefflera* (Araliaceae), and *Diacidia* (Malpighiaceae).

Finally, the Southern Pantepui District comprises the upper sections of the Neblina massif along the border of Brazil with Venezuela. It may also include the botanically unexplored granitic summit of Cerro Tamacuari in the Tapirapecó massif. The Pantepui section of Sierra de la Neblina ranges between 1600 and 3014 m, whereas Cerro Tamacuari reaches about 2350 m elevation. The wide, open summit plains of the quartzitic Cerro Aracamuni and Cerro Avispa to the north of Neblina, as well as some yet unexplored peaks of the mainly granitic Unturán mountain system to the northeast, do not appear to exceed 1600 m elevation; they are probably intermediate between the Central Guayana Province and Pantepui Province, as shown by the floristic list provided by Steyermark and Holst (1989) for Cerro Aracamuni. Future botanical and ecological exploration of this mountain complex may justify the recognition of a western, mainly quartzitic Aracamuni-Neblina Subdistrict between 1500 and 3014 m, and an eastern, mainly igneous Unturán-Tamacuari Subdistrict between 1400 and 2400 m elevation.

The characteristic vegetation types of the Southern Pantepui District on the Neblina massif are extensive shrublands, including *Bonnetia maguireorum* scrub and broad-leaved or tubiform high-tepui meadows. Low forests of other *Bonnetia* species are also relatively frequent in depressions and small valleys of the upper summit plains. The vegetation of the conical peaks of Cerro Tamacuari and Cerro Aratitiyope consists mainly of high-montane, rock-inhabiting plant communities. Floristically, this is the section of Pantepui and of the entire Guayana Region with the highest number of endemic taxa, with one endemic family (Saccifoliaceae) and the endemic genera *Glossarion, Imeria,* and *Neblinaea* (Asteraceae), *Neblinantha* (Gentianaceae), *Pyrrorhiza* (Haemodoraceae), *Comoliopsis* and *Neblinanthera* (Melastomataceae), *Adenarake* (Ochnaceae), *Cephalodendron* and *Neblinathamnus* (Rubiaceae), *Rutaneblina* (Rutaceae), and *Achlyphila* (Xyridaceae).

Western Guayana Province. The flora of the Río Negro basin has long been recognized as being distinct from that of the surrounding lowland Amazon forests. Carl F. von Martius, one of the first botanical explorers of this region in the early 19th century, found many unusual plant species that are adapted to the peculiar white-sand soils common in the area. Since 1940, the distinctiveness of the Río Negro flora has been confirmed by numerous scientific explorations in the area, especially the botanical collections and ecological research carried out around San Carlos de Río Negro in Venezuela and around Araracuara in Colombia (Jordan 1989; Duivenvoorden and Cleef 1994).

The Río Negro basin contains an intricate mosaic of peculiar ecosystems, many of them unique to the northwestern Amazon and growing on extremely nutrient-poor, white, sandy quartzite soils (Fittkau 1983; Goulding et al. 1988). Large tracts of land in this area are flooded during much of the year, but there are also wide, rolling hill-lands and low sandstone table mountains, especially along the upper northwestern tributaries of the Río Negro such as the Río Vaupés and the Río Guainía.

In the past, the entire Río Negro basin was treated as a distinctive part of the Amazon Region phytogeographical unit, also called Hylaea (see Martius 1840–1869; Ducke and Black 1953; Prance 1977). However, both ecological and floristic data from the Río Negro basin now support a closer relationship with the Guayana Region than with the classical Amazon Region, and there seems to be ample justification to designate a separate province of the Guayana Region for the Río Negro basin. First, the area is part of the Guayana Shield, and most of the soils are either derived from the *in situ* decomposition of the underlying granitic parent rocks, or are deposited as fine, bleached sands originating from the intensive weathering processes of the surrounding quartzitic or sandstone mountains (thus producing very nutrient-poor soils). Endemism in the Río Negro flora is high, with a strong representation of typical Guayanan families such as Humiriaceae, Rapateaceae, Tepuianthaceae, Theaceae, and Xyridaceae (see also Chapter 1). Characteristic Guayanan genera such as *Brocchinia* (Plate 65) and *Navia* (Bromeliaceae; Plate 61) also occur in this area.

The Western Guayana Province includes the upper Orinoco lowlands with the Ventuari and Casiquiare peneplains in Amazonas state of southern Venezuela, as well as much of the Río Sipapo basin in western Amazonas; the peneplains, hill-lands and lower mesetas (Cerro Araracuara, Cerro Chiribiquete, Cerro Isibukuri, and Cerro Yapobodá) of southeastern Colombia, between the Inírida and the Caquetá rivers; and the northwestern Amazonas state in Brazil, from the confluence of the Apoporis and Caquetá rivers east to the lower Río Branco and north to the upper Río Demeni. Most of this area consists of macrothermic and ombrophilous lowlands (100–300 m elevation), including extensive floodplains; only the western section in Colombia reaches elevations of almost 1000 m with a submesothermic climate. The main vegetation types of the Western Guayana Province are tall to medium-sized sclerophyllous lowland forests (Rio Negro caatinga) alternating with flooded and nonflooded forests. Shrublands (*campinas* and *banas*) are also widespread, but often occur in isolated patches on dry or intermittently flooded terrain. Herbaceous, savanna-like meadows with dwarf shrubs are particularly frequent in southern Venezuela and adjacent southeastern Colombia. Dense submontane forests predominate on the sandstone mountains of Colombia, and low, often open, sclerophyllous shrublands are frequent near the summits.

Genera endemic to this province are *Duckeanthus, Heteropetalum,* and *Pseudephedranthus* (Annonaceae), *Urospathella* (Araceae), *Aguiaria* (Bombacaceae), *Angostyles, Astrococcus,* and *Chonocentrum* (Euphorbiaceae), *Aste-*

ranthos (Lecythidaceae), *Steyermarkochloa* (Poaceae), *Dendrosipanea* and *Sipaneopsis* (Rubiaceae), *Guacamaya* and *Schoenocephalium* (Rapateaceae; Plate 77), and *Pentamerista* (Tetrameristaceae; Plate 69).

Although the Western Guayana Province is still a relatively poorly explored region, it can be divided into three floristically and ecologically distinct districts. The Upper Río Negro District includes the mainly Brazilian lowlands and floodplains of the upper Río Negro basin, from the mouth of the Rio Branco northward into southern Venezuela and westward to the middle Vaupés and Guainía rivers in Colombia. The most important vegetation types of this district are the sclerophyllous lowland forests (Rio Negro caatinga), as well as a series of peculiar shrublands. Characteristic plant taxa of this district are *Eperua* (Caesalpiniaceae), *Lissocarpa* (Lissocarpaceae), *Compsoneura* (Myristicaceae), and *Astrococcus* and *Micrandra* (Euphorbiaceae) in the caatinga forests, with *Bonnetia* (Theaceae), *Humiria* (Humiriaceae), and *Emmotum* (Icacinaceae) in the shrublands.

The Atabapo-Ventuari District occurs mainly in southern Venezuela, from the middle Río Ventuari southwest to the Río Atabapo and then into eastern Colombia via the Río Guainía. Much of the Río Sipapo basin, with its tributaries, the Río Autana and the Río Guayapo, form a northern extension of the district. Distinctive features of the district are the numerous patches of broad-leaved meadows with dwarf shrubs, as well as isolated shrublands, both growing on extremely bleached white-sand soils and subject to shallow and intermittent flooding. Characteristic taxa of this phytogeographical district are several genera of Rapateaceae (*Schoenocephalium, Guacamaya, Cephalostemon, Monotrema,* and *Saxofridericia*), Araceae (*Urospathella* and *Philodendron*), Arecaceae (*Leopoldinia*), Poaceae (*Steyermarkochloa*), Sapotaceae (*Ecclinusa, Pradosia*), Ochnaceae (*Ouratea, Wallacea, Blastemanthus*), Ixonanthaceae (*Ochthocosmus*), Tetrameristaceae (*Pentamerista*), Theaceae (*Archytaea, Bonnetia*), and Malpighiaceae (*Heteropterys, Tetrapterys*).

The Araracuara District includes the series of low sandstone table mountains in southeastern Colombia, which are also sometimes called collectively Colombian tepuis (Sánchez et al. 1990). The most important mountains are Cerro Isibukuri, Cerro Yapobodá, Cerro Araracuara, and the massif of Chiribiquete that reaches approximately 1000 m elevation. The predominant vegetation types of this district are low sclerophyllous forests covering the lower and middle slopes of the sandstone mountains and sclerophyllous shrublands growing on rocky substrate in the lowlands and on the mountain tops. The key families are similar to those found in the Atabapo-Ventuari District, but with many local endemic species (Estrada and Fuertes 1993).

The Guayanan Periphery

Along the southern and southwestern border of the Guayana Shield, there are many transitional areas with the Amazon Region, making it very difficult to trace a clear line of separation between the two regions. For instance, in the

Manaus Center of Prance (1977), several Guayanan elements are still evident in low forest and scrub types on white sands (Anderson 1981). Another example includes the Rio Branco savannas in Brazil's Roraima state, which have some Guayanan elements but are floristically most similar to the Llanos-type savannas of El Manteco (belonging to the Caribbean Region) in northeastern Bolívar state. There is still a tremendous difference between the southern and eastern limits of the Guayana Region as defined here, and Mori's much broader circumscription of his Lowland Guayana Province, which extends south to the Amazon River and west to the Río Negro. It is hoped that botanists will focus their attention on this area and bring about a better understanding of the interface between two such important phytogeographic regions.

Another transitional or intermediate area occurs in an arc along the northern and northwestern border of the Guayana Shield in Venezuela along the middle Río Orinoco, from the Raudales de Atures in Amazonas state to around Caicara in Bolívar state. In this area, grass savannas alternate with forest islands and gallery forests. While the herbaceous flora of these savannas belongs mainly to the adjacent Llanos Province of the Caribbean Region, some characteristic woody elements are either of Guayanan affinity, such as the common savanna tree *Platycarpum orinocense* (Rubiaceae), or else of Amazonian affinity, such as *Chaunochiton angustifolium* (Olacaceae). Special features of this district are the granitic *lajas* or inselbergs that dot the landscape along the Orinoco, with their own set of over 50 endemic species. Although this whole area is a phytogeographical and ecological transition zone with several indigenous vegetation types, the predominance of Caribbean elements with many deciduous and xerophilous species best places it as a separate Middle Orinoco District in the Llanos Province of the Caribbean Region.

Table 4-8. Endemic genera of the Guayana Shield (138 genera).

Taxon	Number of species	Elevation range (m)	Countries[1]					
Acanthaceae								
Gynocraterium	1	200	•	•	G	•	•	•
Polylychnis	2	100–400	•	•	G	S	F	•
Annonaceae								
Duckeanthus	1	100	•	•	•	•	•	B
Heteropetalum	1	100–200	C	V	•	•	•	B
Pseudephedranthus	1	100–200	•	V	•	•	•	B
Apocynaceae								
Salpinctes	1	1000–1500	•	V	•	•	•	•
Araceae								
Jasarum	1	500–1200	•	V	G	•	•	•
Urospathella	1	100–200	C	V	•	•	•	•
Asteraceae								
Achnopogon	2	1800–2500	•	V	•	•	•	•
Chimantaea	9	1600–2600	•	V	•	•	•	•
Duidaea	4	1000–2800	•	V	•	•	•	•
Glossarion	2	1600–2800	•	V	•	•	•	B
Guayania	6	100–2800	C	V	G	•	•	B
Huberopappus	1	2000–2100	•	V	•	•	•	•
Imeria	1	600–2100	•	V	•	•	•	B
Neblinaea	1	1200–2100	•	V	•	•	•	B
Quelchia	4	1600–2800	•	V	•	•	•	•
Siapaea	1	200–300	•	V	•	•	•	•
Stomatochaeta	6	700–2800	•	V	G	•	•	•
Tuberculocarpus	1	50–200	•	V	•	•	•	•
Tyleropappus	1	2200	•	V	•	•	•	•
Bignoniaceae								
Digomphia	3	100–2500	C	V	G	•	•	B
Potamoganos	1	100–400	•	V	G	S	•	B
Bombacaceae								
Aguiaria	1	100–200	•	•	•	•	•	B
Bromeliaceae[2]								
Brewcaria	2	1500–2600	•	V	•	•	•	•
Brocchinia	19	100–2800	C	V	G	•	•	B
Connellia	5	1300–2800	•	V	G	•	•	•
Lindmania	47	1000–2800	•	V	G	•	•	B
Navia	95	100–2500	C	V	G	S	•	B
Steyerbromelia	3	600–2700	•	V	•	•	•	•
Caesalpiniaceae								
Heterostemon	7	100–300	C	V	G	S	F	B
Clusiaceae								
Neotatea	4	200–2000	C	V	•	•	•	•
Thysanostemon	2	900–1000	•	•	G	•	•	•
Convolvulaceae								
Lysiostyles	1	200–1300	•	V	G	S	•	B
Cyperaceae[3]								
Cephalocarpus	4	600–2700	C	V	G	S	•	B
Didymiandrium	1	400–2400	•	V	G	•	•	B
Exochogyne	1	100–1100	•	V	G	S	F	B
Koyamea	1	500–2000	•	V	•	•	•	B
Mapaniopsis	2	100–1600	•	V	G	•	F	B
Rhynchocladium	1	500–2400	•	V	G	•	•	•

(continued)

Table 4-8. Continued.

Taxon	Number of species	Elevation range (m)	Countries[1]					
Ericaceae								
Ledothamnus	7	1000–2800	•	V	G	•	•	•
Mycerinus	3	1300–2700	•	V	•	•	•	•
Notopora	5	400–2500	•	V	G	•	•	•
Tepuia	7	1500–2600	•	V	•	•	•	•
Eriocaulaceae								
Rondonanthus	6	100–2800	•	V	G	•	•	B
Euphorbiaceae								
Angostyles	1	100	•	•	•	•	•	B
Astrococcus	1	100–200	•	V	•	•	•	B
Celianella	1	1000–1500	•	V	•	•	•	•
Chonocentrum	1	100	•	•	•	•	•	B
Haematostemon	2	100–200	•	V	G	•	•	•
Senefelderopsis	2	100–1600	C	V	•	•	•	•
Euphroniaceae								
Euphronia	3	100–1600	C	V	G	•	•	B
Fabaceae								
Aldina	12	100–1300	C	V	G	•	•	B
Panurea	1	100	C	•	•	•	•	B
Petaladenium	1	100	•	•	•	•	•	B
Spirotropis	2	400–800	•	V	G	S	F	•
Flacourtiaceae								
Euceraea	2	100–2100	C	V	G	S	•	B
Gentianaceae								
Celiantha	3	1500–2800	•	V	G	•	•	B
Chorisepalum	5	800–2400	•	V	G	•	•	•
Neblinantha	2	2200–2600	•	V	•	•	•	•
Sipapoantha	1	1200–2000	•	V	•	•	•	•
Gesneriaceae								
Lembocarpus	1	300–400	•	•	•	S	F	•
Rhoogeton	2	1000–1100	•	V	G	•	•	•
Tylopsacas	1	100–1700	•	V	G	•	•	•
Haemodoraceae								
Pyrrorhiza	1	1800–2100	•	V	•	•	•	•
Hymenophyllopsidaceae								
Hymenophyllopsis	8	700–2800	•	V	G	•	•	B
Ixonanthaceae								
Cyrillopsis	2	100–1100	•	V	•	•	F	B
Lecythidaceae								
Asteranthos	1	100–200	C	V	•	•	•	B
Liliaceae								
Nietneria	2	1000–2800	•	V	G	•	•	B
Malpighiaceae								
Blepharandra	6	100–2200	•	V	G	•	•	B
Diacidia	13	100–2300	C	V	•	•	•	B
Malvaceae								
Uladendron	1	100–200	•	V	•	•	•	•
Melastomataceae								
Acanthella	2	100–1900	C	V	•	•	•	B
Boyania	1	800–1000	•	•	G	•	•	•
Comoliopsis	1	2300	•	V	•	•	•	B
Maguireanthus	1	1000–1500	•	•	G	•	•	•
Mallophyton	1	2000–2500	•	V	•	•	•	•

Taxon	Number of species	Elevation range (m)	Countries[1]					
Neblinanthera	1	1300–1800	•	V	•	•	•	B
Ochthephilus	1	1400–1600	•	•	G	•	•	•
Pachyloma	5–6	50–1200	C	V	G?	•	•	B
Tateanthus	1	800–2000	•	V	•	•	•	B
Tryssophyton	1	1100–1200	•	•	G	•	•	•
Monotaceae								
Pakaraimaea	1	500–1100	•	V	G	•	•	•
Ochnaceae								
Adenanthe	1	1350–2300	•	V	G	•	•	•
Adenarake	1	1500–2800	•	V	•	•	•	B
Blastemanthus	2	100–200	C	V	G	•	•	B
Philacra	4	800–2000	•	V	•	•	•	B
Poecilandra	2	500–2800	C	V	G	•	•	B
Tyleria	12	1200–2500	•	V	•	•	•	B
Wallacea	2	100–200	C	V	•	•	•	B
Olacaceae								
Maburea	1	0–100	•	•	G	•	•	•
Orchidaceae								
Aracamunia	1	1500–1600	•	V	•	•	•	•
Cheiradenia	1	300–900	•	V	G	S	F	B
Degranvillea	1	400–700	•	•	•	•	F	•
Polyotidium	1	100–200	C	V	•	•	•	B
Peridiscaceae								
Whittonia	1	50–100	•	•	G	•	•	•
Poaceae								
Myriocladus	13	700–2700	•	V	•	•	•	B
Steyermarkochloa	1	100–200	C	V	•	•	•	B
Podostemaceae								
Jenmaniella	7	300–1800	•	V	G	•	•	B
Rhyncholacis	25	100–1200?	C	V	G	S	•	B
Weddellina	1	100–400	C	V	G	S	•	B
Rapateaceae								
Amphiphyllum	1	1200–2000	•	V	•	•	•	•
Guacamaya	1	100–200	C	V	•	•	•	•
Kunhardtia	2	100–2100	•	V	•	•	•	•
Marahuacaea	1	2400–2800	•	V	•	•	•	•
Phelpsiella	1	1300–2000	•	V	•	•	•	•
Potarophytum	1	100–400	•	•	G	•	•	•
Schoenocephalium	4	100–800	C	V	•	•	•	B
Stegolepis	35	500–2800	•	V	G	•	•	B
Windsorina	1	50–300	•	•	G	•	•	•
Rubiaceae								
Aphanocarpus	1	1000–2500	•	V	•	•	•	•
Cephalodendron	2	1100–1600	•	V	•	•	•	•
Chalepophyllum	2	400–1600	•	V	G	•	•	•
Coccochondra	1	2000	•	V	•	•	•	•
Coryphothamnus	1	1700–2300	•	V	•	•	•	•
Dendrosipanea	3	100–200	•	V	•	•	•	B
Duidania	1	1200–2600	•	V	•	•	•	•
Henriquezia	3	50–300	C	V	G	•	•	B
Holstianthus	1	1200–1400	•	V	•	•	•	•
Maguireocharis	1	1300–1400	•	V	•	•	•	B

(continued)

Table 4-8. Continued.

Taxon	Number of species	Elevation range (m)	Countries[1]					
Maguireothamnus	3	1200–2800	•	V	G	•	•	•
Merumea	2	1100–1600	•	V	G	•	•	•
Neblinathamnus	4	1200–1900	•	V	•	•	•	B
Neobertiera	1	100–300?	•	•	G	•	F	•
Pagameopsis	2	1100–2500	•	V	G?	•	•	B
Sipaneopsis	7	50–800	C	V	•	•	•	B
Yutajea	1	300–1000	•	V	•	•	•	•
Rutaceae								
Apocaulon	1	600–1300	•	V	•	•	•	•
Decagonocarpus	2	100–1800	C	V	•	•	•	B
Rutaneblina	1	1600–1900	•	V	•	•	•	•
Saccifoliaceae								
Saccifolium	1	2700–2800	•	V	•	•	•	B
Sarraceniaceae								
Heliamphora	5	600–2800	•	V	G	•	•	B
Tepuianthaceae								
Tepuianthus	6	100–2200	C	V	•	•	•	B
Tetrameristaceae								
Pentamerista	1	50–200	C	V	•	•	•	•
Theaceae[4]								
Archytaea	2	50–2000	C	V	G	•	•	B
Xyridaceae								
Achlyphila	1	1700–2300	•	V	•	•	•	•
Aratitiyopea	1	600–1600	C	V	•	•	•	•
Orectanthe	2	500–2800	•	V	G	•	•	B
Total endemic genera per country			C 34	V 118	G 61	S 13	F 10	B 68

[1]C = Colombia, V = Venezuela, G = Guyana, S = Suriname, F = French Guiana, B = Brazil.
[2]The endemic genus *Ayensua* is not included here because it will be synonymized under *Brocchinia* in the flora.
[3]*Mapaniopsis* is likely to be synonymized under *Mapania* in the flora.
[4]The endemic genus *Acopanea* is not included here because it will be synonymized under *Bonnetia* in the flora.

Table 4-9. Genera largely restricted to the Guayana Shield except for one or more outliers.

Taxon	Number of species endemic to Guayana Shield/total species in genus	Distribution outside the Guayana Shield
Apocynaceae		
Galactophora	9/10	One species extends to southern lowland Peru.
Arecaceae		
Leopoldinia	2/3	One species extends into central Brazil.
Asteraceae		
Gongylolepis	14/15	One species occurs in the Andes of Colombia and Venezuela.
Stenopadus	14/15	One species (undescribed) known from southern Ecuador.
Caesalpiniaceae		
Dicymbe	17/18	One species occurs in the middle Amazon basin.
Clusiaceae		
Moronobea	6/7	One species extends into the middle Amazon basin.
Cyperaceae		
Everardia	12/13	One species is disjunct from the Guayana Shield to northern Peru.
Gentianaceae		
Irlbachia	9/10	One species occurs in the middle Amazon basin.
Rogersonanthus	2/3	Two species are high-tepui endemics, the third is endemic to Trinidad.
Ixonanthaceae		
Ochthocosmus	6/7	One species occurs in the central Amazon basin and in Bolivia.
Loganiaceae		
Bonyunia	3/4	One species occurs in central Brazil.
Melastomataceae		
Macairea	16/22	Three species extend from the Guayana Shield to Peru and two to central Amazonian Brazil. Another species is widespread in the Orinoco and Amazon basins, and one species is restricted to the Brazilian Shield.
Macrocentrum	20/22	One species is known only from northern Venezuela, and another species (undescribed) occurs in Colombia and Ecuador.
Phainantha	4/5	One species occurs in Ecuador.
Pteridaceae		
Pterozonium	12/14	One species is disjunct from the Guayana Shield to Amazonian Peru, and a second one is disjunct to Costa Rica, southern Ecuador, and northern Peru.
Rapateaceae		
Monotrema	4/5	One species extends to the southern Amazon basin.
Saxofridericia	8/9	One species occurs in central Brazil.
Spathanthus	0/2	Both species are centered in the Guayana Shield but extend into central Brazil.
Rubiaceae		
Gleasonia	4/5	One species occurs in the southern Amazon basin.
Platycarpum	9/12	One species is disjunct from the Venezuelan Guayana to Amazonian Peru, and two species occur in western and central Amazonian Brazil.

(continued)

Table 4-9. Continued.

Taxon	Number of species endemic to Guayana Shield/total species in genus	Distribution outside the Guayana Shield
Retiniphyllum	15/21	Several species occur in eastern Brazil or the southern Amazon basin, and one extends to Peru.
Rutaceae		
Raveniopsis	18/20	Two species occur in central Brazil.
Theaceae		
Bonnetia	25/28	One species is endemic to Cuba and Puerto Rico, another is endemic to coastal Brazil, and a third is disjunct from the Guayana Shield to the Andes of Venezuela, Ecuador, and Peru.
Thurniaceae		
Thurnia	2/3	One species extends into central Brazil.
Xyridaceae		
Abolboda	15/20	Several species occur south into central Brazil and Paraguay.

Table 4-10. Genera endemic to Pantepui (23).

Taxon	Distribution
Asteraceae	
Achnopogon	Auyán-tepui, Chimantá-tepui
Chimantaea	Aprada-tepui, Auyán-tepui, Cerro El Sol, Macizo del Chimantá, Murisipán-tepui
Glossarion	Sierra de la Neblina
Huberopappus	Sierra de Maigualida
Quelchia	Auyán-tepui, Macizo del Chimantá
Tyleropappus	Cerro Duida
Bromeliaceae	
Brewcaria	Cerro Duida, Cerro Marahuaka
Ericaceae	
Tepuia	Auyán-tepui, Kukenán-tepui, Macizo del Chimantá, Sierra de Lema
Gentianaceae	
Celiantha	Auyán-tepui, Cerro Aracamuni, Ilú-tepui, Kukenán-tepui, Macizo del Chimantá, Roraima-tepui, Serra Pirapucú, Sierra de la Neblina
Neblinantha	Sierra de la Neblina
Haemodoraceae	
Pyrrorhiza	Sierra de la Neblina
Melastomataceae	
Comoliopsis	Sierra de la Neblina
Mallophyton	Macizo del Chimantá
Ochthephilus	Mount Ayanganna
Ochnaceae	
Adenanthe	Macizo del Chimantá, Guyana (Mount Ayanganna)
Adenarake	Sierra de la Neblina
Orchidaceae	
Aracamunia	Cerro Aracamuni
Rapateaceae	
Marahuacaea	Cerro Marahuaka
Rubiaceae	
Coccochondra	Serranía Maigualida, Serranía del Parú, Cerro Yutajé
Coryphothamnus	Auyán-tepui
Rutaceae	
Rutaneblina	Sierra de la Neblina
Saccifoliaceae	
Saccifolium	Sierra de la Neblina
Xyridaceae	
Achlyphila	Sierra de la Neblina

CHAPTER 5

Conservation of the Venezuelan Guayana

Otto Huber

In contrast to most parts of the tropics, much of the vegetation of the Venezuelan Guayana is still relatively undisturbed by human activities, especially in montane areas. Nonetheless, human pressure on the vegetation has increased considerably as the resident population has surged to more than one million inhabitants. Since 1915, several major cities were founded, including Puerto Ayacucho, Ciudad Guayana, and Ciudad Piar. The creation of a special regional development corporation in the early 1960s, the Corporación Venezolana de Guayana (CVG), led to the construction of the largest industrial center in the country at Matanzas, near the confluence of the Caroní and Orinoco rivers in northeastern Bolívar state. Major roads have been built, such as the highway from Ciudad Bolívar to Puerto Ayacucho via Caicara, and the road that cuts through the Gran Sabana from north to south, reaching the Brazilian border near Santa Elena de Uairén. One of the largest hydroelectric power plants in the world, the Represa Raúl Leoni at Guri, was constructed on the lower Río Caroní. Mining, logging, and large-scale agricultural developments have also spread steadily, reaching deep into the Venezuelan Guayana. Finally, small-scale exploitation of natural resources appears to be increasing in many parts of the flora area.

During the economic boom of the late 1960s and 1970s, conservation was not a major concern of governmental institutions. Rather, the emphasis was on developing the abundant natural resources of the region, such as iron, bauxite, gold, diamonds, hydroelectric power, and timber. Starting around 1960, this rapid development surge motivated some conservation-minded citizens to lobby for the establishment of protected areas in southern Venezuela. In the 1980s, concern about the conservation of the Venezuelan Guayana increased dramatically both domestically and internationally, and since the early 1990s, conservation has become a central theme for the future of the Venezuelan Guayana in the nation's economic policies.

The importance given to conservation in the Venezuelan Guayana in the last half of the 20th century is shown by the large number of protected areas in the region (Table 5-1). Since 1962, seven national parks, 29 natural monuments, and two biosphere reserves have been declared. By late 1992, the total surface covered by these protected areas (not including overlapping protection categories) was approximately 142,280 km^2, or almost 31 percent of the Venezuelan Guayana and about 15 percent of the country. These protected areas include huge expanses such as the Alto Orinoco–Casiquiare Biosphere Reserve, which has an area of approximately 83,830 km^2 and is the largest reserve in the tropics.

Protected Areas

There are three main categories of protected conservation areas in the Venezuelan Guayana: national parks, natural monuments, and biosphere reserves. The areas covered by each category are illustrated in Figure 5-1 and treated below in greater detail.

Table 5-1. Officially designated conservation areas in the Venezuelan Guayana.

Conservation area	Year decreed	State	Area (km^2)
National Parks			
Canaima	1962	Bolívar	30,000
Delta del Orinoco	1991	Delta Amacuro	3,310
Duida-Marahuaca	1978	Amazonas	3,737
Jaua-Sarisariñama	1978	Bolívar	3,300
Parima-Tapirapecó	1991	Amazonas	38,290
Serranía La Neblina	1978	Amazonas	13,600
Yapacana	1978	Amazonas	3,200
			95,437
Natural Monuments			
Cerro Autana	1978	Amazonas	0.30
Piedra Cocuy	1978	Amazonas	0.15
Piedra La Tortuga	1992	Amazonas	± 1.00
Piedra Pintada	1992	Amazonas	± 0.50
Tepuyes (25)	1990	Bolívar and Amazonas	10,698
			10,700
Biosphere Reserves			
Alto Orinoco-Casiquiare	1991	Amazonas	83,830
Delta del Orinoco	1991	Delta Amacuro	11,250
			95,080

Source: INPARQUES and personal communications by R. García P. and E. Yerena in December 1993 for national parks and natural monuments; MARNR and O. Huber, unpublished, for biosphere reserves.

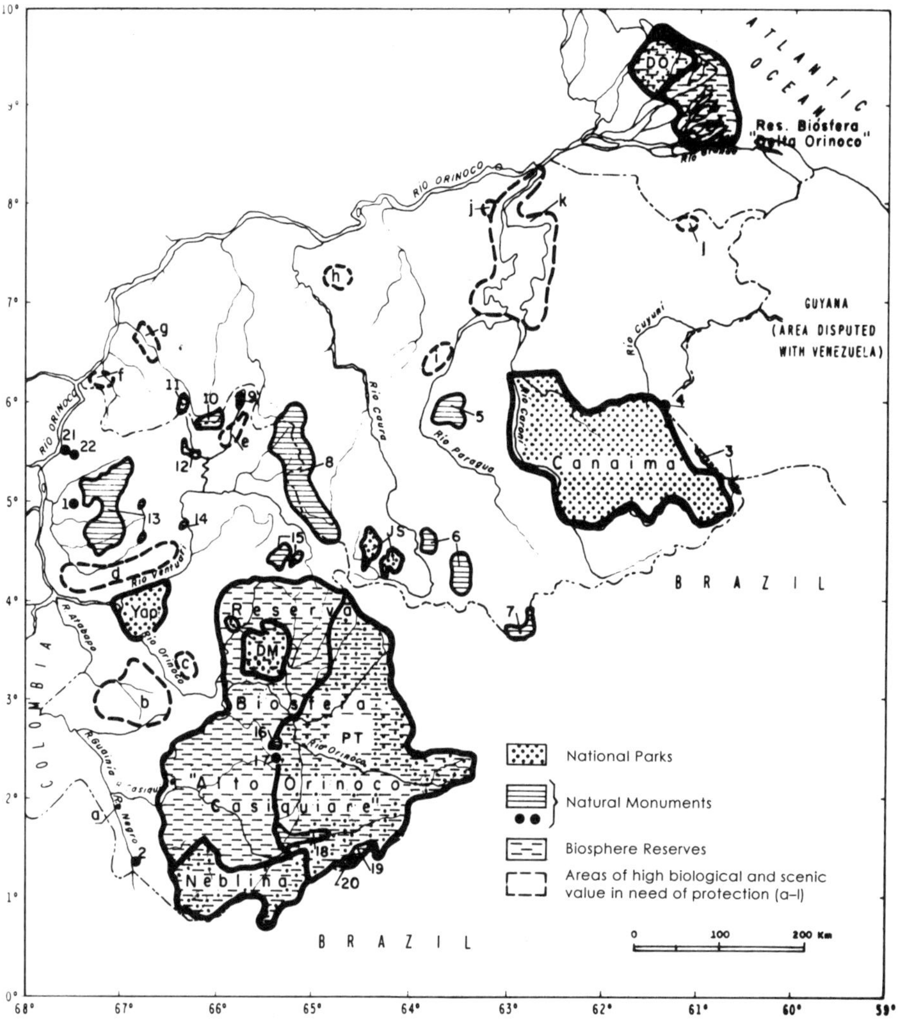

Figure 5-1. Designated conservation areas in the Venezuelan Guayana. Areas recommended for inclusion in protected categories are shown with broken lines. Abbreviations for national parks: DM = Duida-Marahuaca, DO = Delta del Orinoco, JS = Jaua-Sarisariñama, PT = Parima-Tapirapecó, Yap = Yapacana. Numbers refer to natural monuments: 1 = Cerro Autana, 2 = Piedra de Cocuy, 3 = Cadena de Tepuyes Orientales, 4 = Cerro Venamo, 5 = Cerro Guaiquinima, 6 = Cerro Ichúm and Cerro Guanacoco, 7 = Sierra Marutaní, 8 = Sierra Maigualida, 9 = Cerro Yaví, 10 = Serranía Yutajé–Coro Coro, 11 = Cerro Guanay, 12 = Cerro Camani and Cerro Morrocoy, 13 = Macizo Cuao-Sipapo, 14 = Cerro Moriche, 15 = Macizo Parú-Euaja, 16 = Cerro Vinilla, 17 = Cerro Aratitiyope, 18 = Sierra Unturán, 19 = Cerro Tamacuari, 20 = Serranía Tapirapecó, 21 = Piedra La Tortuga, 22 = Piedra Pintada. Areas of high biological or scenic value that still require protection include: a = Rio Negro caatinga (MAB experiment area at San Carlos de Río Negro), b = Rebalse del Atacavi (black-water swamps), c = Río Puruname, d = Guapuchí white-sand meadows, e = Río Parucito flood plains, f = Castillos del Parguaza, g = Lajas del Suapure, h = Serranía del Trueno, i = Montañas de Oris, j = Laja del Elefante, k = deciduous forests around Lago Guri, l = Altiplanicie de Nuria.

National Parks

The seven national parks in the Venezuelan Guayana cover an area of approximately 95,437 km^2. All were established by different presidential decrees and now belong to an administrative unit of the Instituto Nacional de Parques (INPARQUES) called the Subsistema de Parques Nacionales del Sur. Unfortunately, each of these parks is insufficiently staffed, and only Canaima National Park has a management plan for part of the park.

The following section provides additional information for each national park in the Venezuelan Guayana, from the earliest to the most recently declared. The spelling of these parks follows the official decrees, which does not always agree with the toponymy adopted for the *Flora of the Venezuelan Guayana*. More detailed information on these parks can be found in Gondelles et al. (1983) and Gabaldón (1992).

Canaima National Park. Created in June 1962, this was the first national park of the Venezuelan Guayana. It originally covered an area of approximately 10,000 km^2 in southeastern Bolívar state, but was enlarged in September 1975 to the present size of 30,000 km^2. This park includes most of the headwaters of the Río Caroní; its upper tributaries, Río Kukenán, Río Yuruaní, and Río Aponguao; and the middle tributaries, Río Caruay and Río Carrao. The most important mountains in the park are Roraima-tepui, Sierra de Lema, and the large massifs of Auyán-tepui and Chimantá. The Gran Sabana lies entirely in the eastern section of the park.

The main vegetation types of Canaima National Park are dense evergreen submontane and montane forests in the lowlands and uplands; shrublands, savannas, and broad-leaved meadows in the lowlands and uplands; and high-tepui meadows and shrublands on the tepui summits. The summit plateaus of Auyán-tepui, and especially those of the Chimantá massif, have very high degrees of floristic endemicity and ecological diversity.

Places of outstanding scenic beauty that are easily accessible in the park are the Gran Sabana, Angel Falls in the Río Churún valley of Auyán-tepui, and Salto Hacha near the tourist resort of Canaima on the lower Río Carrao. A paved road provides access to the park along its eastern border, and there is commercial airline service to Canaima and towns in the park such as Kamarata, Uonquén, Kavanayén, and Urimán.

Although a preliminary master plan of Canaima National Park was prepared in the early 1970s (Gondelles 1974), the park did not have a management plan until 1991, and then only for the eastern section of the park. A management plan for the western section, which includes Auyán-tepui and the Chimantá massif, was underway in 1994. This park receives the greatest influx of visitors in the Venezuelan Guayana, but it is severely understaffed and is in need of better facilities and logistical support.

In 1989, the Brazilian government established Mount Roraima National Park, covering 1160 km^2 in the adjacent state of Roraima.

Jaua-Sarisariñama National Park. This park, created in December 1978, is located in southwestern Bolívar state in the upper Río Caura drainage basin and covers approximately 3300 km^2. It includes areas of slopes and summit plateaus above 1000 m elevation in the Jaua massif and the Sarisariñama massif farther south. The park can be reached by river along the Río Caura or by small airplane to Canaracuni. The main vegetation types of the park are submontane and montane slope and summit forests, and shrublands and tubiform meadows on the summits. Floristic endemicity is high, especially in the meadows and shrublands.

Near the northeastern edge of Cerro Sarisariñama, there is a system of huge sinkholes (*simas*) that were first explored in 1974 (Brewer-Carías 1976). These unusual chasms and the interesting plant life on the summits of the two tepui massifs were the main reasons this park was established. The park is still largely inaccessible and so far has no management plan.

Yapacana National Park. This park was created in December 1978 and lies south and east of the confluence of the Orinoco and Ventuari rivers in central Amazonas state; it covers an area of approximately 3200 km^2. Most of the park consists of level, lowland landscapes, but it is named after the isolated and imposing Cerro Yapacana, which lies near the southern limit of the park. Access to the park is by boat from the Río Orinoco or the lower Río Ventuari. Small planes can land nearby at Santa Bárbara del Orinoco or at Kanaripó on the northern edge of the park. Yapacana National Park is outstandingly rich floristically, especially in the lowland meadows and submontane forests on the mountain summit. It also has a rich avifauna and an endemic poisonous tree frog, *Minyobates steyermarkii,* which lives on the upper slopes and summit of Cerro Yapacana.

Illegal gold and diamond mining at the base of the mountain and in some of the adjacent forests and meadows has caused considerable impact through repeated burning (Plate 81) and overhunting. Although tourism has increased in and around the park during the past decade, the park has no management plan.

Duida-Marahuaca National Park. Some of the most beautiful mountain scenery in the Venezuelan Guayana is found within this park. It lies along the upper Río Orinoco in the heart of Venezuelan Amazonas and is centered around a spectacular mountain system formed by three tepui massifs, Cerro Duida to the south, Cerro Marahuaka to the northeast, and Cerro Huachamacari to the northwest. The park was created in December 1978, covering approximately 2100 km^2. The lower boundary of the park was originally established at the 1000 m contour line, but there are plans to lower this to the 500 m contour, which would increase the area of the park to more than 3700 km^2. The upper limit of the park reaches a little over 2800 m elevation on the summit of Cerro Marahuaka.

The main vegetation types of this park are extensive lower montane and montane forests, especially on Cerro Duida, as well as many peculiar high-tepui shrublands and meadows. The park is very rich floristically, and the ecological diversity on the large interior plateaus of Cerro Duida and on the summits of Cerro Marahuaka is similarly high.

Duida-Marahuaca National Park can be reached by boat along the upper Río Orinoco or by small airplane to La Esmeralda (on the Río Orinoco) or Culebra (on the Río Cunucunuma).

Serranía La Neblina National Park. This is the southernmost national park of Venezuela, covering an area of approximately 13,600 km^2. It lies in the headwaters of the Yatúa and Baría rivers in Amazonas state and follows part of the border with Brazil. The park was created in December 1978 and is named after Sierra de la Neblina, a large and relatively isolated sandstone massif. This mountain system reaches an elevation of 3014 m, which is the highest point in the Guayana Shield and in South America outside of the Andes (Brewer-Carías 1988). This mountain was not previously reported until Bassett Maguire of the New York Botanical Garden explored it in 1953 (Maguire 1955). From the Venezuelan side, the park is reached by boat along the Siapa, Yatúa, and Baría rivers.

Most of the park is covered by dense, tall, evergreen, basimontane, submontane, and montane forests. On the different summit plateaus of Sierra de la Neblina, there are a variety of shrublands and broad-leaved meadows. Steyermark (1986b) asserted that this mountain system has the highest richness of plant species of all the tepuis, and is the massif with the highest degree of plant species endemism in the Guayana Shield. This was one of the main reasons for establishing the park, which does not yet have a park management plan.

In 1979 the Brazilian government established the Parque Nacional Pico da Neblina, which covers approximately 22,000 km^2 adjoining the Venezuelan park in Brazil's Amazonas state. Subsequently, there has been illegal gold mining in lowland sections of the parks on both sides of the border.

Parima-Tapirapecó National Park. This park was created in June 1991 and is located in the headwaters of the Orinoco and Siapa rivers in southeastern Amazonas state. It covers an area of approximately 38,290 km^2 and is the largest national park in Venezuela; it also lies entirely within the Alto Orinoco–Casiquiare Biosphere Reserve, and most of the park is inhabited by Yanomami Amerindians. It includes most of the Parima uplands and the mountain ranges of Sierra Urucusiro, Sierra Curupira, and Sierra Tapirapecó along the southeastern border of Venezuelan Amazonas adjoining Brazil. The park can be reached by river along the Río Orinoco or by small plane to the settlements of Ocamo, Mavaca, Platanal, and Parima B.

The main types of vegetation in Parima-Tapirapecó National Park are evergreen lowland forests and submontane and montane forests. There are also large areas of mostly secondary savannas in the southern Parima uplands. To date, there is no official management plan for this park. The most severe impacts in the park are caused by illegal gold mining, especially in the area of Cerro Delgado Chalbaud near the sources of the Río Orinoco and by deforestation caused by the Yanomami Amerindians in the southern Parima uplands. Further impact is being caused by unregulated tourism into scattered Yanomami villages in the upper Orinoco and in the Parima uplands.

Delta del Orinoco National Park. This park was created in June 1991 and is located within the Delta del Orinoco Biosphere Reserve in the northeastern section of Delta Amacuro state. It covers an area of approximately 3310 km^2 in the eastern Orinoco Delta and consists mainly of flooded or partially flooded evergreen lowland forests mixed with palm swamps and large areas of floating meadows. Most of the park is inhabited by Warao Amerindians. A management plan has not yet been made for this park.

Natural Monuments

Natural monuments are generally established to provide protection to small areas containing special or unique natural features of very high conservational value, such as caves, waterfalls, and isolated mountains (Gondelles 1992). There are currently 29 natural monuments in the Venezuelan Guayana, but none of them has a management plan prepared by INPARQUES, the government agency responsible for their supervision.

Cerro Autana Natural Monument is a small, towerlike tepui located about 100 km southeast of Puerto Ayacucho in northwestern Amazonas state. It is a major landmark in the area and has a series of ancient caves that cut through the mountain from one side to the other. Piedra de Cocuy Natural Monument is an imposing granitic rock outcrop (inselberg) close to the Río Negro along the border with Brazil in southwestern Amazonas state. These two natural monuments were the first ones established in the area in 1978 because of their striking physical beauty.

Presidential Decree 1223 in November 1990 established 25 new natural monuments in the Venezuelan Guayana to protect the biological and ecological richness of all Venezuelan tepuis that were not previously included in national parks. These monuments are restricted to areas above 800 m elevation, and a few cover large areas that are comparable in size to some of the national parks. A preliminary estimate by INPARQUES of the total area covered by these 25 natural monuments is 10,698.2 km^2. Descriptions of the natural monuments included in the 1990 decree are discussed below in geographical order, starting in southeastern Bolívar state and continuing counterclockwise to southern Amazonas state.

The Cadena de Tepuyes Orientales Natural Monument includes six tepuis along the southeastern border of Bolívar state: Uei-tepui, Kukenán-tepui, Yuruaní-tepui, Wadakapiapué-tepui, Karaurín-tepui, and Ilú-tepui (including Tramen-tepui). Although Roraima-tepui forms part of this same eastern tepui chain, it is not part of this natural monument since it is already included in Canaima National Park. The common base of all these tepuis forms part of the natural monument since it lies above the 800 m contour line.

Cerro Venamo Natural Monument is a mostly forested sandstone mountain at the eastern edge of Sierra de Lema, along the demarcation line of Bolívar state with Guyana. The rich montane cloud forests were the main biological feature that justified establishing this monument.

Cerro Guaiquinima Natural Monument is a large mountain near the lower course of the Río Paragua in east-central Bolívar state. Its summit plateau has a complex mosaic of vegetation types and several endemic plants and animals.

Cerro Ichúm [sic] and Cerro Guanacoco Natural Monuments are formed by the mountain system of Cerro Guanacoco and Cerro Ichún. They are located in southern Bolívar state between the upper Paragua and Caura rivers. Submontane and montane forests are found in both areas, especially on Cerro Ichún. Some unusual high-tepui shrublands and meadows occur on Cerro Guanacoco.

Sierra Marutaní Natural Monument includes the horseshoe-shaped range of Serranía Marutaní, located on the border between southernmost Bolívar state and Brazil. Extensive shrublands and low tepui forests on its summit led to its inclusion as a natural monument.

Sierra Maigualida Natural Monument encompasses a huge, mainly granitic range, the Sierra de Maigualida, that straddles the border of Bolívar and Amazonas states. It is the largest natural monument in the Venezuelan Guayana. A wide range of submontane and montane forests, as well as the peculiar grasslands and meadows on the summits, are major features of this natural monument.

Cerro Yaví Natural Monument, in northeastern Amazonas state, is one of the smallest tepuis in the Venezuelan Guayana. It was declared a natural monument to protect the upper slope forests and the unique summit vegetation of shrublands and pioneer plants.

Serranía Yutajé–Coro Coro Natural Monument is located on the northern border between Amazonas and Bolívar states and includes the large massif formed by the Yutajé and Coro Coro mountains. This area is floristically rich, with many scrub and herbaceous formations. Cerro Guanay Natural Monument also extends along the northern Amazonas-Bolívar border. It was created to protect the diverse shrublands that cover the rocky summit plateau of Cerro Guanay.

Cerro Camani and Cerro Morrocoy Natural Monuments comprise two mountains to the west of San Juan de Manapiare in northern Amazonas state. They were included as natural monuments mostly for their scenic beauty.

Macizo Cuao–Sipapo Natural Monument covers extensive areas of the large mountain system formed by Cerro Cuao, Cerro Sipapo, and Cerro Aracapo in northwestern Amazonas state. The upper region of these mountains includes dense submontane and montane forests, shrublands, and high-tepui meadows, all with high levels of endemism.

Cerro Moriche Natural Monument includes the small mountain of Cerro Moriche near the middle Río Ventuari in north-central Amazonas state. It has several endemic plant species and a high scenic value.

Macizo Parú–Euaja Natural Monument is a large massif in the upper Río Ventuari basin of northeastern Amazonas state. It has extensive shrublands, savannas, and broad-leaved meadows with high plant species endemism.

Cerro Vinilla Natural Monument is a small sandstone mountain in the upper Orinoco region of southeastern Amazonas state; it has several endemic plant species in its upper shrublands and meadows.

Cerro Aratitiyope Natural Monument lies immediately south of Cerro Vinilla and includes one of the most spectacular mountains of the Guayana, the needlelike granitic dome of Cerro Aratitiyope. Although few endemic plant species have been found on its steep slopes, it represents a geographic feature of outstanding scenic beauty (see George 1988; Plate 24).

Sierra Unturán Natural Monument is a large mountain range in the upper Río Mavaca basin and the largest natural monument in southeastern Amazonas state. It is mostly covered by submontane and montane forests, with some unusual montane shrublands.

Cerro Tamacuari and Serranía Tapirapecó Natural Monuments include a spectacular cone-shaped granitic dome, Cerro Tamacuari, and the adjacent Sierra Tapirapecó, both lying along the southern border of Amazonas state with Brazil (FUDECI 1990). Both monuments contain important upland forest types and open rock formations.

In June 1992, two additional natural monuments were declared in the Venezuelan Guayana. Piedra La Tortuga and Piedra Pintada Natural Monuments are both large granitic outcrops (*lajas*) located 10–12 km south of Puerto Ayacucho in northwestern Amazonas state. Besides its peculiar form of a turtle, Piedra La Tortuga has been known since Humboldt's visit as a burial site for the local Amerindians (mainly Piaroas and Guahibos). Piedra Pintada is a granitic boulder with large prehistoric stone carvings.

Biosphere Reserves

The use of biosphere reserves in the conservation policy of Venezuela is very recent. Two biosphere reserves were created by presidential decree in June 1991. Both were designed to offer protection to two important Amerindian groups and their traditional homelands, one at the mouth of the Río Orinoco, the other at its headwaters.

Delta del Orinoco Biosphere Reserve. This reserve protects an area of about 11,250 km^2 in east-central Delta Amacuro state. It covers much of the middle and lower delta of the Río Orinoco, where the main vegetation types are flooded evergreen forests, marshy forests, palm swamps, and mangroves. The entire region is inhabited primarily by the Warao Amerindians, a well-acculturated ethnic group. The administration of the biosphere reserve is assigned to the environmental ministry (MARNR), which is entrusted with formulating a management plan.

Alto Orinoco–Casiquiare Biosphere Reserve. Located in the upper Orinoco region of Amazonas state, this reserve covers an area of about 83,830 km^2. To date, it is the largest biosphere reserve in the tropics and the second largest in the world after the Northeast Greenland Biosphere Reserve (700,000 km^2). It includes the headwaters of the Río Orinoco downstream to the mouth of the Río Cunucunuma and all tributary river basins in between, as well as the area south of the Río Casiquiare from Tamatama to the mouth of the Río Pasimoni. From there, it follows the Pasimoni upstream until it reaches the Brazilian border. The reserve contains two previously declared national parks, Duida-Marahuaca and Serranía la Neblina, as well as five natural monuments (Vinilla, Aratitiyope, Sierra Unturán, Tamacuari, and Sierra Tapirapecó).

At the same time the biosphere reserve was created, Parima-Tapirapecó National Park was declared and included in the reserve. The main purpose of this reserve is to allow Yanomami and Yekwana Amerindians to live within their respective homelands and maintain their traditional lifestyle under relatively undisturbed environmental conditions. The Alto Orinoco–Casiquiare Biosphere Reserve lacks a management plan, but initial actions to remedy this have been taken by the Servicio Autónomo para el Desarrollo Ambiental del Estado Amazonas (SADA-Amazonas), the governmental authority responsible for managing the reserve, and funding has been provided by the European Community.

Other Special Management Areas

Besides national parks, natural monuments, and biosphere reserves, there are several other legal categories that regulate commercial exploitation of natural resources and other forms of land use. Since 1960, large areas have been designated for commercial exploitation. The categories of forest reserve (*reserva forestal*), protected forest area (*área boscosa bajo protección*), and forest lot (*lote boscoso*) are designated for timber exploitation, whereas mineral exploitation is regulated through individually approved mineral concessions. According to a map of areas under special administration (*áreas bajo régimen de administración especial*), published by the environmental ministry (MARNR 1991), the areas designated for logging in the Venezuelan Guayana include four forest reserves, five protected forest areas, and seven forest lots (see Table 5-2).

Table 5-2. Forest reserves and forest lots in the Venezuelan Guayana.

Forest area	Year created	Surface area (km²)
Forest Reserves		
Imataca	1961	32,033
Sipapo	1963	12,155
El Caura	1968	51,340
La Paragua	1968	7,820
		103,348
Forest Lots		
San Pedro	1981	7,574
Sector Caño Blanco	1984	200
El Frío	1985	650
Río Parguaza	1986	657
El Dorado-Tumeremo	1987	790
Fundo Flamerich	1988	192
Paisolandia	1988	81
		10,144

Source: Mapa de Áreas bajo Régimen de Administración Especial (MARNR, Caracas, 1991).

Forest Reserves

Forest reserves generally consist of large areas of natural forests in public lands that are designated for sustainable timber exploitation through specific management and reforestation plans. They are usually established by presidential decree, but some were created previously by simple resolutions of the agriculture ministry. They are now supervised by the Servicio Autónomo Forestal Venezolano (SEFORVEN), part of the environmental ministry (MARNR).

The Imataca Forest Reserve was created by ministerial resolution in 1961 and covers an area of 32,032.6 km^2 in southern Delta Amacuro and northeastern Bolívar states. It includes mainly evergreen and semievergreen lowland forests and is the only forest reserve that is currently being exploited in the Venezuelan Guayana.

La Paragua Forest Reserve includes mainly evergreen lowland forests. It was created by presidential decree in 1968 and covers an area of 7820 km^2 between the lower Paragua and Caroní rivers in east-central Bolívar state. Part of Cerro Guaiquinima Natural Monument lies within this reserve.

El Caura Forest Reserve was created by presidential decree in 1968. It is the largest such reserve in Venezuela, originally covering an area of 51,340 km^2; it includes most of the Río Caura basin, extending south of the parallel 7°30' N, and reaches the border with Amazonas state in the west and the border with Brazil in the southwest. Part of this reserve, particularly the eastern slopes of Sierra Maigualida above 800 m elevation, was declared part of a natural monument in the 1990s. Also, the Jaua-Sarisariñama massif, originally included in the reserve, was designated as Jaua-Sarisariñama National Park in 1978. The rest of this reserve is dominated by evergreen lowland and submontane forests.

The Sipapo Forest Reserve was established in 1963 by a ministerial resolution and originally covered an area of 12,155 km^2. It is located in the Cuao-Sipapo massif in northwestern Amazonas state. In 1990, all land in the massif above 800 m elevation was included in Cuao-Sipapo Natural Monument. Cerro Autana, to the west of Cerro Cuao, was declared a natural monument in 1978. The remaining area of the Sipapo Forest Reserve is covered mostly with evergreen lowland and submontane forests.

Protected Forest Areas

The legal designation of protected forest area (*área boscosa bajo protección*) was created in June 1991 to exploit primary and secondary forest products in large extensions of natural forests on both public and private lands. Other uses, such as agriculture, mining, and tourism are also allowed. Protected forest areas are created by presidential decree and are administered by SEFORVEN.

Three forest areas in Bolívar state and two in Delta Amacuro state were established in June 1991, along with 32 other forest areas in the rest of the country. The protected forest areas in the Venezuelan Guayana include Chivapure-Cuchivero Forest Area, in the upper Río Cuchivero basin of western Bolívar state; San Francisco de la Paragua Forest Area, north of the lower Paragua river in north-central Bolívar; El Choco Forest Area, in the middle Río Yuruari basin of northeastern Bolívar state; Pedernales Forest Area, in northern Delta Amacuro state; and Merejina Forest Area, in east-central Delta Amacuro state, partially overlapping with the Delta del Orinoco Biosphere Reserve. The total area of these protected forest areas has not yet been determined.

Forest Lots

Forest lots are usually small forested areas on public or private lands that are designated for commercial logging. Like the two prior categories, they also require a management plan prior to exploitation and are supervised by SEFORVEN. All forest lots in the Venezuelan Guayana are located in Bolívar state. Six are scattered in northeastern Bolívar state (San Pedro, El Dorado–Tumeremo, Sector Caño Blanco, Fundo Flamerich, El Frío, and Paisolandia), adjacent to the Imataca and Paragua forest reserves, and are under various degrees of exploitation. The seventh, Parguaza Forest Lot, lies in the middle Río Parguaza basin in western Bolívar and has not yet been logged.

Both forest reserves and forest lots originally covered an area of approximately 113,491 km^2, which corresponds to about 25 percent of the surface of the Venezuelan Guayana. Several of the newer national parks and natural monuments, however, lie within former forest reserves (especially in the cases of the La Paragua, El Caura, and Sipapo forest reserves). This has consequently reduced the coverage of the forest reserves to about 100,000 km^2, or roughly 22 percent of the Venezuelan Guayana.

Protective Zone of Southern Bolívar State

Another legal protection category that is widely employed in Venezuela is protective zone (*zona protectora*). This category tries to regulate land use in potentially populated areas by establishing specific management plans. Most of the 48 protective zones in the country lie in river basins or around urban areas. The only such area south of the Río Orinoco is the Zona Protectora Sur del Estado Bolívar (Protective Zone of Southern Bolívar), which was created in 1975 and covers 72,624 km^2. It lies between the Caura and Caroní rivers, mostly south of 7° N to the Brazilian border; to the east it reaches the Icabarú valley and the region around Santa Elena de Uairén just south of Canaima National Park. As of 1994, there is no management plan or other official document that regulates land use in this area.

In summary, by late 1992, 72 percent of the land in Venezuelan Guayana was politically regulated, with 31 percent of the area designated for conservation (national parks, natural monuments, and biosphere reserves), 25 percent of the area designated for exploitation of timber or other forest products (forest reserves, forest lots, and protected forest areas), and 16 percent of the area designated for rational land use (protected zones). The remaining 28 percent of the land area is not subject to special regulations and is where nearly 90 percent of the inhabitants of the Venezuelan Guayana live and work.

Threats to Conservation

With almost three-quarters of its surface under some sort of legal regulation (almost half of this strictly protected), the Venezuelan Guayana should be in an excellent position to ensure the continued protection of the area's natural resources. The extensive network of national parks, natural monuments, and biosphere reserves that was created since 1970 provides a solid base for the development of a powerful and effective conservation policy in the area.

Until the 1990s, most of the Venezuelan Guayana was simply too inaccessible for commercial exploitation, a fact that ensured its preservation despite the arrival of Europeans and their dreams of finding El Dorado. However, in the late 20th century the situation is radically different and there is now a steady flow of people and equipment into even the most remote corners of Guayana.

Although most residents of the Venezuelan Guayana are still concentrated in a few towns and cities along its northern and northeastern edge, the area's population is rapidly increasing and is already causing severe impacts on the local environment. Figure 5-2 indicates the main areas of human impact in the

Venezuelan Guayana. The human activities that produce the most serious environmental degradation in the Venezuelan Guayana are described below.

Mining

When Sir Walter Raleigh visited the Guayana at the end of the 16th century, legends depicted the area as having immense gold and diamond fields deep in its interior (Raleigh 1596). These riches were not readily discovered, but the legends did prove true in a sense centuries later, when gold, diamonds, and other precious and industrial minerals were found in abundance. Gold mining became one of the major economic activities in the Venezuelan Guayana starting in the early 1800s (Gómez 1953).

The first large mining center in the Venezuelan Guayana was established in 1829 around El Callao in Bolívar state, about 220 km southeast of Ciudad Bolívar. The principal mining centers are now widely scattered in Bolívar and to a lesser degree in Amazonas state. The main mines in Bolívar state are the gold mines around El Callao and El Dorado, on the middle and lower Río Yuruari; the gold mines at Kilómetro 88 (Las Claritas; Plate 80), at the northeastern base of Sierra de Lema; the gold mines at Bochinche and Botanamo, on the Río Cuyuní near the border with Guyana; the gold and diamond mines of Icabarú in the central Pakaraima mountain range, west of Santa Elena de Uairén; the diamond mines at San Salvador de Paul on the lower Río Caroní; and the diamond mines of Guaniamo on the Río Guaniamo, a tributary of the Río Cuchivero in south-central Municipio Cedeño. In Amazonas state the major mining centers are the gold mines at the base of Cerro Yapacana in Yapacana National Park, the gold mines in the southern Sierra Parima and on Cerro Delgado Chalbaud in the upper Río Orinoco, and the gold mines at the western and southern base of Sierra de la Neblina in Serranía La Neblina National Park.

Whereas mining is allowed in Bolívar state through government concessions, it was totally banned in Venezuelan Amazonas by a presidential decree in June 1989. This decree was largely a response to the illegal entry into Venezuela by several thousand Brazilian miners (*garimpeiros*) in the headwaters of the Río Orinoco. Although most of these miners were forced out of the country, some small mining camps in parts of southern Sierra Parima in Parima-Tapirapecó National Park are still active. It has been difficult for the Venezuelan government to muster the logistical and financial resources to adequately protect and monitor the parks and national borders in the Guayana.

Although the direct impact caused by gold and diamond mining activity on the vegetation is usually small, the side effects can be severe. Among these are mercury pollution, an increase in the sediment load of the rivers, overhunting, and frequent wildfires. More important are the effects on the indigenous population by the introduction of diseases, the disruption of traditional food supplies, and even the reported killings of villagers.

Illegal mining has become a serious threat in many areas of the Venezue-

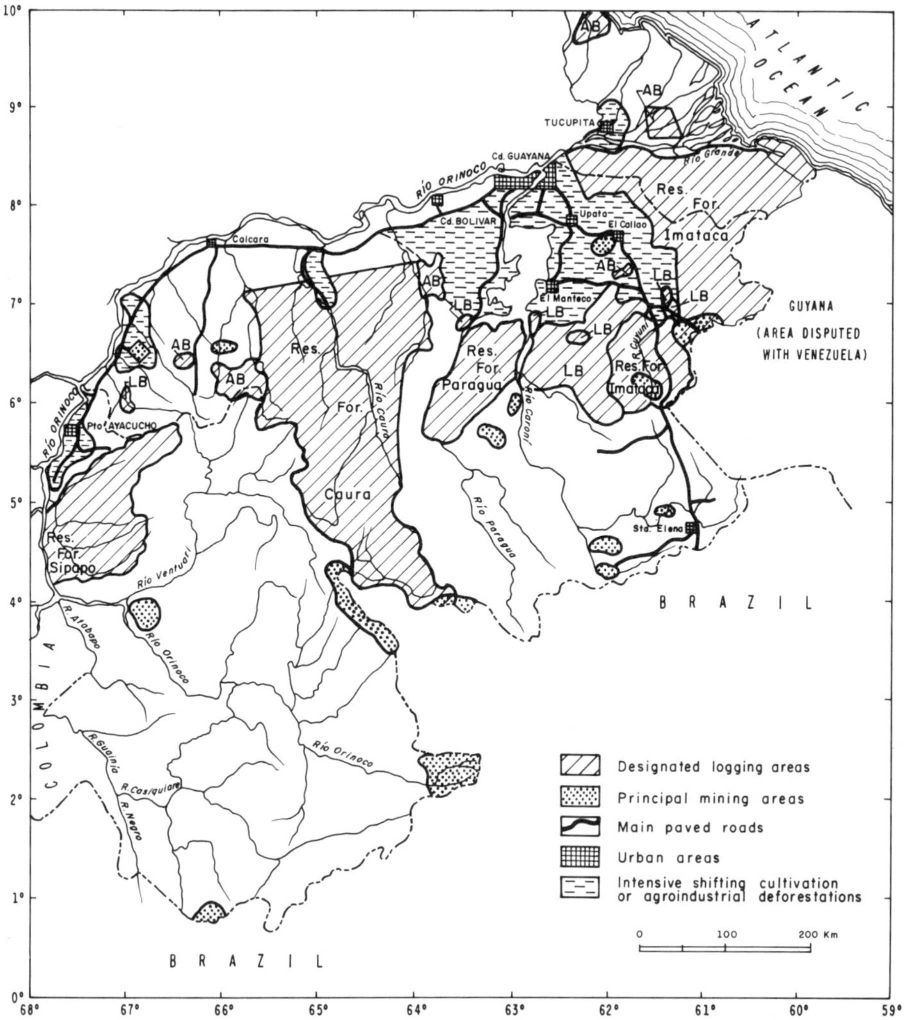

Figure 5-2. Main types of human impact in the Venezuelan Guayana. The three kinds of designated logging areas include the following: AB = *Area boscosa bajo protección* or protected forest area, LB = *Lote boscoso* or forest lot, and Res. For. = *Reserva forestal* or forest reserve. The four principal mining areas in Amazonas state (in the southern third of the map) are unauthorized and are located within existing national parks.

lan Guayana in the 1980s and 1990s, especially due to the large increase in the number of Brazilian miners crossing over the poorly marked border. In the future, legal mining through official concessions is also likely to increase due to the new government policy of generating alternative sources of income as a means of boosting the national economy.

Strip mining for industrial minerals such as iron and bauxite is now causing considerable damage to the environment. The huge iron deposits found on Cerro Bolívar, Cerro San Isidro, Cerro Los Barrancos, and Cerro María Luisa near Ciudad Piar (80 km southeast of Ciudad Bolívar) have been intensively exploited since the early 1940s. This kind of mining destroys not only the original vegetation but eats away at the mountains themselves. Another large-scale deforestation is now occurring at one of the world's largest bauxite mines near Los Pijiguaos, on the lower Río Suapure in northwestern Bolívar state (Huber and Guánchez 1988; Plate 79).

Logging

The traditional economic outlook on the largely intact forest cover of the Venezuelan Guayana still regards it as a vast supply for timber production, once roads are built to make the forests more accessible. Although forest reserves and forest lots in Venezuela are officially regarded as units of both exploitation and protection, they rarely fulfill this dual purpose. Timber concessions in both forest reserves and lots usually require management plans that regulate both the kind and level of timber extraction as well as the restoration of cut-over areas. However, the ecological degradation from the timber activities is usually so severe that restoration or even simple reforestation is a difficult if not impossible task.

During the 1980s, the concept of sustainable forest management was widely introduced into development schemes throughout the tropics in an attempt to achieve a compromise between exploitation and conservation. However, the paucity of proven strategies for sustainable forest management and their poor implementation and monitoring have produced ecological disasters on an unprecedented scale, such as Daniel Ludwig's Jarí ranch experiment in eastern Amazonian Brazil (Palmer 1991).

In the Venezuelan Guayana, commercial logging was prohibited in Amazonas state by Presidential Decree 2552 of January 1978. However, logging in Bolívar and Delta Amacuro states is encouraged through numerous timber concessions. The main areas for timber exploitation are concentrated in the Imataca Forest Reserve and the San Pedro Forest Lot (Plate 84) in northeastern Bolívar and in east-central Delta Amacuro state. New concessions have been approved in the lower Río Paragua basin, in the Chivapure-Cuchivero Forest Area, and in the Parguaza Forest Lot (which directly affects the homeland of the Piaroa Amerindians). Although local timber companies only consider 10–15 species commercially harvestable, forest destruction is high, and reforestation is generally ignored or ineffective. Road construction in new forest areas by logging companies also encourages rapid invasion by squatters and miners, leading to even greater forest degradation (Uhl and Guimarães 1989).

It is rapidly becoming apparent that tropical forests left intact may be worth more than those subject to many kinds of destructive exploitation. This reflects upon the forest's value in the regulation of local, regional, and planetary climate patterns, for the hydrological equilibrium of river basins, in the maintenance and evolution of genetic resources (biodiversity), as a sustainable long-term source of forest products, and as a natural homeland for indigenous populations. International funding to tropical countries that preserve their rain forests is expected to increase dramatically, and Venezuela stands to become one of the main beneficiaries due to the Guayana's huge reserves of intact ecosystems. Rather than logging natural forests in the Venezuelan Guayana, the establishment of tree plantations in secondary vegetation or degraded ecosystems could prove to be an economically viable and ecologically wise alternative.

Hydroelectricity

The many rivers that arise in the mountains of the Guayana Shield provide numerous sites where hydroelectric power plants could be built in Bolívar and Amazonas states. A survey in Amazonas state identified more than 60 potential sites for hydroelectric dams (MARNR 1983–1984).

There is currently only one large hydroelectric power plant in the Venezuelan Guayana, the Represa Raúl Leoni. This dam was built between 1963 and 1986 by Electrificación del Caroní C.A. (CVG-EDELCA) at Guri on the lower Río Caroní. It consists of a huge 162-m high concrete dam and an artificial lake covering over 4250 km^2, with approximately 135 billion m^3 of water (CVG-EDELCA 1985). The area flooded by the lake was previously covered by savannas and by semideciduous and riparian forests. To the south of the dam, the confluence of the Paragua and Caroní rivers was also flooded, together with small extensions of evergreen lowland forest.

Downstream from Guri, three more dams are under construction by CVG-EDELCA on the Río Caroní, at Macagua, Caruachi, and Tocoma (CVG-EDELCA 1985). Although the reservoirs created by these dams will be much smaller (approximately 400 km^2 together) than the reservoir created by the dam at Guri, there will be considerable impact on the riparian vegetation of the lowermost Río Caroní.

Another ambitious program plans to build several new hydroelectric power plants along the middle and upper Río Caroní and on the lower Río Paragua (CVG-EDELCA 1985). One planned at Eutobarima, below the confluence of the Caruay and Caroní rivers, is of environmental concern, because it would severely affect the unusual gallery forests and palm swamps (*morichales*) of the southern Gran Sabana. Part of this dam's artificial lake would also extend into Canaima National Park.

Another large-scale hydroelectric project is planned by CVG-EDELCA in the Río Caura basin. After abandoning the original plan of a huge dam on the lower Caura which would have flooded an immense area of evergreen lowland forests, CVG-EDELCA is considering a new plan that would create a canal to divert part of the water of the Río Caura into the Río Paragua, which feeds into Lago Guri. Besides the considerable on-site impacts produced by the construction of such a canal, the project would cause major hydrological changes in the ecology of the large floodplains in the lower section of the Río Caura valley.

In a few localities, CVG-EDELCA has constructed small, local hydroelectrical power plants that utilize the river flow and do not require large dams. Such small plants, known as *minicentrales,* are in operation near San Ignacio de Yuruaní in the southeastern Gran Sabana and on the lower Río Cuao in Amazonas state. Two more are near completion, one in the headwaters of the Río Aponguao in the northern Gran Sabana, and the other at Salto Hacha alongside the tourist camp at Canaima. These dams provide electricity to small villages or settlements, and their environmental impact is relatively small.

The generation of hydroelectric energy has long been considered a clean, relatively inexpensive energy supply. When there are few dams in a given area, such as the Represa Raúl Leoni in the Venezuelan Guayana, it is relatively easy to control the dam's watershed and promote a strong conservation policy in the area. However, the construction of more dams in the area would probably lead to conflicts with Amerindian groups and other settlers, infringe on existing national parks or monuments, and elicit objections from national and international conservation groups. The costs of maintaining and operating the power plants and their reservoirs, as well as the control of the entire river basin that supplies the dam, have often been underestimated. Today, the construction of large-scale power plants with huge reservoirs is being challenged on the basis of social, financial, and environmental concerns (Fearnside 1989).

Burning

The annual burning of large tracts of land is a common practice in the Venezuelan Guayana, as in other parts of the country and the tropics in general. Almost all fires in the flora area are intentionally lit by Amerindians, miners, or farmers, usually in the dry season. The occurrence of natural fires appears to be negligible compared to the frequency of those started by humans. To date, the areas most heavily affected by fires include open savannas and shrublands of the Gran Sabana uplands (including forested areas in exceptionally dry years); savannas, shrublands, and dry forests of the northern and northwestern lowlands in Bolívar state; savannas and meadows of the northern and west-central lowlands of Amazonas state (Plate 81); and secondary savannas and forests of the southern Parima uplands (Plate 82).

In the Gran Sabana, where CVG-EDELCA has carried out an intensive program of helicopter-supported fire control since the 1980s, approximately 5000 fires are recorded annually, mainly in forested areas designated as high priority for fire control. An additional 5000 fires may be set each year, but these occur in low priority areas or during the rainy season. Most of the fires are set by local Pemón Amerindians, who have numerous explanations for setting the fires, such as "cleaning" the savannas, snake or insect control, hunting, farm clearing, communicating with distant communities, superstitions, and religious reasons (CVG-EDELCA unpublished). Although fires continue to be set in large numbers, the main emphasis in the fire protection program have been to limit the extent of large fires.

Fires in the Gran Sabana are usually started in the open savanna, then spread with the prevailing wind until reaching a forest border or rock outcrop. At times the fire penetrates the forest, where it can kill large numbers of trees by destroying the root mat. Studies on the effect of fire in Gran Sabana forests show a high vulnerability and tree mortality after fire due to the pronounced oligotrophic conditions of the soils (Fölster 1986; Hernández 1987, 1992). The moist meadows, palm swamps, and numerous shrublands on rocky substrate are also affected by the recurrent fires.

The southern Parima uplands are another example of severe deforestation caused by burning (Plate 82). Fires have been set by Yanomami Amerindians and by miners (Huber et al. 1984). The frequency of fires in the lowland savannas, meadows, and shrublands of northern Bolívar and Amazonas states is also increasing, perhaps as a consequence of the increased mobility of the local population for hunting and traveling. During the late dry season (February–March), flying above certain areas can be difficult because the landscape beneath may be hidden by dense layers of smoke.

Givnish et al. (1986) suggested that fires might be a regular feature of high-tepui ecosystems and are capable of producing fire-adapted structures in certain tepui taxa. However, extensive observations argue against the frequent occurrence of fires on tepuis. Numerous scientists have spent altogether thousands of days on top of many tepuis, but so far there is no reliable report witnessing the start of a natural fire, for instance by lightning. This is perhaps more significant considering that most visits to tepuis have been made during the dry season, when fire-prone conditions prevail.

After visits to 41 tepui summits, Otto Huber has seen clear evidence of past fires on only 10 tepuis: Uei-tepui, Sororopán-tepui, Auyán-tepui (northwestern sector), Kurún-tepui, Amurí-tepui, Serranía Yutajé, Cerro Guanay, Cerro Parú, Cerro Duida, and Sierra de la Neblina (Plate 83). Of these, the summits of Uei-tepui, Sororopán-tepui, Kurún-tepui, Cerro Guanay, and Cerro Duida were reached by fires that were set in surrounding lowland areas and climbed up the lower slopes (see Tate in Gleason 1931; Maguire and Deery de Phelps 1951; Mayr and Phelps 1967). A fire on Auyán-tepui was caused in May 1964 by workers accompanying Julian Steyermark during his botanical exploration (Steyermark 1967). Only four tepui summits (Amurí-tepui, Cerro Yutajé, Cerro Parú, and Sierra de la Neblina) show signs of natural fires during modern times. Still, more research is needed to determine if fires in high-tepui environments are only sporadic events occurring at long intervals, or regular ecological features that have molded certain vegetation types.

In most cases, burned high-tepui vegetation recovers very slowly, probably because of the highly oligotrophic environment. Very slow recovery rates for high montane tropical vegetation were also reported in Costa Rica by Janzen (1973). Givnish et al. (1986), however, found evidence for a quick recovery from a fire on the Sierra de la Neblina. Due to the fragility of the vegetation on tepui summits, visitors should refrain from making campfires because of the risk of the fire escaping to the surrounding vegetation.

Shifting Cultivation and Other Agricultural Activities

With the low population density of native Amerindians in the Venezuelan Guayana, the effects of shifting cultivation have been relatively minor. Where religious missions and government-sponsored settlements have induced traditionally semi-nomadic groups to adopt a sedentary lifestyle, for instance around La Esmeralda, San Juan de Manapiare, Puerto Ayacucho, and in the southern Parima uplands of Amazonas state, the method of shifting cultivation has led to a marked increase in forest clearings. The population of these settlements is increasing at a faster rate than that of more nomadic communities, making it continually more difficult to find new areas suitable for forest clearing. The same pattern occurs among the Pemón Amerindians who have settled along the main road from El Dorado to Santa Elena de Uairén.

Large-scale agriculture and cattle raising have taken place mainly in northeastern Bolívar state, in a belt extending from the lower Río Paragua east to the Serranía de Imataca. A number of large agricultural settlements have appeared along the new road from Caicara to Puerto Ayacucho and along the lower Río Caura, near Maripa. They usually consist of extensive clearings of forests or savannas for cattle pasture, or else of medium-sized cultivated fields with maize, cotton, or sorghum as the main crops. In 1985, the total surface area of cultivated land in the Venezuelan Guayana was approximately 3,000,000 hectares (Lairet and Rodríguez 1989), but undoubtedly this has increased steadily since then.

The agricultural advance in northeastern Bolívar state is closely linked to the construction of new logging roads. Since the soils of this region are generally poor, economically viable agriculture is only possible for a few years, after which the land must either be abandoned or heavily fertilized. Pollution of rivers due to excessive use of fertilizers and insecticides has already been observed near La Paragua and between Upata and El Manteco in northeastern Bolívar state.

Population Pressure

The population of Venezuela is growing at approximately 2.5 percent per year, one of the highest rates in Latin America. According to official census data (OCEI 1985, 1992), the country's population increased from approximately 7.5 million inhabitants in 1961 to about 14.5 million in 1981, and then to an estimated 19.3 million in 1990. The number of inhabitants in the Venezuelan Guayana has more than quadrupled from only 259,279 in 1961 to approximately 1,120,000 in 1990. Much of this increase can be attributed to the development of Ciudad Guayana and the adjacent industrial center of Matanzas.

Nearly 90 percent of the inhabitants of the Venezuelan Guayana is now concentrated in a dozen cities and towns along the northern edge of the region. This has caused heightened local demand for land, food, energy, roads, and wood. It can soon be expected that some inhabitants of these crowded towns and cities will move farther south for greater independence and a simpler life style. Before this happens, it is important that sound regional land use plans be made and put into effect to avoid social and environmental conflicts.

Tourism

Since the 1980s, tourism has become one of the most promising economic activities in the Venezuelan Guayana. Because of the numerous places with outstanding scenic beauty and mostly pristine landscapes, the area ranks higher each year as a vacation destination for national and international tourists.

The completion of paved roads through the Gran Sabana and to Puerto Ayacucho, together with daily commercial air routes to Puerto Ordaz, Canaima, and Puerto Ayacucho, has made access to the area much easier. Numerous jungle lodges and camps have opened in remote areas of Bolívar and Amazonas states. Small airplanes are available at many places in the flora area, either for sightseeing trips or for transport to remote areas. Tour operators offer a variety of river trips, especially along the lower Río Caura and the Río Orinoco above Puerto Ayacucho.

The main tourist destinations in Bolívar state currently are the Gran Sabana, along the road from La Escalera to Santa Elena de Uairén, with side trips to Kavanayén and Icabarú; Canaima, a Pemón village and tourist camp near the confluence of the Río Carrao and the Río Caroní, the oldest and largest tourist resort of the Venezuelan Guayana, only accessible by airplane; Kavac, a small resort at the southeastern base of Auyán-tepui, accessible only by airplane; Guri, at the northern end of the lake created for the Raúl Leoni hydroelectric power plant, accessible by paved road; and Salto Pará, on the lower Río Caura, accessible by river.

In Venezuelan Amazonas the main tourist attractions are Puerto Ayacucho and surroundings (including Cerro Autana), with several tourist camps, accessible by road and commercial airline; the Río Manapiare basin, with several tourist camps, accessible mainly by small plane; Santa Bárbara, at the mouth of the Río Ventuari, accessible by boat or small plane; Culebra, on the Río Cunucunuma, at the northern base of Cerro Duida, accessible mainly by small plane; the Yanomami villages of Ocamo, Mavaca, and Platanal on the upper Río Orinoco, accessible by boat or small plane; and the Río Orinoco, with boat trips from Puerto Ayacucho to Ocamo.

Some of the tourist areas listed above that are accessible by plane or boat cater primarily to foreign tourists and wealthy Venezuelans. Most Venezuelan tourists travel en masse by road to the Gran Sabana and Puerto Ayacucho during vacation periods (Christmas to New Year, Carnival week, Easter week, and school vacation period in July and August). During these peak vacation periods, 20,000–30,000 vehicles with an average of 5 passengers each usually visit the Gran Sabana.

Problems caused by excessive tourism are already apparent in several parts of the Venezuelan Guayana and include littering, illegal gathering of plants and animals, and accidental or intentional wildfires. Because of adverse impacts on the indigenous population, a presidential decree in December 1989 restricted tourist activities in areas inhabited by Amerindian populations in Venezuelan Amazonas.

While mass tourism is centered in lowland areas or the uplands of the Gran Sabana, more adventuresome individuals have taken to climbing to the tops of tepuis, particularly Roraima-tepui and Auyán-tepui. The summits of these two mountains are accessible by long, but relatively easy foot trails. During peak periods, several hundred visitors crowd the summit of Roraima-tepui, causing considerable environmental contamination and damage to the naturally sparse vegetation. An elite type of tourism has begun using helicopters to reach the tepui summits. Wealthy Venezuelans and foreigners have hired helicopters for picnics on Cerro Autana or near the famous sinkholes of Cerro Sarisariñama. Other favorite mountains for this kind of tourism are Roraima-tepui, Auyán-tepui, and Cerro Marahuaka.

The increasing presence of tourists on tepui summits poses significant threats to these unique ecosystems. First is the danger of escaped fires. Also, regrowth of trampled plants around campsites is very slow since many plant species of the high-tepui vegetation are very brittle and fragile (Gorzula and Huber 1992). The accidental transfer of either endemic or weedy species from one mountain summit to another may alter the tepui's floristic composition. On rocky summits such as Roraima-tepui, many visitors disfigure or break off and carry out rocks and quartz crystals as souvenirs, or they remove plants for their home garden (especially orchids and bromeliads). Finally, littering of the pristine and fragile landscapes often occurs, even though visitors are expected to remove all their trash. It is not known how excessive wood collecting for campfires might affect the local ecology, nor the effects of eutrophication from human wastes and washing. Because of these dangers, INPARQUES suspended permits to visit all tepui summits in 1989, until a detailed management plan could be made and put into effect.

Commercial Exploitation of Natural Products

The indigenous populations of the Venezuelan Guayana have traditionally made extensive use of natural forest products. Resource overexploitation was probably never a serious threat because of their low population density and migrant life style. This situation, however, changed radically with the arrival of commercially minded colonizers, who quickly began to search for forest products to exploit on a larger scale. One of the main goals of Solano's Expedición de Límites to the upper Orinoco in 1750 was to explore the region's plant resources, especially cocoa, for future trade in Europe (Ramos 1946).

Of the many natural plant products in the Guayanan forests, few have actually played important roles in the modern Venezuelan economy. Those that were important include natural rubber (*caucho,* or *Hevea* spp.); other gums and rubbers (*chicle* and *balatá,* or *Manilkara bidentata* and *Couma* spp.); tonka bean (*sarrapia,* or *Dipteryx odorata*) for perfumes and the tobacco industry; *piassaba* palm fibers (*chiqui-chiqui,* or *Leopoldinia piassaba*) for basketry and brooms (Plate 76); *mamure* (aerial roots of the aroids *Heteropsis flexuosa* and *H. speciosa*) for furniture; *seje* (oil from the palms *Oenocarpus bacaba* and *O. bataua*) for cooking and medicinal use; and palm hearts (*palmito,* or *Euterpe oleracea* and *E. precatoria*).

With the downturn in Venezuela's petroleum-based economy during the 1980s, the exploitation of secondary forest products has been increasing. The palm heart industry has expanded, using the natural palm populations in the swamps of the Orinoco Delta. Harvest of *mamure* and *piassaba* has also grown, and *mamure* has been largely depleted in populated areas such as around Puerto Ayacucho. Lately even the collection of tonka beans in the lower Caura and Cuchivero region has been renewed, since their market price has risen considerably.

The gathering of wild rubber from natural forests was one of the main economic activities in Amazonas state from 1870 until 1930. The boom period ended after 1930, due to the collapse of the world market price for natural rubber, but there was a brief renewal of rubber tapping during World War II. Between 1972 and 1980, the regional development corporation, CODESUR, tried to establish an experimental rubber plantation near Santa Bárbara del Orinoco. This effort was initially promising, but it was later abandoned because of its high economic risks. Another regional development corporation, CVG, began a new experimental rubber plantation near San Fernando de Atabapo. This program's goal is to expand the plantations to over 11,000 hectares in central Amazonas state. The project has been sharply criticized, however, since the plan involves large-scale deforestation (which is prohibited by presidential decree), and the economic viability of the experiment is questionable.

The gathering of secondary or nontimber forest products, commonly called rain forest harvesting, has been advanced by institutions from different countries as an ecologically valid alternative to destructive timber exploitation. In the American tropics, this idea is particularly favored in Brazil and Colombia, where there are several specially designed extractive reserves.

Although this is an interesting new approach to the problem of rational forest management, it has some risks. First, there is very little knowledge available on the reproductive biology of most of the potential forest species, and intensive commercial exploitation of their fruits or seeds may soon lead to decreased conditions for the species' survival. Second, outside commercial managers or groups often take over the trade of these forest products, to the detriment of the indigenous population. Third, many of the secondary forest products only satisfy temporary markets in developed countries. If the product goes out of fashion, the local producers suffer sudden declines in their income, which can further weaken the already fragile local economy. Under these three circumstances, it is clear that exploitation limited to secondary forest products requires important precautions and a certain amount of knowledge for it to be both economically viable and ecologically acceptable.

Demand for unique and interesting plants from the tepuis has also increased. These especially include the carnivorous pitcher plants (*Heliamphora* spp.; Plate 60), bromeliads, and orchids. Even though removing live plants is strictly prohibited in national parks and natural monuments, the danger of illegal plant trade continues to persist because of insufficient staffing in these protected areas.

Climate Change

Because of marked increases in carbon dioxide and other gaseous emissions from the large-scale combustion of fossil fuels in the last half of the 20th century, the specter of major local and global climatic changes has become the focus of both popular concern and serious scientific study. Since the Venezuelan Guayana has such vast expanses of forests and other vegetation types, it stands to play an important role in influencing short- and long-term climatic changes.

Although discussion of this topic is highly speculative, local climate changes directly caused by human activities should be possible to detect over periods measured in years or decades. For instance, the creation of the large lake by the dam at Guri on the lower Río Caroní may well modify the mesoclimate of the region. The large evaporation surface of this lake (more than 4000 km^2) could increase the humidity levels and make the originally subxeric local climate more mesic, but the appropriate meteorological measurements to corroborate this assumption have not been made yet, perhaps due to the short time since the filling of the lake.

Other human activities may alter local climatic conditions in the Venezuelan Guayana. The planned diversion of part of the Río Caura into the Río Paragua is one example. Depending on the size of the river shunt, the floodplains of the lower Río Caura valley may dry out considerably, which would strongly reduce the evapotranspiration rate of their forests and alter the associated micro- and mesoclimates. A second example is the large-scale deforestation in the Serranía de Imataca and in the adjacent Yuruari basin. Reduction in the evapotranspiration rate of the original forest cover may lead to locally drier climatic conditions, especially in the rain shadow zones west of the Imataca range. A third impact on local climatic conditions is the dense layers of smoke caused by vegetation fires during the dry season, especially in northern Amazonas and southeastern Bolívar states. This yearly occurrence could create a local greenhouse effect during the hottest time of the year, leading to even hotter conditions and more intensive damage by the fires.

Estimates of global climate changes in the last half of the 20th century point to periods of increasing warmth and drought in tropical South America. If this is the case, certain kinds of medium- to long-term vegetation changes could be expected to occur in the Venezuelan Guayana. First, the lowland and lower talus slope forests will likely be invaded by a higher number of deciduous tree species, especially in rain shadows. A series of semideciduous and deciduous forests would replace much of the current evergreen forest types. Second, in montane forests, such as cloud forests, the abundance and diversity of epiphytes would decrease, yielding to more drought-resistant bromeliads and ferns. The extent of montane forest types would probably be greatly reduced, both altitudinally and horizontally. Third, on the tepui summits, the process of peat formation would be slowed, causing changes in the associated herbaceous and woody plant cover. This process might eventually promote the speciation of more drought-resistant forms on these sites and in the ecotones. In the meadows and high-tepui grasslands, an increase in grass and sedge species would be expected, displacing more moisture-loving taxa such as Rapateaceae and Xyridaceae. Open rock surfaces would likely increase, and the terrestrial bromeliads might become the predominant element due to their higher photosynthetic adaptive response and greater drought resistance.

A persistent increase in drought would probably cause more significant changes in the vegetation than an increase in humidity, since most of the tepui environments are now well adapted to moist climatic conditions. If the temperature regime were to decrease rather than increase, the frequency of frost would increase on the higher tepui summits. Although some plants such as the woolly Asteraceae appear to have frost adaptations, their actual resistance to prolonged frost periods is still unknown.

Epilogue

To maintain the present floristic and ecological richness of the Venezuelan Guayana, it will be necessary to balance the different types of land use required for the economic benefit of the area's human population with a strong conservation strategy that preserves large tracts of land intact. Both the detailed conservation strategy and the means to enforce it must be put into place by the end of the 20th century, or else increasing demographic and economic pressures will make its implementation more difficult or even impossible.

The future of the Venezuelan Guayana lies in the balance between short-term and long-term uses of the area's natural resources. If greater emphasis is placed on stripping the area of its immediate mineral and botanical wealth (ores, precious metals, timber, and other mostly nonrenewable extractive activities), then the long-term value of the Venezuelan Guayana will be severely diminished. If, on the other hand, activities that involve destructive practices and produce only short-term or one-time profits can be discouraged in those areas with high conservation value, the Venezuelan government should be able to expect a sustained and steadily increasing economic benefit as the number of intact forest ecosystems dwindles in other parts of the tropics, and international agencies rally to preserve what little is left. However, since political mandates are short-lived, there is often little incentive for long-term economic planning. Therefore, intense domestic and international lobbying for the conservation of most of the Venezuelan Guayana must be applied on the Venezuelan government and international funding agencies to counteract the tendency for immediate, uncontrolled exploitation.

Fortunately, the existing system of protected areas in the Venezuelan Guayana includes representatives of almost all major landscape and ecological units that occur there. The only important ecosystems not yet included in protected areas are the periodically flooded areas of the western Casiquiare plains with their unique black-water vegetation and some of the interesting forest-savanna mosaics in northern Bolívar state.

To remain one of the most attractive and biologically significant areas of the world, the Venezuelan Guayana must receive active support from the government and local residents to protect its natural resources. By providing important baseline data on the flora and vegetation of this region, the *Flora of the Venezuelan Guayana* should provide an important tool for understanding and managing the biological wealth of the region.

APPENDIX A

Vascular Plant Families of the Venezuelan Guayana

The following information is based on a working checklist for the *Flora of the Venezuelan Guayana*. Minor changes may occur upon publication of individual treatments.

The angiosperm families follow Cronquist (1981), *An Integrated System of Classification of Flowering Plants*. Whereas some authors recognize a single family of legumes with three subfamilies, this system recognizes three families, the Fabaceae, Caesalpiniaceae, and the Mimosaceae. Cronquist also recognizes certain segregate families that may not be recognized in other systems, such as the Hugoniaceae and the Ixonanthaceae (often placed in the Linaceae) and the Costaceae (sometimes included in the Zingiberaceae). On the other hand, certain traditionally recognized families are subsumed into others, such as the Cochlospermaceae into the Bixaceae, and the Martyniaceae into the Pedaliaceae, and the Liliaceae are treated in an inclusive sense to include segregate families such as the Amaryllidaceae. Changes in family circumscriptions since the publication of Cronquist's system are reflected here, such as the publication of the Euphroniaceae and the placement of the genus *Pakaraimaea* in the Monotaceae instead of the Dipterocarpaceae.

	Genera	Species
Pteridophytes		
Aspleniaceae	1	29
Azollaceae	1	1
Blechnaceae	2	16
Cyatheaceae	3	36
Davalliaceae	1	4
Dennstaedtiaceae	9	45
Dicksoniaceae	1	1
Dryopteridaceae	21	115
Gleicheniaceae	3	12
Grammitidaceae	5	55
Hymenophyllaceae	2	73
Hymenophyllopsidaceae	1	8
Isoëtaceae	1	3
Lycopodiaceae	3	26
Marattiaceae	2	6
Marsileaceae	1	1
Metaxyaceae	1	1
Ophioglossaceae	1	2
Osmundaceae	1	2
Parkeriaceae	1	2
Plagiogyriaceae	1	1
Polypodiaceae	7	46
Psilotaceae	1	1
Pteridaceae	11	63
Salviniaceae	1	1
Schizaeaceae	4	18
Selaginellaceae	1	59
Thelypteridaceae	1	32
Vittariaceae	4	12
Spermatophytes		
Acanthaceae	19	65
Agavaceae	1	1
Aizoaceae	1	1
Alismataceae	2	8
Amaranthaceae	11	24
Anacardiaceae	9	15
Annonaceae	17	102

	Genera	Species
[Spermatophytes]		
Apiaceae	2	3
Apocynaceae	34	162
Aquifoliaceae	1	69
Araceae	20	164
Araliaceae	4	64
Arecaceae	24	71
Aristolochiaceae	1	12
Asclepiadaceae	15	91
Asteraceae	91	257
Balanophoraceae	2	2
Basellaceae	1	1
Begoniaceae	1	11
Bignoniaceae	38	132
Bixaceae	2	4
Bombacaceae	9	38
Boraginaceae	5	47
Brassicaceae	1	1
Bromeliaceae	22	273
Brunelliaceae	1	2
Burmanniaceae	7	24
Burseraceae	7	56
Buxaceae	1	1
Cabombaceae	1	3
Cactaceae	12	14
Caesalpiniaceae	30	203
Campanulaceae	5	8
Cannaceae	1	2
Capparaceae	5	17
Caprifoliaceae	2	3
Caricaceae	1	1
Caryocaraceae	2	8
Caryophyllaceae	3	3
Cecropiaceae	3	29
Celastraceae	4	19
Chenopodiaceae	1	1
Chloranthaceae	1	3
Chrysobalanaceae	7	111
Clethraceae	1	2
Clusiaceae	20	128
Combretaceae	6	38
Commelinaceae	10	16
Connaraceae	4	21
Convolvulaceae	14	74
Costaceae	2	7
Crassulaceae	1	2
Cucurbitaceae	15	39
Cunoniaceae	1	12
Cuscutaceae	1	3
Cyclanthaceae	8	23

	Genera	Species
Cyperaceae	36	243
Cyrillaceae	2	2
Dichapetalaceae	2	8
Dilleniaceae	5	28
Dioscoreaceae	1	28
Droseraceae	1	13
Ebenaceae	1	8
Elaeocarpaceae	2	33
Eremolepidaceae	2	3
Ericaceae	17	69
Eriocaulaceae	7	88
Erythroxylaceae	1	29
Euphorbiaceae	57	237
Euphroniaceae	1	3
Fabaceae	66	319
Flacourtiaceae	12	49
Gentianaceae	20	72
Gesneriaceae	16	44
Gnetaceae	1	6
Haemodoraceae	3	3
Haloragaceae	1	1
Heliconiaceae	1	12
Hernandiaceae	2	2
Hippocrateaceae	11	37
Hugoniaceae	2	5
Humiriaceae	5	22
Hydrocharitaceae	2	2
Hydrophyllaceae	1	2
Icacinaceae	7	14
Iridaceae	4	9
Ixonanthaceae	2	8
Juncaceae	1	1
Krameriaceae	1	2
Lacistemataceae	1	1
Lamiaceae	9	30
Lauraceae	14	143
Lecythidaceae	8	35
Lemnaceae	4	4
Lentibulariaceae	2	53
Liliaceae	10	13
Limnocharitaceae	1	1
Lissocarpaceae	1	2
Loasaceae	1	1
Loganiaceae	5	26
Loranthaceae	8	37
Lythraceae	5	26
Magnoliaceae	1	3
Malpighiaceae	22	123
Malvaceae	16	56
Marantaceae	9	59

	Genera	Species
Marcgraviaceae	3	16
Mayacaceae	1	3
Melastomataceae	50	397
Meliaceae	5	23
Mendonciaceae	1	10
Menispermaceae	12	27
Menyanthyaceae	1	1
Mimosaceae	26	163
Molluginaceae	2	2
Monimiaceae	2	13
Monotaceae	1	1
Moraceae	13	57
Moringaceae	1	1
Musaceae	1	2
Myricaceae	1	1
Myristicaceae	4	20
Myrsinaceae	6	55
Myrtaceae	17	178
Najadaceae	1	1
Nyctaginaceae	6	41
Nymphaeaceae	1	7
Ochnaceae	12	107
Olacaceae	9	30
Oleaceae	1	1
Onagraceae	1	17
Opiliaceae	1	2
Orchidaceae	154	698
Oxalidaceae	2	7
Passifloraceae	3	60
Pedaliaceae	2	2
Peridiscaceae	1	1
Phytolaccaceae	6	10
Piperaceae	2	126
Plantaginaceae	1	1
Plumbaginaceae	1	1
Poaceae	94	420
Podocarpaceae	1	9
Podostemaceae	7	25
Polygalaceae	6	53
Polygonaceae	6	30
Pontederiaceae	2	6
Portulacaceae	2	13
Primulaceae	1	1
Proteaceae	3	17
Quiinaceae	4	17
Rafflesiaceae	1	2
Ranunculaceae	1	1
Rapateaceae	13	70
Rhamnaceae	5	19
Rhizophoraceae	3	8
Rosaceae	3	9
Rubiaceae	83	530
Rutaceae	19	59
Sabiaceae	2	3
Saccifoliaceae	1	1
Santalaceae	1	1
Sapindaceae	16	83
Sapotaceae	10	85
Sarraceniaceae	1	5
Scrophulariaceae	18	42
Simaroubaceae	5	14
Smilacaceae	1	11
Solanaceae	16	79
Sphenocleaceae	1	1
Sterculiaceae	9	40
Strelitziaceae	1	1
Styracaceae	1	8
Symplocaceae	1	10
Taccaceae	1	1
Tepuianthaceae	1	4
Tetrameristaceae	1	1
Theaceae	5	48
Theophrastaceae	2	3
Thurniaceae	1	2
Thymelaeaceae	4	10
Tiliaceae	8	28
Trigoniaceae	1	6
Triuridaceae	3	5
Tropaeolaceae	1	1
Turneraceae	2	21
Typhaceae	1	1
Ulmaceae	3	4
Urticaceae	6	13
Valerianaceae	1	1
Velloziaceae	2	2
Verbenaceae	16	71
Violaceae	8	32
Viscaceae	2	48
Vitaceae	1	8
Vochysiaceae	4	59
Winteraceae	1	1
Xyridaceae	5	95
Zamiaceae	1	3
Zingiberaceae	1	8
Zygophyllaceae	3	3

Summary

	Families	Genera	Species
Pteridophytes	29	92	671
Spermatophytes	(201)	(1694)	(8740)
Gymnosperms	3	3	18
Angiosperms	198	1691	8722
Monocotyledons	(26)	(459)	(2385)
Dicotyledons	(172)	(1232)	(6337)
	230	1786	9411

APPENDIX B

Key to the Families of Spermatophytes in the Venezuelan Guayana

by Paul E. Berry and Bruce K. Holst

Because of the difficulties that both nonspecialists and trained botanists can have with keys that are based upon technical, often difficult-to-assess floral characters, this key attempts to make the initial selections easier by using vegetation and habitat characters, along with leaf position and type. Since many families of plants do not conform to easy-to-see characters such as opposite leaves or palmate leaf venation, however, they may appear more than once in different sections of the key. When in doubt, the key has opted for redundancy rather than excessive parsimony. For optimal use of the key, both flowers and fruits are required. In the last, most difficult section of the key that includes shrubs and trees with alternate, simple leaves, fruiting characters take on greater weight.

Since Volume 2 of the flora includes all the treatments of ferns and fern allies, the key to those families appears at the beginning of that volume. The seed plants, on the other hand, will appear in the other volumes of this series, and their key is included here. Ferns and their allies can be distinguished from seed plants because they reproduce by spores rather than by seeds. Because some fern families, especially the aquatic families, do not appear very "fern-like" and might initially be confused with seed plants, they are also included in appropriate sections of this key.

The basic structure of the key divides the families into eight easily recognizable groups. The first group includes the nonphotosynthetic parasites and saprophytes, that is, plants without chlorophyll. Second are the strict aquatics, third are the typical herbaceous monocots, and fourth is a small group of photosynthetic, nonaquatic plants with leaves lacking or rudimentary. Groups five to eight include the remaining herbs and all woody plants; they are separated by leaf type (simple or compound) and position (opposite or alternate).

The authors were assisted in the preparation of the key by Jason Rauscher (Groups A and B), Jason Bradford (Group G), Ken Olsen (Group H), and Denis Kearns (all groups). We are also grateful to the following colleagues who provided useful suggestions to improve different segments of the key: William Alverson, William Anderson, Gerardo Aymard, Douglas Daly, Rodrigo Duno, Christian Feuillet, Aaron Goldberg, Michael Grayum, Nancy Hensold, Otto Huber, Jacquelyn Kallunki, Ronald Liesner, Mark Olson, John Pruski, James Solomon, Charlotte Taylor, Wayt Thomas, Stephen Tillett, Grady Webster, Anna Weitzman, Henk van der Werff, and John Wurdack.

Not all contributors have had the opportunity to carefully review the key for the groups they are contributing to the flora. Consequently, any errors or omissions are entirely the responsibility of the authors, who would appreciate receiving comments and corrections from any reader to incorporate into future versions of the key.

Key to the Major Groups

1. Nonphotosynthetic plants, the vegetative parts with little or no green pigmentation, saprophytic or parasitic; leaves absent or rudimentary (scale-like or bract-like) . . . . [page 225] **GROUP A**

1. Photosynthetic plants, the vegetative parts with at least some green pigmentation, sometimes partly parasitic; leaves evident during all or part of the year (plants evergreen or deciduous) or, if absent, then stems photosynthetic 2

2(1). Strictly aquatic herbs, floating or entirely or partly submerged, the plant generally not able to keep its leaves or stems upright outside of water . . . . [page 226] **GROUP B**

2. Herbaceous or woody plants generally growing on land, but if in water, usually some parts emergent, and the leaves and stems able to support themselves when out of water (includes some aquatic emergents) 3

3(2). Typical herbaceous monocots: plants without wood or wood-like supporting structure; flowers usually 3-merous; leaves alternate (including rosettes); leaf bases sheathing; venation mainly parallel; stipules absent (Note: arborescent and atypical monocots key out in later groups) . . . . [page 229] **GROUP C**

3. Dicots, gymnosperms, and atypical or arborescent monocots: plants woody or herbaceous; flowers 4-, 5-, or polymerous, sometimes 2- or 3-merous or perianth lacking; leaves alternate, opposite, or whorled; leaf bases usually not sheathing; venation mainly reticulate; stipules present or not; leaves sometimes lacking, but then stems photosynthetic 4

4(3). Leaves absent or rudimentary; stems green and photosynthetic. . . . . [page 232] **GROUP D**

4. Leaves present (if plants deciduous, then leaf scars evident) 5

5(4). Leaves compound 6

5. Leaves simple (including unifoliolate leaves) 7

6(5). Leaves opposite or whorled . . . ["compound-opposite," page 233] **GROUP E**

6. Leaves alternate or in basal rosettes. . . . . ["compound-alternate," page 235] **GROUP F**

7(5). Leaves opposite or whorled . . . . . . ["simple-opposite," page 241] **GROUP G**

7. Leaves alternate or in basal rosettes. . . . . ["simple-alternate," page 255] **GROUP H**

GROUP A: Nonphotosynthetic plants, saprophytic or parasitic, with little or no green coloration, leaves absent or rudimentary (scale-like or bract-like). Includes the following families (monocotyledons and dicotyledons):

Balanophoraceae	Gentianaceae	Rafflesiaceae
Burmanniaceae	Lauraceae	Triuridaceae
Cuscutaceae	Lentibulariaceae	Viscaceae
Eremolepidaceae	Orchidaceae	

1. Slender, herbaceous, erect herbs; plants saprophytic or carnivorous 2
1. Vines, shrubs, subshrubs, inconspicuous stem parasites, root parasites with thick, club-shaped inflorescences, or woody epiphytes; plants parasitic 6
2(1). Rudimentary leaves opposite; flowers with a distinct calyx and a tubular, actinomorphic, sympetalous corolla . . . (*Voyria, Voyriella*) **Gentianaceae**
2. Rudimentary leaves alternate or absent; flowers with 1 perianth whorl or, if 2 whorls present, then corolla strongly zygomorphic 3
3(2). Ovary superior, either with numerous free carpels or with 2 carpels in a 1-locular ovary with free-central placentation 4
3. Ovary inferior, with 1–3 connate carpels 5
4(3). Perianth of 2 distinct whorls (calyx and corolla), strongly zygomorphic; root-like system with small bladder-like traps for capturing microscopic organisms; carpels 2, connate; flowers bisexual . . . . . **Lentibulariaceae**
4. Perianth of 1 whorl, actinomorphic; root system without bladder-like traps; carpels numerous, free; flowers unisexual . . . . **Triuridaceae**
5(3). Flowers actinomorphic or nearly so; stamens 3 or 6, adnate to the perianth; pollen grains not cohering into pollinia . . . . **Burmanniaceae**
5. Flowers zygomorphic; fertile stamen 1, part of a specialized column with the stigma; pollen grains cohering into pollinia . . . . (*Uleiorchis, Wullschlaegelia*) **Orchidaceae**
6(1). Root parasites with subterranean, tuber-like structures; inflorescence emergent, thick-fleshy, club-like . . . . **Balanophoraceae**
6. Plants not as above, either herbaceous vines or stem parasites. 7
7(6). Small, inconspicuous, rootless parasites, imbedded in the stems of a few genera such as *Casearia* (Flacourtiaceae) and some Mimosaceae. . . . . **Rafflesiaceae**
7. Slender herbaceous vines or woody hemiparasites 8
8(7). Plants herbaceous, slender-stemmed vines; stems usually dull yellow or orange; flowers bisexual 9
8. Plants small, branching, and woody; stems usually greenish; flowers unisexual. 10
9(8). Inflorescence glomerular or cymose; perianth of 2 whorls (sepals and petals); petals 5, connate; stamens 5; fruit a capsule . . . . . . **Cuscutaceae**
9. Inflorescence racemose; perianth of 1 whorl; tepals 6, free; stamens 9; fruit a berry . . . . (*Cassytha*) **Lauraceae**
10(8). Leaves (if present) and branches alternate (but opposite in juveniles); anthers 4-locular . . . . (*Eubrachion*) **Eremolepidaceae**
10. Leaves (if present) and branches opposite; anthers 1- or 2-locular . . . . **Viscaceae**

GROUP B: Strictly aquatic herbs, floating or mostly submerged, unable to support themselves outside of water. Includes the following families (monocotyledons, dicotyledons, and pteridophytes):

Alismataceae	Isoëtaceae (fern ally)	Nymphaeaceae
Amaranthaceae	Lemnaceae	Onagraceae
Apiaceae	Lentibulariaceae	Parkeriaceae (fern)
Araceae	Limnocharitaceae	Poaceae
Azollaceae (fern)	Marsiliaceae (fern)	Podostemaceae
Cabombaceae	Mayacaceae	Pontederiaceae
Cyperaceae	Menyanthaceae	Salviniaceae (fern)
Eriocaulaceae	Mimosaceae	Scrophulariaceae
Hydrocharitaceae	Najadaceae	Xyridaceae

1. Plants free-floating, not rooted to a substrate, in still or slow-moving water 2

1. Plants rooted to a substrate (sometimes part of a floating mat, but then usually rooted in a matrix with other plants), in still, slow-moving, or fast-moving water 12

2(1). Plants carnivorous, with small, bladder-like traps borne on modified, submerged, highly dissected leaves; peduncle with a whorl of spongy leaf-like floats .... (*Utricularia benjaminiana, U. breviscapa*) **Lentibulariaceae**

2. Plants not carnivorous, leaves and peduncles not as above 3

3(2). Individual leaf blades or segments of plant < 2 cm long 4

3. Individual leaf blades or segments of plant > 2 cm long 7

4(3). Leaves rhombic-ovate, with an acute base, densely arranged in symmetrical rosettes; petioles 0.5–10 cm long; flowers conspicuous; petals yellow .... (*Ludwigia sedoides*) **Onagraceae**

4. Leaves not shaped as above, neither petiolate nor arranged in rosettes; flowers inconspicuous or plants spore-producing 5

5(4). Leaves and stems not differentiated, each segment 1–6 ¥ 0.5–5 mm .... **Lemnaceae**

5. Leaves and stems differentiated, the leaves evident and stems usually reduced 6

6(5). Leaves 0.5–2 mm long, ovate, without evident veins, papillate but without hairs on upper surface .... (see Pteridophytes) **Azollaceae**

6. Leaves mostly 5–20 mm long, orbicular to ovate, with conspicuous hairs and evident veins on upper surface ...(see Pteridophytes) **Salviniaceae**

7(3). Leaves not lobed, simple 8

7. Leaves pinnately lobed or compound 10

8(7). Leaves opposite, spaced along prostrate or erect stems; stems becoming pneumatophore-like .... (*Alternanthera philoxeroides*) **Amaranthaceae**

8. Leaves in basal rosettes; stems not evident 9

9(8). Leaves sessile, obtriangular, the apex truncate to rounded; plant resembling a floating head of lettuce; inflorescence small and inconspicuous, composed of a spadix surrounded by a whitish or greenish spathe .... (*Pistia*) **Araceae**

9. Leaves with long swollen petioles, the blade orbicular and thick-fleshy; inflorescence paniculate, erect; flowers showy, lavender .... (*Eichhornia*) **Pontederiaceae**
10(7). Leaves 2-pinnate, sensitive to the touch; flowers borne in axillary heads, with conspicuous petaloid filaments... (*Neptunia oleracea*) **Mimosaceae**
10. Leaves clover-like and 4-parted or pinnately lobed or divided, not sensitive to touch; plants without flowers (floating ferns)............... 11
11(10). Leaves arising at nodes along long creeping stems, clover-like, 4-parted, each part < 5 cm long; petioles filiform............................ .... (see Pteridophytes) **Marsiliaceae**
11. Leaves in rosettes, pinnately lobed or divided, the blades > 10 cm long, petioles slightly succulent in rosettes; new plants arising from proliferous buds on the leaves .... (see Pteridophytes) **Parkeriaceae**
12(1). Most leaves floating or emergent, submerged leaves usually not present 13
12. Most or all leaves completely submerged.......................... 23
13(12). Leaf blades simple, completely floating on the surface of the water (no part emergent); petioles elongate 14
13. Leaf blades either 4-parted with long petioles, or leaves emergent and without elongate petioles 19
14(13). Leaves peltate, the margins crenate; inflorescences simple axillary umbel .... (*Hydrocotyle umbellata*) **Apiaceae**
14. Leaves not peltate, the margins entire, dentate, or crenulate; inflorescences various... 15
15(14). Leaf venation palmate (primary veins not parallel to main vein) 16
15. Leaf venation parallel (the primary veins parallel to main vein except for diverging basal veins in *Sagittaria guyanensis*, Alismataceae) 17
16(15). Petals 5, margins fimbriate; inflorescences umbel-like, cymose clusters emerging from petiole near base of leaf; flowers small .. **Menyanthaceae**
16. Petals 8, margins entire; inflorescences of solitary flowers, emerging from the base of the plant; flowers large .... **Nympheaceae**
17(15). Leaf blades sagittate, basal veins not parallel to main vein .... (*Sagittaria guyanensis*) **Alismataceae**
17. Leaf blades orbicular, veins parallel or curved-convergent 18
18(17). Flowers borne in an umbel-like inflorescence on a long scape, bisexual; plants with milky juice .... **Limnocharitaceae**
18. Flowers solitary, unisexual (plants monoecious); plants without milky juice .... (*Limnobium*) **Hydrocharitaceae**
19(13). Leaves clover-like, 4-parted, with filiform petioles arising along long creeping stems .... (see Pteridophytes) **Marsiliaceae**
19. Leaves not 4-parted nor with filiform petioles 20
20(19). Leaves with parallel (or curved-convergent) venation; flowers either 3-merous or without an obvious perianth 21
20. Leaves with reticulate venation; flowers 4- or 5-merous, with a conspicuous perianth.. 22
21(20). Leaves orbicular, with long, swollen petioles; corolla showy .... **Pontederiaceae**
21. Leaves linear, without swollen petioles; flowers extremely reduced, borne

in complex spikelets grouped into spike-like or panicle-like inflorescences (grasses) . . . . **Poaceae**

22(20). Leaves opposite; petals connate and forming a tube with stamens attached to it, white or bluish; ovary superior . . . . **Scrophulariaceae**

22. Leaves partly or totally alternate; petals free, yellow or rarely white; ovary inferior . . . . **Onagraceae**

23(12). Leaves basal; stem short, usually unbranched 24

23. Leaves borne on elongate, sometimes branching stems 29

24(23). Leaves usually strongly divided or lobed; plants growing on rocks in fast-flowing streams and rivers (resembling algae), flowering when rivers drop in the dry season and the plant or inflorescence becomes exposed . . . . **Podostemaceae**

24. Leaves not divided nor lobed; plants not resembling algae. 25

25(24). Leaves petiolate; blades linear to lanceolate 26

25. Leaves sessile; blades filiform or linear 27

26(25). Petioles usually more than twice as long as the leaf blades; flowers conspicuous, usually in racemes or panicles; perianth evident; pistils 6–many, free . . . . (*Echinodorus*) **Alismataceae**

26. Petioles shorter than leaf blades; flowers inconspicuous, borne on a spadix surround by a spathe; perianth lacking, pistil 1 . . . . . (*Jasarum*) **Araceae**

27(25). Leaves linear and quill-like, tightly clustered, leaf base swollen with a spore-bearing structure . . . . (see Pteridophytes) **Isoëtaceae**

27. Leaves not quill-like, without swollen leaf bases; flowering plants with inflorescence borne on an elongate peduncle 28

28(27). Leaf blades filiform, flaccid, equitant, distichous; leaf sheaths with margins scarious at base and prominent ligules; inflorescences imbricate-bracted spikes; flowers bisexual; petals conspicuous on live plants, yellow . . . . (*Xyris aquatica*) **Xyridaceae**

28. Leaf blades filiform and densely clustered or linear; leaf sheaths poorly defined and without scarious margins or ligules; inflorescences involucrate heads; flowers unisexual, very small; petals inconspicuous, never yellow . . . . **Eriocaulaceae**

29(23). Plants with specialized leaves forming small bladder-like traps, carnivorous . . . . (*Utricularia*) **Lentibulariaceae**

29. Plants without bladder-like traps, not carnivorous 30

30(29). Leaf margins dentate. 31

30. Leaf margins entire or leaves absent 32

31(30). Leaves with a distinct sheath; flowers unisexual, sessile or short-pedunculate, without an evident perianth; fruit 1-seeded . . . . **Najadaceae**

31. Leaves without a sheath; flowers usually bisexual, long-pedunculate, sepals 3, petals 3; fruit 3–8-seeded . . . . (*Elodea*) **Hydrocharitaceae**

32(30). Leaves entire or apex bifid, sessile 33

32. Leaves (or stem, if without leaves) dissected and/or petiolate; stem often highly branched 34

33(32). Individual flowers inconspicuous, densely grouped in involucrate heads; leaves entire . . . . **Eriocaulaceae**

33. Individual flowers small but readily visible, solitary; leaves often bifid at apex . . . . **Mayacaceae**

34(32). True leaves absent; stem finely branched .... (*Egleria*, *Websteria*) **Cyperaceae**
34. True leaves present 35
35(34). Leaves finely pinnately dissected; petals connate and forming a tube .... (*Benjaminia*) **Scrophulariaceae**
35. Leaves of 2 types, the submerged ones palmately dissected into 3–7 dichotomously or trichotomously branched, linear segments, the petioles to 4 cm long, the floating leaves entire (usually present when flowering); petals free .... **Cabombaceae**

GROUP C: Typical herbaceous monocots: plants without wood or wood-like supporting structure; flowers usually 3-merous; leaves alternate (including rosettes); leaf bases sheathing; veins mainly parallel; stipules absent (Note: arborescent and atypical monocots key out in later groups). Includes the following families (all monocotyledons):

Agavaceae
Alismataceae
Araceae
Bromeliaceae
Cannaceae
Commelinaceae
Costaceae
Cyclanthaceae
Cyperaceae
Dioscoreaceae
Eriocaulaceae
Haemodoraceae
Heliconiaceae
Iridaceae
Juncaceae
Liliaceae
Marantaceae
Orchidaceae
Poaceae
Rapateaceae
Smilacaceae
Strelitziaceae
Taccaceae
Thurniaceae
Typhaceae
Velloziaceae
Xyridaceae
Zingiberaceae

1. Carpels free, 6–many; plants emergent aquatics; milky sap often present in the leaves or stems .... (*Echinodorus*, *Sagittaria*) **Alismataceae**
1. Carpels connate, 2–many or solitary; plant habit various 2
2(1). Flowers numerous, crowded, small, aggregated on a spadix and subtended by 1 or more spathes 3
2. Flowers few to numerous, small to large and showy, never aggregated on a spadix and without subtending spathes 4
3(2). Leaves often plicate and mostly bifid or 2-parted, occasionally entire; venation parallel; spadix subtended by 2–several large, deciduous spathes .... **Cyclanthaceae**
3. Leaves simple or compound but not bifid, 2-parted, or plicate; venation mostly reticulate; spadix subtended by a single, usually persistent spathe .... **Araceae**
4(2). Leaves broad, ± reticulate-veined, without a basal sheath; petiole well-defined; flowers actinomorphic 5
4. Leaves mostly narrow and parallel-veined, with a basal sheath (sheath absent if leaf emerges from the top of a pseudobulb); petiole usually absent, occasionally present; or if leaves reticulate-veined and petiolate, then flowers zygomorphic....................................... 7
5(4). Stemless herbs; primary venation of leaves palmate or pinnate, the veins

not converging; flowers bisexual, borne in an umbel-like inflorescence on top of a scape, with long, filiform, drooping bracts; tepals conspicuous, dark-colored . . . . **Taccaceae**

5. Vines; primary venation of leaves palmate, with 3–13 main converging veins and reticulate veinlets of higher orders; flowers unisexual, borne in axillary inflorescences without long bracts; perianth rather inconspicuous. 6

6(5). Petioles bearing a pair of tendrils; flowers borne in umbels; ovary superior . . . . **Smilacaceae**

6. Petioles without tendrils; flowers borne in racemes, spikes, or panicles; ovary inferior . . . . **Dioscoreaceae**

7(4). Flowers greatly reduced, the perianth inconspicuous and often chaffy; ovary superior 8

7. Flowers generally well developed, the perianth (at least some parts) usually petaloid; ovary superior or inferior 13

8(7). Inflorescences of 1 or more dense, often whitish, hemispheric to globose, involucrate heads on ribbed peduncles, each subtended by a tubular sheathing leaf; flowers almost always unisexual . . . . **Eriocaulaceae**

8. Inflorescences not as above; flowers unisexual or bisexual 9

9(8). Inflorescences dense, either a thick cylindric scapose spike, or a terminal globose head subtended by spreading leafy bracts; plants colonial and emergent from shallow water 10

9. Inflorescences not as above; plants either emergent from water or more commonly terrestrial, colonial or not 11

10(9). Inflorescences dense, elongate, cylindric, scapose spikes; scapes > 1 m tall; plants usually in still water . . . . **Typhaceae**

10. Inflorescences dense, terminal, globose heads subtended by spreading, leafy bracts; plants usually in or along moving water (Note: some *Scleria*, Cyperaceae, may also key out here) . . . . **Thurniaceae**

11(9). Perianth biseriate, chaffy, small but evident; flowers not borne in spikes or spikelets; fruits capsular, 1–3-locular, ovules 3–many . . . . . . **Juncaceae**

11. Perianth not as above; flowers sessile, borne in the axis of chaffy or scale-like bracts (spikes or spikelets); fruits indehiscent, 1-locular, ovule 1 12

12(11). Stems usually solid and triangular, rarely cylindrical and hollow; leaf sheath usually closed; seed coat free from the pericarp; flowers spirally or less often distichously arranged on the axis of the spike or spikelet, each one subtended by a single scale . . . . **Cyperaceae**

12. Stems usually cylindrical, never triangular, hollow to occasionally solid; leaf sheath usually open; seed coat usually adnate to the pericarp; flowers distichously arranged on the axis of the spikelet, each one subtended by a pair of scales . . . . **Poaceae**

13(7). Ovary superior 14

13. Ovary inferior 19

14(13). Leaves usually spread along an evident stem, often somewhat succulent, the sheath closed around the stem and the nodes often swollen . . . . **Commelinaceae**

14. Leaves all basal or grouped in a generally tight rosette, generally not suc-

culent, the sheath usually not closed around the stem 15
15(14). Stamens 3; leaves equitant or forming rosettes 16
15. Stamens 6; leaves not equitant, frequently in rosettes 17
16(15). Inflorescences branched; flowers zygomorphic, the inner and outer tepals not strongly differentiated . . . . **Haemodoraceae**
16. Inflorescences cylindric heads or dense spikes, usually scapose; flowers actinomorphic, the perianth strongly differentiated into sepals and petals . . . . **Xyridaceae**
17(15). Tepals generally all petaloid, not strongly differentiated; leaves generally lasting less than 1 year, growing from an underground rhizome, bulb, or corm . . . . **Liliaceae**
17. Perianth strongly differentiated into sepals and petals; leaves generally persisting from year to year, not growing from an underground bulb or corm, but sometimes from a stout, partly emergent rhizome 18
18(17). Anthers dehiscing by apical pores; leaves usually distichous, the lamina base oblique; leaf sheaths folded; inflorescence a multiflowered head or spike (sometimes 1 or a few flowers) borne at the summit of a scape, composed of spikelets with a single terminal flower subtended by a series of imbricate bractlets (plants less often with spirally arranged leaves with nonfolded sheaths and without an evident rhizome) . . . . . . **Rapateaceae**
18. Anthers dehiscing by longitudinal slits; leaves in short-stemmed rosettes, the lamina base even; leaf sheaths not folded; inflorescences various, but not as above . . . . **Bromeliaceae**
19(13). Fertile stamens 1, 2, or 5; flowers strongly zygomorphic or irregular . . . 20
19. Fertile stamens 3, 6, or rarely more; flowers zygomorphic or actinomorphic 26
20(19). Leaves generally lacking prominent midribs (but see *Epistephium*); venation, when visible, parallel; stamens 1 or 2, adnate to the style in a column; pollen aggregated into pollinia or distinct sticky masses; fruits 1(3)-locular; placentation parietal; seeds numerous, minute. . . . . **Orchidaceae**
20. Leaves with prominent midribs; venation pinnate; fertile stamens 1 or 5, not adnate to the style; pollen free; fruits usually 3-locular; placentation axile; seeds neither minute nor very numerous 21
21(20). Fertile stamens 5; leaves distichous 22
21. Fertile stamen 1; leaves distichous or spirally arranged. 23
22(21). Plants small to large herbs; fruit separating into 2 or 3 generally blue, 1-seeded drupes; seeds without arils . . . . **Heliconiaceae**
22. Plants tall, woody-stemmed; fruit a loculicidal capsule; seeds numerous, arillate . . . . **Strelitziaceae**
23(21). Fertile stamen symmetrical, not petaloid, with a complete, 2-chambered anther; flowers zygomorphic; sepals connate at base 24
23. Fertile stamen asymmetrical, petaloid, with a 1-chambered anther; flowers irregular; sepals free 25
24(23). Leaves distichous; aromatic oils present; staminodes 2, small or petaloid . . . . **Zingiberaceae**
24. Leaves spirally arranged; aromatic oils absent; staminodes absent . . . . **Costaceae**

25(23). Leaves ± distichous; petiole with an apical pulvinus; ovules 1 per carpel; seeds arillate .... **Marantaceae**
25. Leaves spirally arranged; petiole without a pulvinus; ovules several per carpel; seeds without arils .... **Cannaceae**
26(19). Stamens 3; perianth of 6 petaloid segments; leaves usually equitant..... .... **Iridaceae**
26. Stamens 6 or rarely more; perianth either as above or sepals and petals strongly differentiated; leaves not or rarely equitant 27
27(26). Flowers solitary, long-pedicellate, borne in leaf axils near branch tips; plants perennial, with evident stems, sometimes shrubby; leaves firm, leaf sheaths or old leaves persisting on stem .... **Velloziaceae**
27. Flowers generally grouped in inflorescences, if solitary, then not long-pedicellate and plant not as above 28
28(27). Perianth usually well differentiated into petals and sepals; trichomes multicellular, scale-like or hair-like; plants terrestrial or epiphytic .... **Bromeliaceae**
28. Perianth with 1 or 2 whorls of petaloid tepals; trichomes lacking or, if present, never scale-like; plants terrestrial 29
29(28). Leaves fleshy and thick, with an apical spine and recurved spines on the margins; inflorescence a panicle several meters tall .... **Agavaceae**
29. Leaves membranous to coriaceous, lacking spines; inflorescence umbellate, racemose or, if paniculate, then < 1 m tall .... **Liliaceae**

GROUP D: Leaves absent or rudimentary; stems green and photosynthetic; plants terrestrial, epiphytic, or hemiparasitic. Includes the following families (all dicotyledons):

Cactaceae	Lentibulariaceae	Santalaceae
Lauraceae	Polygalaceae	

1. Plants herbaceous, filamentous vines, hemiparasitic .... (*Cassytha*) **Lauraceae**
1. Plants trees, shrubs, or herbs, erect or scandent, terrestrial or epiphytic 2
2(1). Plants usually succulent, terrestrial, climbing, or epiphytic; stems flattened, round in cross-section, or ribbed, often with clusters (areoles) of spines; flowers with many perianth segments and stamens, usually quite showy and > 10 mm long; ovary inferior; fruits fleshy .. **Cactaceae**
2. Plants not succulent, terrestrial; stem generally round in cross-section, without spines; flowers with 10 or fewer perianth segments and stamens, generally < 10 mm long; ovary superior or inferior; fruits dry .. 3
3(2). Flowers actinomorphic; perianth with 1 whorl, tepals 5, connate and forming a short tube; fruit indehiscent, nut-like, often 10-ribbed, the perianth often persistent .... **Santalaceae**
3. Flowers zygomorphic; perianth with 2 whorls (calyx and corolla); fruit capsular... 4
4(3). Calyx 2- or 5-lobed; corolla bilabiate, lower lip usually conspicuously

spurred, 2–5-lobed; stamens 2; fruit a 1-locular, many-seeded, globose capsule; carnivorous herbs of wet or damp places, the underground or submerged portions bearing small bladder-like traps or forked, tubular traps .... **Lentibulariaceae**

4. Calyx of 5 sepals, the lower 2 connate and the lateral 2 larger and petaloid; corolla not bilabiate or spurred, with 5 petals, typically adnate to the 5 stamens and forming a common tube; fruit a 2-locular, 2-seeded, flattened capsule; autotrophic herbs of wet to dry places, without traps ... . (*Polygala*) **Polygalaceae**

GROUP E: Leaves compound, opposite or whorled. Includes the following families (all dicotyledons):

Asteraceae	Crassulaceae	Ranunculaceae
Bignoniaceae	Cunoniaceae	Rutaceae
Brunelliaceae	Fabaceae	Verbenaceae
Caprifoliaceae	Mimosaceae	Zygophyllaceae
Caryocaraceae	Quiinaceae	

1. Plants herbaceous or suffruticose 2
1. Plants woody 5
2(1). Leaves ternately compound and glandular-punctate; inflorescence terminal; flowers white .... (*Ertela*) **Rutaceae**
2. Leaves pinnate or, if ternate, then not glandular-punctate; inflorescence, if terminal, then not with white flowers 3
3(2). Plants procumbent; leaves even-pinnate; petals free; fruit separating into 5–12 mericarps .... (*Kallstroemia, Tribulus*) **Zygophyllaceae**
3. Plants erect; leaves odd-pinnate or ternate; petals connate; fruit not as above 4
4(3). Stem and leaves succulent; leaflets strongly crenate; inflorescences paniculate; carpels 4, free; ovary superior .... **Crassulaceae**
4. Stem and leaves not succulent; leaflets crenate; inflorescences heads subtended by bracts; carpels 2, united; ovary inferior.................. .... (*Bidens, Cosmos*) **Asteraceae**
5(1). Vines.. 6
5. Trees or shrubs 7
6(5). Petals connate, zygomorphic; fruit capsular; seeds winged or unappendaged; carpels united .... **Bignoniaceae**
6. Petals free, actinomorphic; fruit an achene with an elongate, plumose style; carpels free .... **Ranunculaceae**
7(5). Leaves ternately or palmately compound 8
7. Leaves pinnately compound... 12
8(7). Leaves glandular-punctate and/or pellucid-punctate .... (*Raputia, Raveniopsis*) **Rutaceae**
8. Leaves neither glandular-punctate nor pellucid-punctate 9
9(8). Corolla actinomorphic; petals free; stamens 8, 10, or 60–750, the filaments free or connate basally 10

9. Corolla zygomorphic; petals connate; stamens 4, adnate to the petals . . 11
10(9). Stamens 8 or 10, the filaments free; fruit a bifid capsule; inflorescence a many-flowered raceme or panicle; in montane areas over 1000 m elevation . . . . **Cunoniaceae**
10. Stamens 60–750, the filaments connate basally and deciduous with the petals; fruit a drupe; inflorescence a few- to many-flowered, short, terminal raceme or corymb; mostly in lowland areas, but a few species in montane areas over 1000 m elevation . . . . . . (*Caryocar*) **Caryocaraceae**
11(9). Flowers 5–15(–30) mm long; corolla tube ± cylindrical, usually blue, violet, purple, or rarely white, sometimes with a yellow lip; fruits fleshy, 4-celled drupes; trichomes, when present, simple . . . . (*Vitex*) **Verbenaceae**
11. Flowers usually larger than above; corolla tube usually more open or compressed, usually yellow, pink, or white, rarely bluish; fruit capsular; trichomes usually stellate or of lepidote scales . . . . (*Godmania, Tabebuia*) **Bignoniaceae**
12(7). Corolla ± zygomorphic; petals connate and forming a tube; stamens 4 or 5, adnate to the corolla 13
12. Corolla actinomorphic or zygomorphic; petals free (if connate, then only so at base, and corolla actinomorphic); stamens > 5, generally not adnate to the corolla 14
13(12). Corolla only slightly zygomorphic, < 10 mm long, white; stamens 5; fruit a berry; inflorescences compound, umbellate cymes that emerge above the leaves . . . . (*Sambucus*) **Caprifoliaceae**
13. Corolla strongly zygomorphic, usually much larger than 10 mm long, usually in colors other than white; stamens 4; fruit a capsule; inflorescences various but not as above . . . . **Bignoniaceae**
14(12). Leaves 2-pinnate; inflorescence a head . . . . (*Parkia*) **Mimosaceae**
14. Leaves pinnate; inflorescence a raceme, cyme, panicle, or fascicle 15
15(14). Leaves even-pinnate; leaflets in 2 or 3 opposite pairs; flowers blue or purple, in fascicles near stem apices . . . . (*Guaiacum*) **Zygophyllaceae**
15. Leaves odd-pinnate or if seemingly even-pinnate, the leaflets alternate on the rachis; flowers without the above combination of characters 16
16(15). Flowers zygomorphic, medium-sized; petals yellow-orange or purple-lilac; fruit a flattened, 1-seeded legume; leaflets with a pulvinulus . . . . (*Platycarpum, Platymiscium, Taralea*) **Fabaceae**
16. Flowers actinomorphic, small, petals yellowish, whitish, or absent; fruit a berry, capsule, or follicle; leaflets without a pulvinulus 17
17(16). Leaflets 9–15, glabrous or finely pubescent; secondary veins > 15 per side, straight and parallel; lowland areas . . . (*Froesia, Touroulia*) **Quiinaceae**
17. Leaflets 3–many, glabrous to densely pubescent, secondary veins < 15 per side and/or not notably straight and parallel; montane areas over 1000 m 18
18(17). Flowers unisexual; petals absent; inflorescences large, dichotomously branched, cymose; fruit of (2–)5 follicles; leaflets > 8 cm long; rachis not winged . . . . **Brunelliaceae**
18. Flowers bisexual; petals small, whitish; inflorescences racemes or small, few-branched cymes; fruits bifid capsules; leaflets usually < 8 cm long; rachis often winged . . . . **Cunoniaceae**

GROUP F: Leaves compound, alternate. Includes the following families (monocotyledons, dicotyledons, and gymnosperms):

Anacardiaceae	Cecropiaceae	Passifloraceae
Araceae	Connaraceae	Proteaceae
Araliaceae	Convolvulaceae	Rosaceae
Arecaceae	Cucurbitaceae	Rutaceae
Bixaceae	Dioscoreaceae	Sabiaceae
Bombacaceae	Euphorbiaceae	Sapindaceae
Brassicaceae	Fabaceae	Simaroubaceae
Burseraceae	Meliaceae	Solanaceae
Caesalpiniaceae	Mimosaceae	Sterculiaceae
Capparaceae	Moringaceae	Vitaceae
Caryocaraceae	Oxalidaceae	Zamiaceae

1. Nonclimbing herbs (sometimes slightly woody at the base) 2
1. Vines (herbaceous or woody), erect or scandent shrubs, or trees 10
2(1). Leaves 1–few per plant, emerging from a corm or an underground rhizome; flowers grouped on a spadix subtended by a spathe. . . . . (*Anaphyllopsis, Dracontium, Xanthosoma*) **Araceae**
2. Leaves generally more numerous, not emerging from a corm; flowers not grouped as above 3
3(2). Leaves 2-pinnate . . . . (*Mimosa, Neptunia*) **Mimosaceae**
3. Leaves 1-pinnate or palmately compound 4
4(3). Leaves pinnately compound 5
4. Leaves palmately compound or 3-foliolate 7
5(4). Flowers zygomorphic; fruit a legume . . . . **Fabaceae**
5. Flowers actinomorphic; fruit a 2-valved or 5-parted capsule 6
6(5). Leaves basal or spread out along the stem; petals 4; stamens 6; fruit a 2-valved capsule with a thin central partition . . . . **Brassicaceae**
6. Leaves clustered at the shoot apex; petals 5; stamens 10; fruit a 5-parted capsule . . . . (*Biophytum*) **Oxalidaceae**
7(4). Leaves congested on short stems, long-petiolate, 3-foliolate, pellucid-punctate; inflorescence axillary, a long-pedunculate dichasium with a central flower and 2 racemose branches . . . . (*Apocaulon*) **Rutaceae**
7. Leaves and inflorescences not as above 8
8(7). Leaves 3-foliolate, each leaflet emarginate and folding along the midvein at night; petals 5, flowers actinomorphic . . . . (*Oxalis*) **Oxalidaceae**
8. Leaves 3- or more-foliolate, if 3-foliolate, then not as above; petals 4 or, if 5, then flowers zygomorphic. 9
9(8). Leaves 3- or usually more-foliolate; petals 4, free; flowers slightly zygomorphic; placentas 2 or more . . . . (*Cleome*) **Capparaceae**
9. Leaves 3-foliolate; petals 5, lower 2 connate and forming a keel; flowers strongly zygomorphic; placentas 1 . . . . **Fabaceae**
10(1). Vines or scandent shrubs 11
10. Self-supporting shrubs or trees 28
11(10). Leaves pinnately compound or biternate (leaflets > 3) 12
11. Leaves palmately or pedately compound (leaflets 3–many-foliolate) . . . 17
12(11). Upper leaflets modified into retrorse hooks that help the plant climb; flowers

borne on a thick rachis; inflorescence enveloped in bud by a spathe . . . . (*Desmoncus*) **Arecaceae**

12. Upper leaflets not modified into retrorse hooks; flowers and inflorescence not as above 13

13(12). Leaves 2-pinnate; stamens exserted and forming the most conspicuous part of the inflorescence; fruit a usually flattened, dehiscent legume . . . . (*Mimosa, Piptadenia*) **Mimosaceae**

13. Leaves 1-pinnate or, less commonly, biternate or 2-pinnate; stamens included and not particularly conspicuous; fruit not as above 14

14(13). Plants usually with axillary tendrils, often with milky sap; flowers zygomorphic; petals with petaloid appendages on the inner surface; stamens 8; fruit of (1–)3 connate carpels . . . . **Sapindaceae**

14. Plants without tendrils or milky sap; flowers actinomorphic or zygomorphic; petals without appendages; stamens 3–5 or 10; fruit a follicle, of 2–5 separate follicles, a legume, or a berry 15

15(14). Fruit a berry; stamens 3–5 . . . . (*Picramnia*) **Simaroubaceae**

15. Fruit a legume or of 1–5 follicles; stamens 10 16

16(15). Flowers strongly zygomorphic; fruit a legume; seeds without arils . . . . **Fabaceae**

16. Flowers actinomorphic or nearly so; fruit a follicle or of 2–5 separate follicles; seeds arillate . . . . **Connaraceae**

17(11). Stems entirely herbaceous; flowers much reduced and grouped on a thick spadix subtended by a spathe . . . . (*Anthurium, Philodendron, Syngonium*) **Araceae**

17. Stems herbaceous or woody; flowers not as above 18

18(17). Scandent shrubs with recurved prickles on stems and petioles; leaflets 3(5), finely serrate . . . . (*Rubus*) **Rosaceae**

18. Vines or lianas without recurved prickles; leaflets ≥ 3, the margins various 19

19(18). Tendrils usually present 20

19. Tendrils absent 23

20(19). Tendrils strictly axillary 21

20. Tendrils lateral to the leaf axil or leaf-opposed 22

21(20). Flowers generally in ± dense inflorescences, small (mostly 5–10 mm diameter); petals 4 or 5, whitish, free, with petaloid appendages on the inner surfaces; stamens 8; tendrils helically coiled; stems often with milky sap and/or with unusual xylem patterns in cross-section . . . . . **Sapindaceae**

21. Flowers generally solitary, large, showy; petals connate at base, with 5 spreading lobes, colors various; stamens 5; tendrils spirally coiled; stems without milky sap or unusual patterns in cross-section . . . . **Passifloraceae**

22(20). Tendrils lateral to the petiole, usually tightly coiled; leaf nodes generally not swollen or jointed; flowers unisexual; inflorescences racemes, panicles, fascicles, or flowers solitary; ovary inferior . . . . **Cucurbitaceae**

22. Tendrils emerging directly opposite the leaves, not tightly coiled; leaf nodes generally swollen or jointed; flowers bisexual; inflorescences corymbose; ovary superior . . . . **Vitaceae**

23(19). Petiolules cylindrically thickened for their whole length; petiole base com-

monly pulvinate; fruit a follicle, of 2–5 separate follicles, or a legume 24
23. Petiolules not as above; petiole base various; fruit of 2 or more connate carpels 25
24(23). Flowers strongly zygomorphic; fruit a legume; seeds without an aril . . . . . . **Fabaceae**
24. Flowers actinomorphic or nearly so; fruit a follicle or of 2–5 separate follicles; seeds arillate . . . . . . **Connaraceae**
25(23). Flowers bisexual; petals 5, connate and usually showy; fruit a berry or a 2-locular capsule 26
25. Flowers unisexual; petals absent or tepals 6 in 2 series; fruit a 3-locular capsule 27
26(25). Sepals overlapping; anthers dehiscing by longitudinal slits, free; fruit a capsule . . . . . . **Convolvulaceae**
26. Sepals not overlapping; anthers pressed together around the style, dehiscing by terminal pores; fruit fleshy . . . . . . (*Solanum*) **Solanaceae**
27(25). Ovary superior; perianth absent or of much reduced sepals; flowers much reduced and borne in a common structure (pseudanthium) subtended by 2 large bracts; capsule 3-lobed; seeds round, 1 per locule . . . . . . (*Dalechampia*) **Euphorbiaceae**
27. Ovary inferior; perianth of 6 tepals in 2 series; flowers borne in spikes, racemes, or panicles; capsule 3-angled or 3-winged; seeds ± flattened, > 1 per locule . . . . . . **Dioscoreaceae**
28(10). Leaves palmately compound, with 3 or more leaflets 29
28. Leaves either 2-foliolate, clearly pinnate with 3 or more leaflets, or 2- or more-pinnate 41
29(28). Leaves large, to many meters long, divided into many segments that usually remain partially joined at the base; flowers grouped on a thick, often branched rachis, the whole structure subtended by a spathe in bud; solitary or cespitose palms with the leaves clustered at the top of the shoot . . . . . . **Arecaceae**
29. Without the above combination of characters (i.e., not palms) 30
30(29). Leaves and usually parts of the flower with glandular-punctate and/or pellucid-punctate dots (generally visible on the lower surface or in some cases when held against a light source) . . . . . . **Rutaceae**
30. Leaves and flowers (except flowers of *Connarus*, Connaraceae) without pellucid-punctate dots 31
31(30). Leaves generally with a broad, sheathing petiole base; flowers small, < 1 cm long, grouped in simple or compound umbels or heads; ovary inferior; stamens 4 or 5 . . . . . . (*Schefflera*) **Araliaceae**
31. Without the above combination of characters 32
32(31). Flowers conspicuous, often large; sepals connate; petals free or connate only at the base; stamens numerous, the filaments conspicuously exserted and connate basally to form a staminal tube that divides into separate stamens or bundles of stamens, anthers 1-locular; fruits capsular; plants usually with stellate or lepidote trichomes . . **Bombacaceae**
32. Without the above combination of characters 33
33(32). Leaflets > 3 34

33. Leaflets 3 36
34(33). Petals large, showy, and yellow; fruit a 1-locular capsule with parietal placentas and hairy seeds . . . . (*Cochlospermum*) **Bixaceae**
34. Petals absent or, if present, small, hooded, and purplish; fruit not as above 35
35(34). Stipules connate, enveloping the terminal bud; flowers strongly reduced, unisexual, borne in spikes or heads; fruit dry or fleshy, < 4 cm long. . . . . (*Cecropia*, *Pourouma*) **Cecropiaceae**
35. Stipules free, not enveloping the terminal bud; flowers not strongly reduced, bisexual, borne in fascicles or cauliflorous; fruit a deeply ridged, hispid pod > 4 cm long . . . . (*Herrania*) **Sterculiaceae**
36(33). Fruit a large, globose, glabrous, coriaceous berry 4–6 cm in diameter; flowers 4-merous; ovary elevated on an elongated stipe. . . . . (*Crataeva*) **Capparaceae**
36. Fruit smaller, not as above; flowers various; ovary not elevated on a stipe 37
37(36). Stipules connate and enveloping the terminal bud; flowers greatly reduced, numerous, and aggregated on spikes or heads . . . . (*Pourouma*) **Cecropiaceae**
37. Stipules, if present, free and not enveloping the terminal bud; flowers not as densely aggregated nor as reduced as above 38
38(37). Flowers strongly zygomorphic, usually red or orange; fruit a legume. . . . . (*Erythrina*) **Fabaceae**
38. Flowers actinomorphic or weakly zygomorphic, usually not reddish; fruit a drupe or capsule. 39
39(38). Flowers 4-merous; petals with petaloid appendages on the inner surface; stamens 8; carpels 2, only 1 developing into a drupaceous fruit . . . . (*Allophyllus*) **Sapindaceae**
39. Flowers 3- or 5-merous; petals absent or if present, then without appendages; stamens 3–many; carpels 3–many; fruits capsular 40
40(39). Flowers bisexual, 5-merous; petals and sepals present; stamens grouped into 5 fascicles; fruit a globose 8–20-locular capsule; plants without latex . . . . (*Anthodiscus*) **Caryocaraceae**
40. Flowers unisexual, 3- or 5-merous, petals absent; stamens not grouped into fascicles; fruit a 3- or 5–20-lobed and -locular capsule, explosively dehiscent; plants with or without milky latex . . . . **Euphorbiaceae**
41(28). Leaves 2- or more-pinnate 42
41. Leaves 1-pinnate 47
42(41). Petioles swollen and ± sheathing at the base; stipules absent; inflorescence terminal and umbellate; leaves 3- or 4-pinnate . . . . (*Sciadodendron*) **Araliaceae**
42. Petioles and inflorescence not as above; leaves 2- or 3-pinnate 43
43(42). Petioles with a basal pulvinus 44
43. Petioles without a pulvinus 45
44(43). Flowers actinomorphic; petals connate at least basally; stamens often > 10 and forming the conspicuous part of the inflorescence. . . **Mimosaceae**
44. Flowers moderately zygomorphic; petals free or only the lower 2 connate; stamens mostly 10, the petals forming the conspicuous part of the inflo-

rescence .... **Caesalpiniaceae**

45(43). Leaflets with an unequal base and deeply serrate margins; inflorescence a terminal panicle; native trees, usually far removed from towns or urban centers .... (*Dilodendron*) **Sapindaceae**

45. Leaflets with a ± equal base and entire or shallowly serrate margins; inflorescences axillary, not strictly terminal; usually cultivated trees near towns and cities, occasionally escaped.......................... 46

46(45). Leaflets serrate; flowers pink, blue, or lilac-colored, actinomorphic; stamens 10, the filaments connate and forming a conspicuous staminal tube; fruit a fleshy drupe .... (*Melia*) **Meliaceae**

46. Leaflets entire; flowers cream-colored, zygomorphic; stamens 5 and free; fruit an elongate, 1-locular, 3-valved capsule .... **Moringaceae**

47(41). Palms or palm-like plants; leaves generally large, long-lived, always clustered at the shoot apex; stem unbranched and either solitary, cespitose (with several stems from a common base), or without an above-ground stem 48

47. Without the above combination of characters (not palms or cycads).... 49

48(47). True palms, with flowers grouped on a spadix subtended by a spathe .... **Arecaceae**

48. Cycads, producing cones or strobili near the shoot apex .... **Zamiaceae**

49(47). Dwarf trees or shrubs < 1 m tall; leaves tightly clustered at the shoot apex .... (*Biophytum*) **Oxalidaceae**

49. Larger shrubs and trees; leaves either clustered or well spaced along the stems 50

50(49). Leaf blades glandular-punctate and/or pellucid-punctate (or if not, as in *Spathelia*, fruit a samara); fruit a (4)5-locular capsule, a few- to several-seeded berry or hesperidium, a drupe, a samara, or of 1–5 mericarps with ventral, loculicidal dehiscence and often with separating, elastic, bony endocarp; seeds 1–several per carpel .... **Rutaceae**

50. Leaf blades not glandular-punctate or pellucid-punctate (large glands may be present on rachis between the leaflets)........................ 51

51(50). Legumes; ovary with 1 carpel (rarely 2 or more free carpels), in fruit commonly dry and dehiscent along both sutures, but sometimes indehiscent or breaking transversely into 1-seeded segments; stamens usually 10, sometimes 9, or fewer or more numerous, usually connate by their filaments to form an open or closed sheath around the ovary; leaves usually stipulate; petiole base pulvinate and whole length of petiolules usually cylindrically thickened................................ 52

51. Nonlegumes; ovary of various kinds, but not forming leguminous fruits described above; stamens usually 3–12, free or sometimes connate and forming a tube; leaves with or without stipules; petiole base and petiolules pulvinate or not 54

52(51). Leaves even-pinnate with glands between each pair of leaflets; flowers actinomorphic; petals connate and forming a tube; stamens numerous and collectively forming the conspicuous part of the inflorescence .... (*Inga*) **Mimosaceae**

52. Leaves without the above combination; flowers slightly or strongly zygomorphic; petals free or only the basal 2 connate, forming the conspicuous

part of the inflorescence; stamens usually 10, 9, or fewer 53

53(52). Flowers usually moderately zygomorphic; corolla not papilionaceous, upper petal borne internally to the lateral petals and usually smaller than them; sepals mostly free .... **Caesalpiniaceae**

53. Flowers usually strongly zygomorphic; corolla papilionaceous, with 3 types of petals, upper petal (the standard) borne externally to the others and generally the largest, the 2 lateral free petals (the wings), and the 2 lower, innermost petals mostly connate to form a keel enfolding the stamens and ovary (the keel); sepals mostly connate and tubular at the base .... **Fabaceae**

54(51). Stipules absent; flowers bisexual, perianth of 4 petaloid sepals, the petals absent; stamens 4, inserted opposite each perianth lobe; ovary 1-locular, with 2 ovules .... **Proteaceae**

54. Stipules present or absent; flowers bisexual or unisexual, without the above combination of perianth and stamens; ovary 1- or more-locular, with 1, 2, or more ovules per locule 55

55(54). Pulvinae present at petiole bases and along entire length of the cylindrically thickened petiolules; leaflets entire; flowers 5-merous; stamens 10 in 2 series; carpels 1–5, free; ovules 2 per carpel; seeds arillate..... .... **Connaraceae**

55. Pulvinae absent or, if present, not on all leaflets or comprising just part of the petiolules (these thickened only at base); otherwise without the above combination of characters............................... 56

56(55). Flowers (4)5-merous; stamens usually 8, occasionally 5–10; petals usually with internal scales or petaloid appendages; annular nectary disk usually present and located outside the stamens .... **Sapindaceae**

56. Flowers various, but only rarely with 8 stamens; petals usually without internal scales or appendages; disk, if present, located inside the stamens.. 57

57(56). Leaflets ca. 30 × 12 cm, thickly coriaceous; petiolule to 8 mm thick with dense ferrugineous scales; flowers rose-red; only 1 species known from flora area in southwestern Amazonas along Brazilian border at ca. 1250 m elevation .... (*Ophiocaryon*) **Sabiaceae**

57. Leaflets and flowers not as above 58

58(57). Terminal leaflets with a pulvinulate petiolule, or the fruit a dehiscent drupe; ovary and fruit with 2 ovules per locule; trunk and branches usually with a translucent to milky resin, smelling strongly of turpentine and drying hard or tacky and often whitish around trunk wounds .... **Burseraceae**

58. Terminal leaflets without a pulvinulate petiolule; fruit with 1 or more ovules per locule; trunk and branches usually not as above 59

59(58). Fruits drupes; inflorescences pendent racemes or panicles .... (*Picramnia*) **Simaroubaceae**

59. Fruits capsules, drupes or schizocarps; inflorescences various 60

60(59). Leaves simple, but branches with numerous small leaflets arranged distichously and resembling a pinnately compound leaf; flowers axillary, solitary or in fascicles, usually pendulous; fruit a 3-locular capsule with 2 seeds per locule .... (*Phyllanthus*) **Euphorbiaceae**

60. Leaves truly pinnately compound; flowers not solitary and axillary nor fruits as above 61
61(60). Stamen filaments 8–12, connivent or connate and forming a tube (except *Cedrela*, with 4–6 stamens adnate to an androgynophore and a thick, lobed columella); bark usually with a sweet smell, without resinous sap; carpels connate; leaves with a terminal leaflet "bud" in *Guarea* . . . . **Meliaceae**
61. Stamen filaments 3–10, free, not connivent; bark either bitter tasting or with resinous sap; carpels connate or free; leaves never with terminal "buds" 62
62(61). Ovary with 2–5 free to connate carpels; styles free or connate; fruit schizocarpic, separating into separate mericarps, or apocarpous and often with flattened drupaceous carpels; stamen filaments usually appendaged; plants never resinous . . . . **Simaroubaceae**
62. Ovary with 1 or 2–5 connate carpels; styles free or connate; fruit a single drupe per floral receptacle, not flattened; stamen filaments never appendaged; plants usually resinous . . . . **Anacardiaceae**

GROUP G: Leaves simple, opposite or whorled (including unifoliolate leaves).

Key to the Subgroups

1. Nonclimbing herbs (mainly terrestrial, but some epiphytic) . . . . [page 241] **SUBGROUP G-1**
1. Shrubs, trees, vines, or lianas 2
2(1). Climbing plants (herbaceous vines or woody lianas) . . . . [page 246] **SUBGROUP G-2**
2. Upright woody plants (trees and shrubs, including nonclimbing woody epiphytes and hemiparasites) . . . . [page 249] **SUBGROUP G-3**

SUBGROUP G-1: Nonclimbing herbs (mainly terrestrial, but some epiphytic), with simple, opposite or whorled leaves. Includes the following families (all dicotyledons):

Acanthaceae	Gesneriaceae	Pedaliaceae
Aizoaceae	Haloragaceae	Piperaceae
Amaranthaceae	Lamiaceae	Polygalaceae
Apocynaceae	Lentibulariaceae	Portulacaceae
Asclepiadaceae	Loasaceae	Primulaceae
Asteraceae	Loganiaceae	Rubiaceae
Caryophyllaceae	Lythraceae	Scrophulariaceae
Clusiaceae	Melastomataceae	Urticaceae
Crassulaceae	Molluginaceae	Verbenaceae
Euphorbiaceae	Nyctaginaceae	Violaceae
Gentianaceae		

1. Leaf margins dentate, serrate, or crenate 2
1. Leaf margins entire or at most undulate 16

2(1). Perianth of 1 whorl, greenish or scarious, or highly reduced or absent . . 3
2. Perianth of 2 distinct whorls; corolla variously colored, not greenish or scarious 5
3(2). Ovary inferior; fruits nut-like with 4–8 well-defined ridges . . . . **Haloragaceae**
3. Ovary superior; fruits achenes or 3-parted capsules 4
4(3). Leaves with linear, fusiform, or spherical cystoliths (small crystalline inclusions best seen with 10× magnification, on dry leaves sometimes resembling appressed hairs); venation palmate; ovary 1-locular; fruit an achene; plants without milky latex . . . . . .(*Boehmeria, Pilea*) **Urticaceae**
4. Leaves without cystoliths; venation palmate or pinnate; ovary 3-locular; fruit a 3-parted capsule, explosively dehiscent and falling apart, not enclosed by the perianth; plants with abundant milky latex . . . . (*Chamaesyce, Euphorbia*) **Euphorbiaceae**
5(2). Petals free 6
5. Petals connate and forming a corolla tube 9
6(5). Leaves palmately veined or pliveined; stamens usually twice the number of petals; anthers dehiscing by terminal pores, connective prolonged below the anther and usually appendaged; placentation axillary . . . . **Melastomataceae**
6. Leaves pinnately veined; stamens 4, 5, or opposite the petals in groups of 4 or 5; anthers dehiscing by longitudinal slits, connective not appendaged; placentation parietal or apical 7
7(6). Flowers zygomorphic; ovary superior; stamens 5; fruit a 3-valved capsule . . . .(*Hybanthus*) **Violaceae**
7. Flowers actinomorphic; ovary inferior; stamens 4 or opposite the petals in groups of 4 or 5; fruit nut-like and ridged or a 4-valved capsule 8
8(7). Flowers bisexual; stamens 16 or 20, opposite the petals in groups of 4 or 5; fruit a 4-valved capsule with barb-tipped trichomes . . . . . . **Loasaceae**
8. Flowers unisexual; stamens 4; fruit nut-like, glabrous, with well-defined ridges . . . . **Haloragaceae**
9(5). Ovary inferior; flowers borne in heads subtended by bracts; fruit an achene usually crowned with the persistent calyx modified into a pappus composed of bristles, awns, or scales; corolla tubular, at least basally so; stamens attached to the corolla, the anthers connate and forming a tube, stigmas with 2 branches . . . . **Asteraceae**
9. Plants without the above combination of characters 10
10(9). Carpels free; calyx long-tubular and somewhat inflated; corolla actinomorphic; leaves thick and fleshy . . . . **Crassulaceae**
10. Carpels connate; calyx not as above; corolla usually zygomorphic; leaves not thick and fleshy or, if so, then plants epiphytic 11
11(10). Indumentum either viscid-villous or of simple and short-stalked trichomes mixed with multicellular, head-shaped, glandular trichomes (use 10× magnification); fruit an oblong capsule or a drupe with a curved beak . . . . **Pedaliaceae**
11. Indumentum of 1 kind of trichome, usually not viscid-villous, or plants glabrous; fruit not as above 12
12(11). Ovary consisting of 4 distinct segments, or deeply lobed to appear so; style usually borne at the base of the ovary; leaves often aromatic; stem cross-

section often square; calyx often 10-veined .... **Lamiaceae**
12. Ovary not deeply 4-lobed; style attached to apex of ovary; leaves generally not aromatic; stem cross-section usually round; calyx usually 5-veined 13
13(12). Ovule 1 per locule; fruits drupes or schizocarps .... **Verbenaceae**
13. Ovules 2 or more per locule; fruits capsules or berries 14
14(13). Anthers connivent in pairs or 4 together; ovary 1-locular; placentation parietal; plants frequently with long, soft hairs; corollas often pouch-like or swollen on 1 side at the base .... **Gesneriaceae**
14. Anthers free; ovary 2-locular; placentation axile; plants glabrous or, if pubescent, then generally without long and soft hairs; corollas usually not pouch-like or swollen on 1 side at the base 15
15(14). Leaves frequently with linear cystoliths (small crystalline inclusions best seen with 10× magnification, on dry leaves resembling appressed hairs); seeds generally flattened and circular, relatively large, usually attached to the placenta by slender, hook-like stalks; fruits explosively dehiscent .... **Acanthacea**e
15. Leaves without cystoliths; seeds generally small and angled or elongated (pyramidal, conical, oblong or linear), not attached to the placenta as above; fruits not explosively dehiscent .... **Scrophulariaceae**
16(1). Petals or tepals scarious, greenish, brownish, thin, and translucent, or absent; individual flowers usually inconspicuous 17
16. Petals or tepals petaloid, white or shades of red, blue, or yellow; individual flowers usually conspicuous.................................. 24
17(16). Inflorescence a dense, rarely lax spike borne opposite the leaves; flowers numerous, minute; perianth absent; leaves usually succulent; petioles sheathing at base and often with wing-like margins; fruits small 1-seeded drupes .... *(Peperomia)* **Piperaceae**
17. Plants without the above combination of characters 18
18(17). Opposing leaves usually unequal in size, bearing long cystoliths (small crystalline inclusions best seen with 10× magnification), most perpendicular to the midrib .... *(Pilea microphylla)* **Urticaceae**
18. Leaves opposite and equal in size or leaves whorled 19
19(18). Flowers unisexual; stamens 1; plants with milky latex; fruit a 3-locular capsule, explosively dehiscent and falling apart, leaving a central column; seeds 1 or 2 per locule.................................. .... *(Chamaesyce, Euphorbia)* **Euphorbiaceae**
19. Flowers bisexual; stamens 2–5; plants without milky latex; fruit not as above 20
20(19). Leaves whorled.. 21
20. Leaves opposite... 22
21(20). Stipules scarious or thin and translucent; plants pubescent with simple hairs .... *(Polycarpaea, Polycarpon)* **Caryophyllaceae**
21. Stipules absent; plants glabrous or pubescent with stellate hairs........ .... **Molluginaceae**
22(20). Leaves palmately veined .... *(Drymaria)* **Caryophyllaceae**
22. Leaves pinnately veined 23
23(22). Leaves sessile; flowers solitary and axillary, present at nearly all leafy nodes .... *(Rotala)* **Lythraceae**

23. Leaves petiolate; flowers borne in capitate or spicate inflorescences, not present at nearly all leafy nodes .... **Amaranthaceae**

24(16). Flowers borne in heads subtended by bracts; ovary inferior; fruit an achene usually crowned with the persistent calyx modified into a pappus composed of bristles, awns, or scales; corolla tubular, at least basally so; stamens attached to the corolla, the anthers connate and forming a tube, stigmas with 2 branches .... **Asteraceae**

24. Plants without the above combination of characters 25

25(24). Leaves palmately veined or pliveined with major veins connected by conspicuous lower order venation, connecting veins with ± equal spacing from each other and perpendicular to the major veins; ovary mostly inferior, or, if superior, surrounded by the hypanthium and appearing inferior; petals free; stamens often with elbowed filaments and appendaged anthers; anthers dehiscing by terminal pores .... **Melastomataceae**

25. Plants without the above combination of characters 26

26(25). Petals free or absent 27

26. Petals connate 37

27(26). Leaves whorled 28

27. Leaves opposite 29

28(27). Flowers zygomorphic; fruit a flattened, 2-valved capsule .... (*Polygala*) **Polygalaceae**

28. Flowers actinomorphic; fruit a spherical or ovoid, 3–5-valved capsule.... .... **Molluginaceae**

29(27). Leaves succulent 30

29. Leaves not succulent 32

30(29). Stems 4-angled; petals 4 .... (*Crenea*) **Lythraceae**

30. Stems round in cross-section or irregularly angled; petals or tepals 5 .. 31

31(30). Petioles of opposing leaves connate at base and surrounding the stem; flowers solitary in axils of leaves; ovary 3-locular; perianth 1-whorled; tepals 5, connate at base; seashore plants .... **Aizoaceae**

31. Petioles of opposing leaves free at base; flowers not solitary or, if so, borne only on stem terminus; ovary 1-locular; perianth 2-whorled; sepals 2, often deciduous; petals usually 5, free; plants not of seashores...... .... **Portulacaceae**

32(29). Flowers zygomorphic 33

32. Flowers actinomorphic 35

33(32). Calyx forming an elongated, striate, zygomorphic tube; petals borne on the edge of the calyx tube; stipules usually present as several small linear appendages at the petiole base .... (*Cuphea*) **Lythraceae**

33. Calyx and petals not as above; stipules absent 34

34(33). Plants densely or sparsely viscid-pubescent; leaf margins undulate; fruit an obovoid, 3–5-angled achene; inflorescence lax .... (*Boerhavia*) **Nyctaginaceae**

34. Plants mostly glabrous or at least not viscid-pubescent; leaf margins entire; fruit a flattened, 2-valved capsule; inflorescence usually a dense spike or panicle of spikes .... (*Polygala*) **Polygalaceae**

35(32). Leaves pellucid-punctate; petals bright yellow ... (*Hypericum*) **Clusiaceae**

35. Leaves not pellucid-punctate; petals white 36
36(35). Leaves palmately veined; sepals parted to the base; flowers pedicellate . . . . (*Drymaria*) **Caryophyllaceae**
36. Leaves pinnately veined; sepals connate; flowers sessile . . . . (*Rotala*) **Lythraceae**
37(26). Flowers zygomorphic 38
37. Flowers actinomorphic 44
38(37). Leaves all basal; corolla not tubular, the lower lobe with a long spur or rarely merely saccate; roots modified to trap small organisms, either tubular or with small bladder-like structures . . . . **Lentibulariaceae**
38. Leaves borne along the stem; corolla tubular; roots not as above 39
39(38). Pubescence of simple trichomes mixed with short-stalked multicellular capitate trichomes; fruit an oblong capsule . . . . (*Sesamum*) **Pedaliaceae**
39. Pubescence of 1 kind of trichome, or plants glabrous; fruit not as above 40
40(39). Ovary consisting of 4 distinct segments, or deeply lobed to appear so; style usually gynobasic (borne at the base of the ovary); leaves often aromatic; stem cross-section often square; calyx often 10-veined . . . . . . **Lamiaceae**
40. Ovary not deeply 4-lobed; style attached to apex of ovary; leaves usually not aromatic; stem cross-section usually round; calyx usually 5-veined 41
41(40). Anthers connivent in pairs or 4 together; plants frequently with long, soft hairs; corollas usually pouch-like or swollen on 1 side near base . . . . **Gesneriaceae**
41. Anthers free or, if paired, then not connivent; plants glabrous or, if pubescent, then generally without long and soft hairs; corollas usually not pouch-like or swollen on 1 side near base 42
42(41). Seeds generally flattened and circular, relatively large, usually attached to the placenta by slender, hook-like stalks; fruits explosively dehiscent, stipitate; leaves frequently with linear cystoliths (small crystalline inclusions best seen with 10× magnification, on dry leaves resembling appressed hairs) . . . . **Acanthaceae**
42. Plants not as above 43
43(42). Interpetiolar line present or petioles connate at base and surrounding the stem; stems 4-angled (seen best in young growth or on peduncles); plants glabrous; leaves palmately veined or pliveined (rarely venation obscured) . . . . **Gentianaceae**
43. Plants not as above (or, if so, then stems weak and sprawling as in *Micranthemum umbrosum* or *Bacopa gratioloides,* which also has glandular-punctate leaves) . . . . **Scrophulariaceae**
44(37). Ovary inferior 45
44. Ovary superior 46
45(44). Interpetiolar stipules absent; opposing leaves connate at base; leaves succulent; seashore plants . . . . **Aizoaceae**
45. Interpetiolar stipules present, sometimes well developed and the leaves appearing whorled; opposing leaves free; leaves not succulent; plants of various habitats . . . . **Rubiaceae**
46(44). Leaves with linear cystoliths (small crystalline inclusions best seen with

10× magnification, on dry leaves resembling appressed hairs) . . . . (*Mirabilis*) **Nyctaginaceae**

46. Leaves without cystoliths 47

47(46). Inflorescence terminal, secund (flowers all oriented on 1 side of the inflorescence), usually helicoid and unbranched . . . . (*Spigelia*) **Loganiaceae**

47. Inflorescence not as above. 48

48(47). Interpetiolar line present or petioles connate at base and surrounding the stem; stems 4-angled (seen best in young growth or on peduncles); plants glabrous (pubescent in *Macrocarpaea rugosa*); leaves palmately veined or pliveined (rarely venation obscure or pinnate: see *Coutoubea* with spicate inflorescences; *Macrocarpaea* with greenish or pale yellow flowers; *Symbolanthus* with strongly overlapping calyx lobes); ovaries with parietal placentation . . . . **Gentianaceae**

48. Plants without the above combination of characters 49

49(48). Corolla yellow; fruit a circumscissile, rounded capsule; plants without milky latex . . . . **Primulaceae**

49. Corolla not yellow; fruit a longitudinally dehiscent, usually elongate capsule; plants usually with milky latex 50

50(49). Corolla without an evident tube; pollen grains aggregated in specialized pollen sacs (pollinia) joined in pairs; stamen filaments and stigma head fused to form a complex pentagonal central columnar structure (the gynostegium) . . . . **Asclepiadaceae**

50. Corolla with an evident tube; pollen grains free; anthers borne on evident filaments; stigma borne on an evident style . . . . **Apocynaceae**

SUBGROUP G-2: Climbing plants (herbaceous vines or woody lianas) with simple, opposite or whorled leaves. Includes the following families (monocotyledons, dicotyledons, and gymnosperms):

Acanthaceae	Dioscoreaceae	Mendonciaceae
Amaranthaceae	Gesneriaceae	Rubiaceae
Apocynaceae	Gnetaceae	Scrophulariaceae
Asclepiadaceae	Hippocrateaceae	Thymeleaceae
Asteraceae	Loasaceae	Trigoniaceae
Bignoniaceae	Loganiaceae	Valerianaceae
Clusiaceae	Malpighiaceae	Verbenaceae
Combretaceae	Melastomataceae	

1. Inflorescences 4-flowered heads subtended by 4 bracts; ovary inferior; fruit an achene crowned by numerous setae . . . . (*Mikania*) **Asteraceae**

1. Inflorescences cymose or racemose, or flowers solitary; ovary superior or inferior; fruit a berry, drupe, capsule, or naked achene 2

2(1). Leaves palmately veined or pliveined. 3

2. Leaves pinnately veined 8

3(2). Tendrils usually present, thick, spiral; fruit a hard-shelled, subglobose or ovoid berry . . . . (*Strychnos*) **Loganiaceae**

3. Tendrils absent; fruit a drupe, capsule, or soft-shelled berry 4

4(3). Calyx tubular; petals connate and showy; ovary superior. 5

4. Calyx not tubular; perianth parts free or, if petals connate, then individual

flowers fairly inconspicuous and ovary inferior 6
5(4). Petiole as long as or longer than the leaf blade, narrowly winged; corolla orange-yellow with a purple throat; fruit a 2-locular capsule .... (*Thunbergia alata*) **Acanthaceae**
5. Petiole much shorter than the leaf blade, not winged; corolla red; fruit a drupe .... (*Clerodendron*) **Verbenaceae**
6(4). Flowers 3-merous, unisexual, borne in axillary racemes; fruit a 3-winged or 3-angled capsule .... **Dioscoreaceae**
6. Flowers 4- or 5-merous, bisexual, borne in terminal cymes, racemes, or panicles; fruit an achene, capsule, or berry........................ 7
7(6). Petals connate and forming a tube; corolla ± spurred at base; stamens fewer than petals; anthers dehiscing longitudinally; fruit an achene........ .... (*Valeriana scandens*) **Valerianaceae**
7. Petals free; stamens twice as many as petals; anthers dehiscing by terminal pores; fruit a capsule or berry................................ .... (*Leandra*, *Phainantha myrteoloides*) **Melastomataceae**
8(2). Leaf margins noticeably serrate 9
8. Leaf margins entire or lobed, generally not serrate.................. 12
9(8). Flowers actinomorphic; petals free, generally < 5 mm long or wide 10
9. Flowers zygomorphic; petals connate, generally showy and > 5 mm long or wide 11
10(9). Woody lianas, often with bent tendril-like lateral branches; ovary and fruit glabrous; ovary superior; flowers with exposed nectary disk; stamens 3, dehiscing by transverse slits; inflorescence axillary .... **Hippocrateaceae**
10. Herbaceous vines, branches never tendril-like; ovary and fruit covered with minutely barb-tipped trichomes; ovary inferior; flowers without a nectary disk; stamens opposite petals in groups of 4 or 5, dehiscing by longitudinal slits; inflorescence terminal .... **Loasaceae**
11(9). Leaf blades nearly sessile, elliptic-cordate, with 4–8 very prominent glandular teeth per side; anthers free; occurring above 2000 m elevation... .... (*Vellosiella*) **Scrophulariaceae**
11. Leaf blades usually petiolate, not cordate, with > 8 teeth per side; anthers usually joined in pairs or 4 together .... **Gesneriaceae**
12(8). Leaf nodes conspicuously thickened and jointed, the continuing stem or lateral branchlet apparently articulate just above the node; plant a dioecious gymnosperm, without true flowers; male reproductive branches resembling a string of beads, female parts developing into oblong-ellipsoid drupe-like "fruits" 3–5 cm long .... **Gnetaceae**
12. Leaf nodes variously continuous or somewhat jointed, but never articulate as above; true flowering plants, flower and fruit types various...... 13
13(12). Petals free or absent 14
13. Petals connate or hypanthium present............................. 19
14(13). Flowers minute, arranged in whitish, head-like clusters in branching inflorescences; floral parts scarious, only tepals and bracteoles present.. .... (*Iresine*, *Pfaffia*) **Amaranthaceae**
14. Individual flowers clearly distinguishable, not in tight, scarious clusters as above; both calyx and corolla present 15
15(14). Styles 3; sepals with (1)2 large glands externally; petals clawed; indument

often sericeous, the hairs 2-branched . . . . **Malpighiaceae**
15. Styles 1 or absent; sepals without glands; petals not clawed; indument, if present, not sericeous nor of 2-branched hairs 16
16(15). Flowers zygomorphic, leaves often white-sericeous on the lower surface; seeds hairy . . . . **Trigoniaceae**
16. Flowers actinomorphic; leaves not white-sericeous on the lower surface; seeds not hairy 17
17(16). Flowers unisexual; stamens few to many; plants with white, yellow, or orange latex . . . . **Clusiaceae**
17. Flowers bisexual; stamens either 3 or else twice as many as petals; plants without latex or latex red 18
18(17). Ovary inferior; fruit 1-locular and 1-seeded; stamens twice as many as the petals, dehiscing longitudinally, nectary when present borne inside the hypanthium tube . . . . (*Combretum*) **Combretaceae**
18. Ovary superior; fruit generally with > 1 seed and locule; stamens usually 3, dehiscing by transverse slits; nectary present as a prominent extrastaminal disk . . . . **Hippocrateaceae**
19(13). Stipules interpetiolar; ovary inferior; corolla actinomorphic . . . **Rubiaceae**
19. Stipules absent or leaf node sometimes with an interpetiolar line; ovary superior or, if inferior, then the corolla zygomorphic 20
20(19). Leaves with the lower order pinnate venation very finely parallel (without evident reticulation); inflorescence terminal, 10–16-flowered, narrowly corymbose, subtended by several pairs of pink bract-like leaves; hypanthium narrow, with 5 calyx lobes alternating with vestigial, hairy petals . . . . (*Lophostoma*) **Thymeleaceae**
20. Leaves generally more evidently reticulate-veined, without very finely parallel lower order venation; inflorescence without pairs of pink bract-like leaves; hypanthium absent, the sepals and petals forming 2 distinct and separate whorls 21
21(20). Plants usually with abundant milky latex; anthers as many as the corolla lobes, usually 5; flowers actinomorphic (sometimes slightly zygomorphic 22
21. Plants without latex; anthers fewer than the corolla lobes, usually in pairs (2 or 4); flowers usually zygomorphic, but sometimes barely so 23
22(21). Corolla without an evident tube; pollen grains aggregated in specialized pollen sacs (pollinia) joined in pairs; stamen filaments and stigma head fused to form a complex pentagonal central columnar structure (the gynostegium) . . . . **Asclepiadaceae**
22. Corolla with an evident tube; pollen grains free; anthers borne on evident filaments; stigma borne on an evident style . . . . **Apocynaceae**
23(21). Flowers subtended by 2 large, spathe-like bracts; fruit a 1-locular drupe with 1 or 2 seeds . . . . **Mendonciaceae**
23. Flowers not subtended by 2 large bracts; fruit various, but if a drupe usually with > 1 locule or seed 24
24(23). Anthers connivent in pairs or 4 together; ovary inferior or superior, 1-locular; fruit a loculicidal capsule or less often a berry; seeds numerous, unwinged . . . . **Gesneriaceae**
24. Anthers not connivent in pairs or 4 together; ovary superior, 2- or more-

locular; fruit indehiscent or, if dehiscent, then with winged seeds . . . 25
25(24). Fruit a 2–4-seeded fleshy drupe or dry and indehiscent with calyx lobes persistent as wings; seeds not winged. . . . . (*Aegiphila*, *Petrea*) **Verbenaceae**
25. Fruit a 2-locular, dehiscent capsule with winged seeds or berry-like with a cupular calyx and many small seeds (*Schlegelia*) . . . . **Bignoniaceae**

SUBGROUP G-3: Upright woody plants (trees and shrubs, including nonclimbing woody epiphytes and hemiparasites) with simple, opposite or whorled leaves. Includes the following families (all dicotyledons):

Acanthaceae	Euphorbiaceae	Oleaceae
Apocynaceae	Gentianaceae	Proteaceae
Asclepiadaceae	Gesneriaceae	Quiinaceae
Asteraceae	Hippocrateaceae	Rhamnaceae
Bignoniaceae	Lamiaceae	Rhizophoraceae
Buxaceae	Lauraceae	Rubiaceae
Caprifoliaceae	Loganiaceae	Rutaceae
Celastraceae	Loranthaceae	Sapotaceae
Chloranthaceae	Lythraceae	Trigoniaceae
Clusiaceae	Malpighiaceae	Verbenaceae
Combretaceae	Melastomataceae	Violaceae
Cunoniaceae	Monimiaceae	Viscaceae
Elaeocarpaceae	Myrtaceae	Vochysiaceae
Ericaceae	Nyctaginaceae	

1. Stipules present, especially on young leaves near the stem apex, often conspicuous, sometimes enveloping the terminal bud or surrounding the stem, sometimes quickly deciduous but then stipular scars usually evident at the leaf nodes 2
1. Stipules absent 20
2(1). Stipules interpetiolar and centered between the opposing petiole bases (sometimes also fused intrapetiolarly and forming a sheath around the leaf node), or reduced to an interpetiolar line 3
2. Stipules lateral to the petiole base, on the petiole and more or less covering the axillary bud (intrapetiolar), or sometimes present as nectariferous glands, but never interpetiolar and centered between adjacent petiole bases 13
3(2). Stipules interpetiolar and centered between the opposing petiole bases; ovary inferior; leaves entire; petals connate . . . . **Rubiaceae**
3. Plants without the above combination of characters 4
4(3). Leaf node surrounded by a sheath formed from the connate petiole bases or connate stipules 5
4. Leaf node not surrounded by a sheath; petioles free; stipules free or reduced to an interpetiolar line. 6
5(4). Leaves serrate; flowers unisexual; petals lacking; ovary inferior . . . . **Chloranthaceae**
5. Leaves entire; flowers bisexual; petals present; ovary superior . . . .

. . . . (*Pagamea*) **Rubiaceae**

6(4). Petals free, 5, the lower 2 forming a keel; fruit a 3- or 4-locular capsule; seeds round and covered with hairs; flowers zygomorphic . . . . **Trigoniaceae**

6. Petals free or connate, if petals 5, then not forming a keel; fruit a berry, nut, few-seeded indehiscent nut-like capsule with a bony endocarp, or a 2–many-locular dehiscent capsule; seeds various; flowers actinomorphic or zygomorphic (some Malpighiaceae) 7

7(6). Petals connate and forming a tube, 4, 5, or 10; stamens as many as the petals 8

7. Petals free, 4–6; stamens mostly twice as many as the petals or more . . . 9

8(7). Fruit a flattened capsule, corollas zygomorphic; ovary initially inferior, then superior in fruit . . . . (*Henriquezia, Platycarpum*) **Rubiaceae**

8. Fruit a berry or capsule round in cross-section; corollas actinomorphic . . . . **Loganiaceae**

9(7). Sepals each with 2 large glands externally . . . . **Malpighiaceae**

9. Sepals without glands 10

10(9). Fruit a 2-locular capsule, with 2 distinct persistent styles; seeds numerous, small and winged . . . . **Cunoniaceae**

10. Fruit a berry or a 3–6-locular capsule, styles usually not persistent; seeds few, either without appendages or sometimes arillate, hairy, or winged 11

11(10). Stipules usually long and separate; fruit berry-like, occasionally dehiscent; seeds 1–4, hairy . . . . **Quiinaceae**

11. Stipules either large and sheathing the terminal bud or short and connate, sometimes only an interpetiolar line evident; fruit a capsule or a berry with either a single large seed or numerous, nonhairy ones 12

12(11). Plants usually with abundant colored latex; stipules mostly evident as an interpetiolar line; fruit a berry . . . . (*Vismia*) **Clusiaceae**

12. Plants without latex; stipules either short and connate or large and sheathing the bud; fruit either a 1-seeded berry with a semi-inferior ovary or a capsule with arillate or winged seeds . . . . **Rhizophoraceae**

13(2). Stipules intrapetiolar or epipetiolar, free or connate 14

13. Stipules borne lateral to the petiole, always free; 2-branched hairs generally absent; sepals without glands 15

14(13). Sepals each with (1)2 large glands externally (some populations of *Byrsonima* are eglandular); petals free; 2-branched hairs usually present at least on young growth; styles 3 . . . . **Malpighiaceae**

14. Sepals lacking glands; petals connate; hairs when present simple; style 1 . . . . (*Capirona, Eleagia, Gleasonia*) **Rubiaceae**

15(13). Leaves verticillate, the blade broadly ovate and usually as long as or shorter than the petiole; stipules small and quickly deciduous, but usually leaving a round gland-like scar next to the petiole base or between the 2 adjacent petioles; plant with abundant milky latex; fruit a 3-locular capsule, explosively dehiscent and falling apart, leaving a central column . . . . (*Euphorbia cotinifolia*) **Euphorbiaceae**

15. Leaves mostly opposite, the blade longer than the petiole; plants without milky latex; fruit not as above 16

16(15). Leaves with a pair of large glands at the base of the leaf blade above the

petiole, sometimes on a small flap extending slightly from the lower blade margin; flowers borne in pairs of dichotomously branched axillary cymes .... (*Colubrina*) **Rhamnaceae**
16. Leaves without glands on the lower margin 17
17(16). Petiole apex pulvinate; petals absent, anthers dehiscing by apical pores; fruits capsular, with 4 or 5 broadly spreading, usually spiny lobes .... (*Sloanea*) **Elaeocarpaceae**
17. Petiole not pulvinate; petals present, anthers dehiscing longitudinally or by transverse slits; fruits without broadly spreading, spiny lobes...... 18
18(17). Flowers actinomorphic, borne in cymose inflorescences; anthers usually 3, dehiscing by transverse slits; nectary disk present; fruit a drupe.... .... **Hippocrateaceae**
18. Flowers zygomorphic or nearly actinomorphic, usually borne in racemes or panicles; anthers 1 or 5, dehiscing longitudinally; nectary disk absent; fruit a capsule or 1-seeded, indehiscent and generally winged 19
19(18). Stipule bases thick or forming nectariferous glands; sepals connate, 1 of them spurred at base; petals 1 or 3; fertile anther 1; placentation axile .... **Vochysiaceae**
19. Stipules not as above; sepals free, not spurred; petals 5; anthers 5; placentation parietal .... **Violaceae**
20(1). Trees growing in shallow salt or brackish water (mangroves)......... 21
20. Nonmangrove trees and shrubs.................................... 22
21(20). Leaves pubescent on lower surface, with conspicuous salt-excreting glands; ovary superior .... (*Avicennia*) **Verbenaceae**
21. Leaves glabrous on both surfaces, without salt glands; petiole with 2 glands near apex; ovary inferior .... (*Laguncularia*) **Combretaceae**
22(20). Flowers borne in heads subtended by bracts; ovary inferior; fruit an achene usually crowned with the persistent calyx modified into a pappus composed of bristles, awns, or scales; corolla tubular, at least basally so; stamens attached to the corolla; anthers connate and forming a tube; stigmas with 2 branches .... **Asteraceae**
22. Flowers not borne in heads; without the above combination of characters 23
23(22). Leaves palmately veined, with 1 or more pairs of veins arcuately parallel to the midvein and running to the leaf apex; lower order venation perpendicular to the main veins 24
23. Leaves pinnately veined; lower order venation not as above 25
24(23). Flowers bisexual; calyx and corolla present; petals free, usually brightly colored; anthers usually twice as many as the petals, dehiscing by terminal pores; connective often lobed or spurred ...... **Melastomataceae**
24. Flowers unisexual; only dull-colored sepals present; anthers 4, dehiscing longitudinally; connective not as above .... **Buxaceae**
25(23). Leaves with numerous, small glandular punctations (use 10× magnification) 26
25. Leaves not punctate or rarely with a few large glands (Malpighiaceae)... 31
26(25). Ovary inferior; fruit generally a berry 27
26. Ovary superior; fruit capsular or with free mericarps or, if a berry, then

tepals 6 or plants with abundant latex 28
27(26). Flowers actinomorphic; petals free; stamens numerous . . . . **Myrtaceae**
27. Flowers zygomorphic; petals connate, golden yellow; stamens as many as the corolla lobes . . . . (*Gaiadendron*) **Loranthaceae**
28(26). Flowers somewhat zygomorphic; petals partly connate; fruits of 1–5 mericarps . . . . **Rutaceae**
28. Flowers actinomorphic; petals free; fruit a capsule or a berry 29
29(28). Hypanthium cup-like; petals small, attached between the calyx lobes. . . . . (*Adenaria, Pehria*) **Lythraceae**
29. Hypanthium absent, petals emerging from the basal receptacle of the flower 30
30(29). Perianth of 6 tepals; stamens dehiscing by valves or flaps; fruit a 1-seeded berry . . . . **Lauraceae**
30. Perianth of 5 sepals and 5 petals; stamens dehiscing by longitudinal slits; fruit a capsule or multiseeded berry . . . (*Hypericum, Vismia*) **Clusiaceae**
31(25). Plants hemiparasitic, epiphytic, without roots but with specialized organs (haustoria) which penetrate the host plant; ovary inferior; fruit usually sticky 32
31. Plants not hemiparasitic; roots present 33
32(31). Perianth of 1 whorl; inflorescence determinate . . . . **Viscaceae**
32. Perianth of 2 whorls (calyx and corolla); inflorescence indeterminate . . . . **Loranthaceae**
33(31). Perianth of 1 whorl (either calyx or corolla or tepals), perianth segments free, or perianth absent 34
33. Perianth of 2 whorls (calyx and corolla or tepals) or, if just calyx present, then sepals connate and forming a petaloid tube 38
34(33). Leaves with secondary venation fine, parallel, and closely spaced; latex yellow . . . . (*Calophyllum brasiliensis*) **Clusiaceae**
34. Leaves with secondary venation reticulate or, if parallel, then not fine and closely spaced; latex absent 35
35(34). Petioles pulvinate; anthers dehiscing by apical pores; fruits capsular, with 4 or 5 broadly spreading, usually spiny valves . . . . **Elaeocarpaceae**
35. Petioles not pulvinate; anthers dehiscing by slits; fruits drupaceous or, if splitting open, then somewhat fleshy. 36
36(35). Hypanthium campanulate to urceolate; inflorescences short, axillary cymes; carpels 4–many, free; fruit of separate drupes either enclosed in an accrescent hypanthium (*Siparuna*) or borne on a reflexed undulate hypanthium (*Mollinedia*) . . . . **Monimiaceae**
36. Hypanthium absent; inflorescences racemose or paniculate; carpels 1; fruit a solitary drupe or woody, tardily dehiscent, and globose or ellipsoid 37
37(36). Tepals 6, not recurved; stamens 3, 6, or 9, dehiscing by valves or flaps; fruit a 1-seeded berry, usually subtended by a cupule, or fruit seated on a woody, slightly swollen pedicel . . . . **Lauraceae**
37. Sepals 4, recurved; stamens 4, dehiscing by longitudinal slits; fruit woody, tardily dehiscent, globose or ellipsoid, never with a basal cupule . . . . (*Panopsis*) **Proteaceae**
38(33). Petals free or connate only basally 39

38. Petals or petaloid sepals connate for most their length 46
39(38). Leaves small, linear, and with a ventral groove that conceals the stomata, apparently veinless; fruit a capsule; small shrubs < 3 m tall, mostly endemic to highlands above 1500 m elevation . . . (*Ledothamnus*) **Ericaceae**
39. Leaves larger and broader than above, the lower order venation usually evident; generally trees or shrubs, most common below 1500 m elevation 40
40(39). Perianth of 2 whorls of 3 greenish tepals; stamens 3, 6, or 9, dehiscing by valves or flaps; fruit a 1-seeded berry, usually subtended by a cupule or fruit seated on a woody, slightly swollen pedicel . . . . (*Persea*) **Lauraceae**
40. Perianth of sepals and usually 4 or 5 petals; stamens 2–many, usually dehiscing by longitudinal slits; fruit various 41
41(40). Petals 4, white, long, narrow, very shortly connate basally; anthers 2; fruit a drupe; inflorescence a terminal panicle with opposite branching . . . . **Oleaceae**
41. Petals (3)4 or 5(–12), variously colored but not long and narrow; anthers 3 or more; fruits various; inflorescence axillary or terminal 42
42(41). Sepals each with (1)2 large glands externally; petals clawed . . . . **Malpighiaceae**
42. Sepals without conspicuous glands; petals not clawed 43
43(42). Inflorescences axillary, dichotomously branched cymes, borne on opposite sides of the stem; flowers < 5 mm wide; either leaves crenate and with drupaceous fruits (*Elaeodendron*), or entire with winged fruits (*Zinowiewia*) . . . . **Celastraceae**
43. Without the above combination of characters 44
44(43). Ovary inferior; flowers bisexual; stamens 8 or 10 (twice the number of petals), dehiscing by terminal pores; connective asymmetrical. . . . . (*Mouriri*, *Votomita*) **Melastomataceae**
44. Ovary superior; flowers bisexual or unisexual; stamens 3, 5 or, if more, usually more than twice the number of petals, dehiscing by longitudinal or transverse slits; connective symmetrical 45
45(44). Plants with white, yellow, or orange latex; flowers often unisexual; stamens numerous; anthers dehiscing by longitudinal slits; fruits drupes, berries, or capsules . . . . **Clusiaceae**
45. Plants without latex or, if present, then latex red; flowers bisexual; stamens 3(5); floral nectary disk present; anthers dehiscing by transverse slits or, if dehiscing by longitudinal slits, then fruit a pruinose drupe; fruits drupes or 3-lobed capsules . . . . **Hippocrateaceae**
46(38). Perianth of 1 whorl, forming a petaloid tube; flowers unisexual, plants dioecious; inflorescence usually terminal, cymose; ovary superior; stamens 5–10; fruit berry-like, 1-seeded . . . . (*Neea*, *Guapira*, *Pisonia*) **Nyctaginaceae**
46. Perianth of 2 whorls (calyx and corolla); flowers bisexual; inflorescence various, if terminal and cymose, then ovary inferior; stamens 4 or 5; fruits various 47
47(46). Flowers numerous, small (< 5 mm long), whitish, in terminal or axillary corymbiform inflorescences; ovary inferior; fruit a 1-seeded drupe . . . . (*Viburnum*) **Caprifoliaceae**

47. Without the above combination of inferior ovaries and corymbiform inflorescences; fruit not a 1-seeded drupe 48
48(47). Flowers actinomorphic; stamens as many as the corolla lobes; plants with or without abundant milky latex 49
48. Flowers slightly to strongly zygomorphic; stamens 2 or 4; corolla lobes 5; plants without latex 53
49(48). Shrubs usually < 8 m tall, without latex in stems or leaves; fruit a berry or capsule with parietal placentation, or a drupe usually with an accrescent calyx. 50
49. Shrubs or trees, usually with abundant white or colored latex in stems and leaves; fruit of follicles (usually paired), a berry with axile placentation, or a drupe without an accrescent calyx 51
50(49). Fruit a 2-celled berry or 2-locular capsule with parietal placentation; calyx not accrescent . . . . **Gentianaceae**
50. Fruit mostly a 3–5-celled drupe, usually with an accrescent calyx . . . . **Verbenaceae**
51(49). Flowers < 5 mm long, borne in axillary fascicles; trichomes, when present, 2-branched and forming a brownish, sericeous indument on lower surface of leaves, flowers, or fruits; fruit a berry or drupe; seeds shiny and with long scars . . . . (*Pouteria arcuata, Pradosia*) **Sapotaceae**
51. Flowers > 5 mm long, borne in various kinds of inflorescences; hairs if present, not as above; fruit follicular or, if a berry, seeds not as above . . 52
52(51). Corolla without an evident tube; pollen grains aggregated in specialized pollen sacs (pollinia) joined in pairs; stamen filaments and stigma head fused to form a complex pentagonal central columnar structure (the gynostegium); small to medium-sized shrubs . . . . **Asclepiadaceae**
52. Corolla with an evident tube; pollen grains free; anthers borne on evident filaments; stigma borne on an evident style; small shrubs to large trees . . . . **Apocynaceae**
53(48). Seeds generally flattened and circular, relatively large, usually attached to the placenta by slender, hook-like stalks; fruits explosively dehiscent, stipitate; leaves frequently with linear cystoliths (small crystalline inclusions best seen with 10× magnification, on dry leaves resembling appressed hairs) . . . . **Acanthaceae**
53. Plants not as above 54
54(53). Fruit a capsule or a many-seeded berry; placentation parietal 55
54. Fruit of 4 nutlets or a 1–4-seeded drupe or capsule; placentation basal-axile 56
55(54). Leaves coriaceous, sclerophyllous, always entire; stigma forked, the stigmatic lobes unfolding to expose the inner receptive surface as the flower matures; calyx lobes well developed and elongated, often partly connate; fruit a 2-valved capsule; seeds winged; corollas without a pouch at base; shrubs to large trees . . . . **Bignoniaceae**
55. Leaves membranous or succulent, entire or serrate; stigma 2-lobed (but not distinctly forked); calyx lobes mostly free to base; fruit a berry or 2-valved capsule; seeds not winged; corollas often forming a pouch at base; small shrubs, frequently with long, soft hairs . . . . . . **Gesneriaceae**
56(54). Leaves serrulate, scabrous on both surfaces; flowers bluish lavender; sta-

mens 4; panicles axillary, compact; calyx tubes elongated with 5 tooth-like lobes; fruit composed of 4 nutlets, enclosed in the persistent calyx tube . . . . (*Hyptidendron*) **Lamiaceae**

56. Without the above combination of characters; 1 ovule per locule; fruit a drupe or capsule . . . . **Verbenaceae**

GROUP H: Leaves simple, alternate (including unifoliolate leaves or those with a pair of leaves at the same node on one side of the stem).

Key to the Subgroups

1. Nonclimbing herbs (mainly terrestrial, but some epiphytic) . . . . [page 255] **SUBGROUP H-1**

1. Shrubs, trees, vines, or lianas 2

2(1). Climbing plants (herbaceous vines or woody lianas) . . . . [page 260] **SUBGROUP H-2**

2. Upright woody plants (trees and shrubs, including nonclimbing woody epiphytes and hemiparasites) . . . . [page 267] **SUBGROUP H-3**

SUBGROUP H-1: Nonclimbing herbs or succulents (mainly terrestrial, but some epiphytic), with simple, alternate leaves or leaves in a rosette. Includes the following families (dicotyledons and monocotyledons):

Amaranthaceae
Apiaceae
Araceae
Asteraceae
Begoniaceae
Boraginaceae
Burmanniaceae
Campanulaceae
Capparaceae
Chenopodiaceae
Convolvulaceae
Droseraceae
Euphorbiaceae
Fabaceae
Gesneriaceae
Haloragaceae
Hydrophyllaceae
Lentibulariaceae
Malvaceae
Melastomataceae
Moraceae
Ochnaceae
Onagraceae
Pedaliaceae
Phytolaccaceae
Piperaceae
Plantaginaceae
Plumbaginaceae
Polygalaceae
Polygonaceae
Portulacaceae
Primulaceae
Rubiaceae
Santalaceae
Sarraceniaceae
Scrophulariaceae
Solanaceae
Sphenocleaceae
Sterculiaceae
Taccaceae
Tiliaceae
Turneraceae
Urticaceae
Violaceae

1. Petals or tepals (or bracts of pseudanthia in Euphorbiaceae) connate, occasionally only basally so, usually forming a conspicuous corolla or perianth 2

1. Petals or tepals free, conspicuous or inconspicuous, or absent. 25

2(1). Ovary inferior 3

2. Ovary superior 10

3(2). Flowers pedicellate. 4

3. Flowers sessile 6

4(3). Leaf blade margins serrate or glandular; flowers zygomorphic, calyx and corolla forming distinct whorls; corollas red, purple, violet, pink, or orange, rarely green or white; plants usually with milky latex.......... .... **Campanulaceae**
4. Leaf blades, or major lobes of leaf blades, with entire margins; flowers actinomorphic, calyx and corolla not forming distinct whorls; flowers white, cream, greenish, or dull purple; plants never with milky latex 5
5(4). Leaf venation strongly parallel; leaf blades entire; inflorescence subtended by small bracts with inconspicuous venation .... **Burmanniaceae**
5. Leaf venation pinnate; leaf blades frequently deeply lobed; inflorescence subtended by a whorl of large, foliaceous, parallel-veined bracts .... **Taccaceae**
6(3). Flowers borne in dense or lax spikes or panicles of spikes 7
6. Flowers borne in dense heads.. 8
7(6). Leaves reduced to scales, plants appearing leafless; inflorescence a lax spike or panicle of spikes .... **Santalaceae**
7. Leaves not reduced to scales; inflorescence a dense spike .... **Sphenocleaceae**
8(6). Inflorescences subtended by 2 large, pink bracts that exceed the flowers; leaves distichous .... (*Didymochlamys*) **Rubiaceae**
8. Inflorescence not as above, usually subtended or surrounded by small, greenish or brownish bracts that do not exceed the flowers; leaves polystichous 9
9(8). Leaf venation parallel; fruit a dehiscent capsule .... **Burmanniaceae**
9. Leaf venation pinnate, reticulate, or 3-veined; fruit an achene .... **Asteraceae**
10(2). Leaves in a basal rosette.. 11
10. Leaves borne along the stem 12
11(10). Leaves entire; stamens 2; corolla not tubular, the lower lobe with a long spur or rarely merely saccate; plants with highly specialized, sometimes subterranean structures modified to trap tiny organisms, these structures of leaf or stem origin, but may appear root-like, either Y-shaped or bladder-like .... **Lentibulariaceae**
11. Leaves serrate or dentate; stamens 4; anthers connivent in 2 pairs or 4 together; corolla tube elongated, without a long spur, at most gibbous at the base; without specialized traps as above .... **Gesneriaceae**
12(10). Pubescence of simple trichomes mixed with short-stalked multicellular capitate trichomes (use 10× magnification); fruit a 4-parted, oblong, initially apically dehiscent capsule .... (*Sesamum*) **Pedaliaceae**
12. Pubescence of 1 kind of trichome; fruit not as above 13
13(12). Leaves serrate, dentate, or conspicuously lobed..................... 14
13. Leaves entire, not lobed 15
14(13). Leaves lobed; flowers > 5 mm long................................. .... (*Datura*, *Physalis*, *Solanum*, *Witheringia*) **Solanaceae**
14. Leaves serrate or dentate; flowers < 5 mm long .... (*Laportea*, *Phenax*) **Urticaceae**
15(13). Stipules present 16

15. Stipules absent 17

16(15). Stipules connate, forming an ochrea (a sheath surrounding the stem), with apical filiform appendages; fruit an achene; plants never with milky latex .... (*Polygonum*) **Polygonaceae**

16. Stipules free, without apical appendages; fruit a 3-lobed capsule, 3- or 6-seeded; plants often with milky latex .... (*Euphorbia*, *Pedilanthus*) **Euphorbiaceae**

17(15). Calyx with conspicuous stalked glands; petioles clasping the stem; corollas blue .... **Plumbaginaceae**

17. Calyx glabrous, pubescent or spiny; petioles, if present, not clasping the stem; corollas white, yellow, pink, red, greenish, or purplish 18

18(17). Leaves and stems completely glabrous............................ 19

18. Leaves and/or stems with some trichomes......................... 21

19(18). Leaves palmately veined, sessile, with the basal lobes clasping the stem; stems narrowly winged; fruit a circumscissile capsule .... **Primulaceae**

19. Leaves pinnately veined, petiolate; stems not winged; fruit a berry, longitudinally dehiscent capsule, or dry and indehiscent 20

20(19). Calyx of 5 unequal, strongly overlapping sepals that are free to the base . ... (*Ipomoea schomburgkii*) **Convolvulaceae**

20. Calyx of 5 equal, nonoverlapping sepals that are partially or completely connate .... (*Schwenkia*, *Solanum*) **Solanaceae**

21(18). Flowers subtended by red or scarlet bracts .... (*Castilleja arvensis*) **Scrophulariaceae**

21. Flowers not subtended by bracts or, if subtended by bracts, then bracts green 22

22(21). Plants covered with gland-tipped hairs (use 10× magnification); usually occurring in moist places .... **Hydrophyllaceae**

22. Plants with nonglandular hairs; occurring in a variety of habitats..... 23

23(22). Sepals partially or completely connate, frequently forming a 5-lobed or 5-dentate cup at the base of the flower .... **Solanaceae**

23. Sepals free 24

24(23). Corolla strongly 5-lobed; inflorescences scorpioid cymes (flowers axillary and solitary in *Heliotropium lagoense*); fruits splitting into 2 or 4 nutlets .... (*Heliotropium*) **Boraginaceae**

24. Corolla entire or subentire; inflorescences 1–few-flowered axillary dichasia or dense, terminal or lateral spicate heads; fruits capsular .... (*Evolvulus*, *Ipomoea argentea*) **Convolvulaceae**

25(1). Petals or tepals scarious, greenish, brownish or rarely colored, or absent. 26

25. Petals or tepals petaloid, white or colored 39

26(25). Leaves in a basal rosette... 27

26. Leaves borne along the stem 29

27(26). Flowers borne in a disk-like or cup-like receptacle; leaf margins crenate . ... (*Dorstenia*) **Moraceae**

27. Flowers borne on dense spikes; leaf margins or margins of major leaf lobes entire 28

28(27). Spikes usually thick and fleshy, subtended and sometimes surrounded by a usually persistent spathe; fruit a berry .... **Araceae**

28. Spikes relatively thin, not fleshy, not subtended by a spathe; fruit a circumscissile capsule .... **Plantaginaceae**
29(26). Leaves not lobed, entire 30
29. Leaves lobed and/or dentate or serrate............................. 34
30(29). Petioles usually sheathing at base; inflorescences dense spikes, subtended by bracts or individual flowers subtended by peltate scales; perianth absent; fruit generally a berry or drupe............................ 31
30. Petioles without sheaths; inflorescences spikes, panicles, or racemes, without subtending bracts; individual flowers not subtended by peltate scales; perianth present or, if absent, then fruit a 3-parted, dehiscent capsule; otherwise fruit a utricle or achene 32
31(30). Spike (spadix) usually thick and fleshy, terminal, subtended and sometimes surrounded by a usually persistent, often colored bract (spathe); individual flowers not subtended by peltate scales; leaves fleshy, but thin .... **Araceae**
31. Spike usually thin, borne opposite the leaves, without subtending bracts; individual flowers subtended by peltate scales; leaves succulent .... (*Peperomia*) **Piperaceae**
32(30). Flowers unisexual; ovary 3-carpellate; fruit an explosively dehiscent capsule; plants often with milky latex or abundant clear or colored sap ... . (*Euphorbia, Phyllanthus*) **Euphorbiaceae**
32. Flowers usually bisexual; ovary 1- or 6–15-carpellate; fruit a utricle or a spiny or tuberculate achene; plants never with milky or abundant clear-colored sap 33
33(32). Fruit a spiny or tuberculate achene; flowers white or greenish white .... (*Microtea*) **Phytolaccaceae**
33. Fruit a utricle; seeds lenticular, shiny; flowers greenish, brownish, purple, yellow, or red, rarely white .... **Amaranthaceae**
34(29). Leaves lobed, sometimes also dentate or serrate 35
34. Leaves never lobed, dentate, or serrate 36
35(34). Petioles with sheaths, at least basally; inflorescence a dense, usually thick and fleshy spike (spadix) that is subtended and sometimes surrounded by a spathe; fruit a berry .... **Araceae**
35. Petioles without sheaths; inflorescence not as above; fruit a capsule .... (*Cnidoscolus, Croton, Jatropha, Ricinus*) **Euphorbiaceae**
36(34). Fruit a 3-parted, explosively dehiscent capsule; plants sometimes with milky or clear-colored and abundant latex .... (*Acalypha, Caperonia, Croton, Sebastiana*) **Euphorbiaceae**
36. Fruit an achene, utricle, or nut-like; plants without milky latex....... 37
37(36). Leaves palmately veined .... (*Laportea, Phenax*) **Urticaceae**
37. Leaves pinnately veined 38
38(37). Ovary inferior; calyx lobes 4, persistent; fruits nut-like, with well-defined ridges; ovules 4, pendent .... **Haloragaceae**
38. Ovary superior; calyx lobes 5, not persistent; fruits utricles, smooth; ovules solitary, basal .... **Chenopodiaceae**
39(25). Leaves forming a hollow tube with a hood-like or thickened appendage at the apex, interior of tube with stiff, downward-pointing hairs (pitcher plants) .... **Sarraceniaceae**
39. Leaves not as above 40

40(39). Leaves club-shaped, forming a basal rosette, covered with conspicuous gland-tipped hairs, plants frequently reddish (sundews) .... **Droseraceae**
40. Leaves flat or succulent and round in cross-section 41
41(40). Leaves dentate, spinose, crenate, or lobed 42
41. Leaves entire and not lobed 50
42(41). Ovary inferior 43
42. Ovary superior 45
43(42). Fruit a winged capsule; inflorescences cymose; leaf blades asymmetric basally, succulent; stipules present; stamens numerous ... **Begoniaceae**
43. Fruit nut-like or a schizocarp; inflorescences dense spikes or glomerules; leaf blades symmetric basally, not succulent; stipules absent; stamens 4 or 5 44
44(43). Flowers 5-merous; inflorescences dense spikes, subtended by small or large bracts; flowers bisexual; stamens 5; fruits schizocarps with 2 tuberculate mericarps; petioles sheathing the stems .... (*Eryngium*) **Apiaceae**
44. Flowers 4-merous; inflorescences glomerulate, not subtended by bracts; flowers unisexual; stamens 4; petioles not sheathing the stems .... **Haloragaceae**
45(42). Flowers unisexual; perianth usually strongly reduced; petals and/or sepals often absent; plants often with milky latex or abundant clear or colored sap and/or a pair of leaf glands at the junction of petiole and blade; fruit a (2)3(–20)-locular capsule, explosively dehiscent and falling apart, leaving a central column, or rarely drupaceous or berry-like; seeds 1 or 2 per locule, sometimes with a fleshy apical appendage ...... **Euphorbiaceae**
45. Flowers bisexual; perianth usually evident; plants without latex (but may have mucilage); otherwise without the above combination of characters 46
46(45). Stipule margins laciniate or deeply ciliate-glandular; staminodes present .... (*Sauvagesia*) **Ochnaceae**
46. Stipules, if present, entire or ciliate at most; staminodes absent 47
47(46). Stamens 5; placentation parietal 48
47. Stamens > 5; placentation axillary 49
48(47). Flowers actinomorphic; sepals connate at least basally; styles 3 .... **Turneraceae**
48. Flowers zygomorphic; sepals free; style 1 .... **Violaceae**
49(47). Stamen filaments connate and forming a long tube around the style; fruit less than twice as long as wide .... **Malvaceae**
49. Stamen filaments free or only connate at the base, not forming a tube; fruit more than twice as long as wide .... (*Corchorus*) **Tiliaceae**
50(41). Ovary inferior 51
50. Ovary superior 53
51(50). Leaf venation palmate ...... (*Bertolonia venezuelensis*) **Melastomataceae**
51. Leaf venation pinnate or the veins inconspicuous 52
52(51). Leaves succulent, round in cross-section or flat; leaf axils frequently with abundant, long white hairs; capsules circumscissile; low prostrate herbs .... (*Portulaca*) **Portulacaceae**
52. Leaves membranous, flat; leaf axils without white hairs; capsules irregularly dehiscent or indehiscent; erect herbs .... **Onagraceae**

53(50). Flowers zygomorphic 54
53. Flowers actinomorphic 55
54(53). Stipules present; fruit a 1-locular capsule (legume), generally longer than wide . . . . (*Crotalaria, Desmodium, Indigofera, Phaseolus*) **Fabaceae**
54. Stipules absent; fruit a 2-locular capsule, generally about as long as wide . . . . (*Polygala*) **Polygalaceae**
55(53). Stipules forming an ochrea (a sheath surrounding the stem), with apical filiform appendages . . . . (*Polygonum*) **Polygonaceae**
55. Stipules, if present, not as above 56
56(55). Stamens numerous with the filaments connate and forming a tube around the style . . . . (*Cienfugosia, Sida linifolia*) **Malvaceae**
56. Stamens less than twice the number of petals or tepals or, if numerous, then the filaments free 57
57(56). Fruit a 3-parted, explosively dehiscent capsule, falling apart and leaving a central column; plants with milky latex; flowers unisexual . . . . (*Euphorbia, Pedilanthus*) **Euphorbiaceae**
57. Fruit not as above; plants without milky latex; flowers mostly bisexual 58
58(57). Leaves stipulate, linear; petals yellow; fruit an oblong, laterally compressed capsule . . . . (*Cleome guianensis*) **Capparaceae**
58. Leaves without stipules, not linear; petals white, cream, greenish yellow, or pink; fruit not as above 59
59(58). Sepals 2; leaves succulent . . . . (*Talinum*) **Portulacaceae**
59. Sepals > 2; leaves not succulent 60
60(59). Inflorescences small axillary cymes; sepals valvate in bud; flowers cream or greenish yellow; stems strongly angled . . . . (*Byttneria genistella*) **Sterculiaceae**
60. Inflorescences racemes, spikes, or a panicle of racemes or spikes; sepals imbricate in bud; flowers white or purplish; stems round in cross-section or longitudinally ridged . . . . (*Microtea, Petiveria, Phytolacca, Rivina*) **Phytolaccaceae**

SUBGROUP H-2: Climbing plants (herbaceous vines or woody lianas) with simple, alternate leaves. Includes the following families (dicotyledons and monocotyledons):

Amaranthaceae
Araceae
Aristolochiaceae
Asteraceae
Basellaceae
Boraginaceae
Caesalpiniaceae
Campanulaceae
Convolvulaceae
Cucurbitaceae
Dichapetalaceae
Dilleniaceae
Dioscoreaceae
Ericaceae
Euphorbiaceae
Fabaceae
Hernandiaceae
Icacinaceae
Marcgraviaceae
Melastomataceae
Menispermaceae
Nyctaginaceae
Olacaceae
Passifloraceae
Phytolaccaceae
Piperaceae
Plumbaginaceae
Poaceae
Polygalaceae
Polygonaceae
Rhamnaceae
Smilacaceae
Solanaceae
Sterculiaceae
Tropaeolaceae
Ulmaceae
Violaceae
Vitaceae

1. Plants with herbaceous tendrils or some branches coiled or transformed into hooks 2
1. Plants without tendrils or branches transformed into hooks (these may have epidermal or stipular thorns, prickles, or spines), climbing by other means or with stems running along the ground. 13
2(1). Tendrils emerging from the leaf nodes either directly opposite the petiole or at a 90° angle to the base of the petiole 3
2. Tendrils emerging directly from the leaf axil, the lower part of the petiole, or from the tip of the inflorescence; otherwise lateral branches or basal parts of the inflorescence modified into hooks or woody tendrils. 4
3(2). Tendrils lateral to the petiole, usually tightly coiled; leaf nodes generally not swollen or jointed; flowers unisexual; inflorescences racemes, panicles, fascicles, or flowers solitary; ovary inferior . . . . **Cucurbitaceae**
3. Tendrils emerging directly opposite the leaves, not tightly coiled; leaf nodes generally swollen or jointed; flowers bisexual; inflorescences corymbose; ovary superior . . . . **Vitaceae**
4(2). Leaves bifid with a central sinus, resembling a cow's hoof . . . . (*Bauhinia*) **Caesalpiniaceae**
4. Leaves not bifid. 5
5(4). Tendrils occurring in pairs, derived from the stipule and emerging from the lower petiole; flowers small, unisexual, borne in umbels . . . . **Smilacaceae**
5. Tendrils emerging directly from the stem, the leaf axil, or part of the inflorescence, not in stipule-derived pairs; flowers never borne in umbels 6
6(5). Tendrils or hooks emerging from the tip of the inflorescence 7
6. Tendrils emerging from leaf axils, the main stem, or the basal part of the inflorescence 8
7(6). Tendrils elongate and strongly coiled; leaves ± cordate at the base; stipules forming an ochrea (a sheath surrounding the stem, which upon abscision leaves a circular scar at the node); flowers borne in racemes . . . . (*Antigonon*) **Polygonaceae**
7. Tendrils short, ± woody, hook-like; leaves obovate; stipules usually small and deciduous; flowers solitary or borne in a few-flowered dichasium . . . . (*Ancistrothyrsus*, *Dilkea*) **Passifloraceae**
8(6). Tendrils helical or spiral, emerging from the leaf axils or the basal part of the inflorescence. 9
8. Occasional branches transformed either into curved hooks or somewhat woody tendrils without strong coiling 10
9(8). Tendrils helical, emerging only from the leaf axils; glands present on the petioles and/or leaf blade; flowers generally conspicuous, 5-parted, with coronal filaments and a stalked ovary; fruit a unilocular berry. . . . . **Passifloraceae**
9. Tendrils spiral; glands usually absent or only a pair at the junction of the blade and petiole; flowers inconspicuous, borne in elongate racemes; fruit 3-angled capsules . . . . (*Gouania*) **Rhamnaceae**
10(8). Leaves palmately veined; flowers inconspicuous; fruit a drupe or large berry. 11

10. Leaves pinnately veined; flowers with small but noticeable petals; fruit dry, flattened or winged 12
11(10). Leaves with a pair of conspicuous glands present at the junction of the blade and petiole; inflorescence an alternately branching panicle; often with red or orange-red latex; fruit 3-seeded ....(*Omphalea*) **Euphorbiaceae**
11. Leaves without glands; inflorescence strongly dichotomously branched; without colored latex; fruit 1-seeded ....(*Sparattanthelium*) **Hernandiaceae**
12(10). Fruit usually with the seed centrally located, with or without wings ....(*Dalbergia*) **Fabaceae**
12. Fruit usually with the seed at one end and a wing at the other (samara) or an oblanceolate, flattened capsule with an apical notch ... **Polygalaceae**
13(1). Plants with spines, thorns, or prickles 14
13. Plants without spines, thorns or prickles.......................... 20
14(13). Petals hooded, connected to the staminal tube; stamens 5; fruit subspherical, a 5-seeded capsule usually dehiscent into distinct segments...... ....(*Byttneria*) **Sterculiaceae**
14. Petals not hooded, not connected to a staminal tube; fruits not as above 15
15(14). Leaves finely serrate; leaf nodes of young branches often bearing a single, recurved spine ....(*Celtis*) **Ulmaceae**
15. Leaves entire; leaf nodes bearing various kinds of thorns or prickles... 16
16(15). Spines or thorns usually single at the leaf nodes or scattered along the stems 17
16. Some leaf nodes bearing pairs of recurved spines.................... 18
17(16). Stems with single straight or slightly curved thorns at leaf nodes; inflorescence with brightly colored bracts ..... (*Bougainvillea*) **Nyctaginaceae**
17. Stems or midveins on lower surface of leaves with scattered prickles; inflorescences without brightly colored bracts .... (*Solanum*) **Solanaceae**
18(16). Stamens and tepals numerous .... (*Pereskia aculeata*) **Cactaceae**
18. Stamens, petals, or corolla lobes 5 or fewer........................ 19
19(18). Flowers borne in heads subtended by bracts; fruits achenes crowned with bristles; ovary inferior .... (*Dasyphyllum*) **Asteraceae**
19. Flowers borne in ± loose panicles; fruits samaras; ovary superior........ ....(*Seguieria*) **Phytolaccaceae**
20(13). Plants herbaceous or barely woody vines.......................... 21
20. Plants clearly woody vines or large lianas......................... 31
21(20). Individual flowers reduced (< 5 mm long), without easily visible perianth parts.. 22
21. Individual flowers with readily visible perianth parts, the petals (or tepals) always connate and forming a tube 26
22(21). Flowers tightly aggregated on a thickened spike with a subtending spathe .... **Araceae**
22. Flowers ± loosely scattered over the inflorescence axis, without a subtending spathe, or with 2 cordate, colored bracts subtending the condensed inflorescence.. 23
23(22). Fruit a 3-lobed capsule, explosively dehiscent and falling apart, leaving a central column; plants with stinging trichomes (*Tragia*), or with 2 cor-

date, colored bracts subtending the condensed inflorescence (*Dalechampia*) . . . . **Euphorbiaceae**
23. Fruit a berry, utricle, or 3-winged capsule; plants without stinging hairs or inflorescences subtended by colored bracts 24
24(23). Leaves orbicular to ovate-cordate, often peltate; inflorescence often with round, leaf-like bracts; fruits berry-like. . . . . (*Cissampelos*) **Menispermaceae**
24. Leaves cordate to acute at the base, acute to acuminate at the apex, never peltate; inflorescence without conspicuous leafy bracts; fruits utricles or capsules 25
25(24). Leaves ± succulent; flowers bisexual, 5-merous; fruit a flattened utricle with lateral wings . . . . **Basellaceae**
25. Leaves membranous; flowers unisexual, 3-merous; fruit a 3-winged capsule . . . . **Dioscoreaceae**
26(21). Flowers zygomorphic 27
26. Flowers actinomorphic 29
27(26). Plants with milky latex; leaves with a strongly serrate margin; fruit an inflated (hollow) berry . . . . (*Burmeistera*) **Campanulaceae**
27. Plants without latex; leaves entire to moderately serrate; fruit a solid berry or dry when mature 28
28(27). Leaves peltate; flowers reddish orange, hanging on a long pedicel, with a long tubular sepaline spur and small, fringed petals . . . . **Tropaeolaceae**
28. Leaves basifixed; flowers purple-red or brownish, without petals, the modified calyx tube inflated at the base, followed by a narrow tube and an expanded, sometimes lobed limb . . . . **Aristolochiaceae**
29(26). Calyx narrowly tubular, with conspicuous stalked glands; corolla salverform, light blue . . . . **Plumbaginaceae**
29. Calyx and corolla not as above 30
30(29). Sepals overlapping, free to the base; plants sometimes with latex . . . . **Convolvulaceae**
30. Sepals not overlapping, connate at least basally; plants without latex . . . . **Solanaceae**
31(20). Individual flowers (or flower-like inflorescences) generally showy or at least easily visible (usually each one more than 8–10 mm long), and part of the flower (corolla, sepals, or bracts) usually tubular or with evidently connate parts; if flowers are smaller than above but still tubular, then the inflorescence of scorpioid cymes or a short spike with round drupaceous fruits > 2 cm diameter 32
31. Individual flowers generally smaller than above and inconspicuous, usually grouped in a ± dense inflorescence, the perianth parts various or absent 46
32(31). Flowers (including associated bracts) noticeably zygomorphic 33
32. Flowers actinomorphic or nearly so 39
33(32). Leaves bifid with a central sinus, resembling a cow's hoof . . . . (*Bauhinia*) **Caesalpiniaceae**
33. Leaves entire or lobed, but not bifid 34
34(33). Bracts 1 or 2, constituting the showiest part of the flower or inflorescence 35
34. Bracts not forming the showiest part of the flower 36

35(34). Inflorescence subtended by 2 large, ± heart-shaped bracts; flowers tiny, unisexual; leaves mostly membranous . . (*Dalechampia*) **Euphorbiaceae**

35. Inflorescence not subtended by large bracts, but bearing saccate, spurred, or hollow, usually reddish or greenish, nectariferous bracts; flowers bisexual; leaves succulent-coriaceous . . . . **Marcgraviaceae**

36(34). Leaves palmately veined; conspicuous stipules often present; flowers solitary; perianth of 1 whorl (connate sepals) . . . . **Aristolochiaceae**

36. Leaves pinnately veined; stipules absent or inconspicuous; flowers borne in inflorescences; perianth of 2 whorls (calyx and corolla) 37

37(36). Lower petal white and developed into a long, saccate spur; fruit an ovoid capsule . . . . (*Corynostylis*) **Violaceae**

37. Lower petal not developed into a long spur; fruit a flattened legume, oblanceolate capsule, samara, or large drupe 38

38(37). Fruit usually with the seed centrally located, with or without wings . . . . (*Dalbergia*) **Fabaceae**

38. Fruit usually with the seed at one end and a wing at the other (samara), an oblanceolate capsule with an apical notch, or a large drupe. . . . . **Polygalaceae**

39(32). Inflorescences heads or scorpioid cymes. 40

39. Inflorescences not as above or flowers solitary 41

40(39). Flowers borne in heads subtended by bracts; ovary inferior; fruit an achene; anthers connate and forming a tube . . . . (*Pentacalia*) **Asteraceae**

40. Flowers borne in scorpioid racemes or spikes grouped in dichotomous panicles; ovary superior; fruit a drupe; anthers free . . . . (*Tournefortia*) **Boraginaceae**

41(39). Flowers unisexual; perianth of 1 whorl (at least in pistillate flowers); white or colored latex usually present . . . . . (*Croton, Manihot*) **Euphorbiaceae**

41. Flowers bisexual; perianth of 2 distinct whorls (calyx and corolla); latex lacking 42

42(41). Leaves usually coriaceous, often palmately veined (3 or 5 main veins arching from the base towards the tip) or pliveined; flowers firm-fleshy or with separate, thin petals; stamens > 5, twice as many as the corolla lobes 43

42. Leaves membranous or coriaceous, mostly pinnately veined; flowers either thin and soon wilting or, if firmer, with 5 or fewer stamens. 44

43(42). Petals connate, firm-fleshy; stamens only slightly asymmetrical; fruit a berry . . . . **Ericaceae**

43. Petals free, thin; stamens strongly asymmetrical; fruit a capsule. . . . . (*Phainantha*) **Melastomataceae**

44(42). Sepals strongly overlapping; plants sometimes with latex . . . . **Convolvulaceae**

44. Sepals not overlapping; plants without latex 45

45(44). Stamens 8, the filaments connate and forming a tube; leaves thick-coriaceous, obovate, and drying yellow-green . . . . (*Moutabea*) **Polygalaceae**

45. Stamens 4 or 5, the filaments free; leaves not as above . . . . **Solanaceae**

46(31). Leaves parallel-veined, the bases sheathing the stem; plants climbing bamboos with hollow internodes and thickened nodes . . . . **Poaceae**

46. Leaves pinnately or palmately veined or pliveined; plants not as above 47
47(46). Petioles usually sheathing at base; inflorescences dense spikes; perianth absent; inflorescences subtended by bracts or individual flowers subtended by peltate scales; fruit generally a berry or drupe 48
47. Petioles without sheaths; inflorescences various, if a spike then perianth present; inflorescences without subtending bracts and flowers not subtended by peltate scales; fruits various 49
48(47). Spike usually thick and fleshy (spadix), terminal, subtended and sometimes surrounded by a usually persistent, often colored bract (spathe); individual flowers not subtended by peltate scales; adult plants with long, hanging, aerial roots . . . . . **Araceae**
48. Spike usually thin, borne opposite the leaves, without subtending bracts; individual flowers subtended by peltate scales; roots growing in ground . . . . . **Piperaceae**
49(47). Leaves palmately veined. 50
49. Leaves pinnately veined 55
50(49). Leaves with a pair of conspicuous glands near the base of the blade; leaf margins entire; plants often with milky latex or abundant clear or colored sap . . . . . (*Plukenetia*) **Euphorbiaceae**
50. Leaves without a pair of conspicuous glands near the base or, if so, then leaf margins crenate or serrulate; plants without latex. 51
51(50). Inflorescence strongly dichotomously branched; fruit a single globose or ellipsoid drupe. 52
51. Inflorescence not strongly dichotomously branched, spicate-racemose, paniculate, or flowers few and axillary; fruit dehiscent and angled, spiny, or an apocarpous group of commonly 3 drupes or nuts 53
52(51). Leaf blades firmly membranous, ovate-elliptic; petiole ca. 1/3 the length of the blade; fruits oblong-elliptic, 5-ribbed at apex. . . . . (*Sparattanthelium*) **Hernandiaceae**
52. Leaf blades coriaceous, (narrowly) oblong-elliptic; petiole less than 1/5 the length of the blade; fruits subglobose . . . (*Ampelozizyphus*) **Rhamnaceae**
53(51). Leaf margin often crenate or serrulate; inflorescence spicate; fruit a 3-angled capsule . . . . . (*Gouania*) **Rhamnaceae**
53. Leaf margin entire; inflorescence in branched thyrses or abbreviated cymes; fruit a spiny 5-parted capsule or apocarpous and usually with 2 or 3 drupes or nuts on a gynophore 54
54(53). Flowers unisexual, mostly on branched thyrses but sometimes few and axillary; fruit apocarpous, the monocarps either drupes or nuts. . . . . **Menispermaceae**
54. Flowers bisexual, on abbreviated cymes; fruit a spiny, dehiscent capsule. . . . . (*Byttneria*) **Sterculiaceae**
55(49). Plants with abundant milky latex; leaf margins serrulate; inflorescence like a bottlebrush, with mostly male flowers and 1–4 larger female flowers at the base . . . . . (*Mabea*) **Euphorbiaceae**
55. Plants without abundant milky latex; leaf margins entire or (some Dilleniaceae) dentate or serrate; inflorescence not as above 56
56(55). Inflorescence narrowly brush-like, densely covered with chaffy, tan-whitish

bracts and bracteoles, individual flowers mostly hidden ....
.... (*Chamissoa*) **Amaranthaceae**

56. Inflorescence without dense chaffy bracts and bracteoles, the individual flowers small but not hidden by other parts of the inflorescence..... 57

57(56). Inflorescence strongly dichotomously branched, often emerging from the petiole or the leaf blade itself, sometimes axillary .... **Dichapetalaceae**

57. Inflorescence axillary or terminal, not emerging from the petiole or leaf blade and not strongly dichotomously branched 58

58(57). Stipules forming an ochrea (a sheath surrounding the stem, which upon abscision leaves a circular scar at the node) .... (*Coccoloba*) **Polygonaceae**

58. Stipules, if present, often deciduous or small and inconspicuous, never forming an ochrea 59

59(58). Leaf blades serrate or entire, smooth or noticeably asperous (sandpaper-like) to the touch, characteristically with regularly spaced, closely parallel secondary veins; fruit dry to ± fleshy, dehiscent; seeds arillate ...
.... **Dilleniaceae**

59. Leaf blades usually entire, not sandpapery to the touch, the secondary veins not parallel nor regularly spaced; fruit various; seeds not arillate
.... 60

60(59). Petals ± hairy; styles 2, stigmas globose; fruit an ellipsoid nut ....
.........................(*Dicranostyles, Lysiostyles*) **Convolvulaceae**

60. Petals or tepals glabrous; style simple; fruit a drupe, berry, spiny capsule, or flat, 1-seeded legume 61

61(60). Petals purplish, hooded; fruit a 5-seeded spiny capsule separating into 1-seeded cocci .... (*Byttneria*) **Sterculiaceae**

61. Flowers not as above; fruit indehiscent, either fleshy or a coriaceous, flattened legume 62

62(61). Flowers zygomorphic, tightly clustered in short axillary racemes or panicles; fruit a coriaceous, flattened, 1-seeded legume ....
.... (*Dalbergia*) **Fabaceae**

62. Flowers actinomorphic, inflorescence various; fruit fleshy, a drupe or a berry... 63

63(62). Sepals completely connate and forming either a free open collar around the fruit or a sheath tightly enclosing the fruit; inflorescence an axillary fascicle or a short raceme or panicle usually < 5 cm long ..... **Olacaceae**

63. Sepals separate, either broad and spreading or small and inconspicuous, not enclosing the fruit; inflorescence usually > 5 cm long, an axillary raceme or panicle... 64

64(63). Sepals broad and spreading, usually > 4 mm long; fruit a many-seeded berry .... **Phytolaccaceae**

64. Sepals ± inconspicuous, usually < 4 mm long; fruit a many-seeded drupe.
.... **Icacinaceae**

SUBGROUP H-3: Upright woody plants (trees and shrubs, including nonclimbing woody epiphytes and hemiparasites) with simple, alternate leaves or leaves in a rosette. Because this subgroup forms the largest section of the key, important couplets are in bold type to facilitate its use. Includes the following families (monocotyledons, dicotyledons, and gymnosperms):

Anacardiaceae
Annonaceae
Apocynaceae
Aquifoliaceae
Araceae
Araliaceae
Arecaceae
Asteraceae
Bignoniaceae
Bixaceae
Bombacaceae
Boraginaceae
Burseraceae
Cactaceae
Caesalpiniaceae
Campanulaceae
Capparaceae
Caricaceae
Cecropiaceae
Celastraceae
Chrysobalanaceae
Clethraceae
Clusiaceae
Combretaceae
Convolvulaceae
Cyrillaceae
Dichapetalaceae
Dilleniaceae
Ebenaceae
Elaeocarpaceae
Eremolepidaceae
Ericaceae
Erythroxylaceae
Euphorbiaceae
Euphroniaceae
Fabaceae
Flacourtiaceae
Hernandiaceae
Hugoniaceae
Humiriaceae
Icacinaceae
Ixonanthaceae
Krameriaceae
Lacistemataceae
Lauraceae
Lecythidaceae
Lissocarpaceae
Loranthaceae
Magnoliaceae
Malvaceae
Menispermaceae
Monotaceae
Moraceae
Musaceae
Myricaceae
Myristicaceae
Myrsinaceae
Myrtaceae
Ochnaceae
Olacaceae
Onagraceae
Opiliaceae
Passifloraceae
Peridiscaceae
Piperaceae
Poaceae
Podocarpaceae
Polygalaceae
Polygonaceae
Proteaceae
Rhamnaceae
Rosaceae
Rutaceae
Sabiaceae
Saccifoliaceae
Sapindaceae
Sapotaceae
Simaroubaceae
Solanaceae
Sterculiaceae
Strelitziaceae
Styracaceae
Symplocaceae
Tepuianthaceae
Tetrameristaceae
Theaceae
Theophrastaceae
Thymeleaceae
Tiliaceae
Turneraceae
Ulmaceae
Urticaceae
Velloziaceae
Verbenaceae
Violaceae
Winteraceae

1. Shrubby or tree-like monocots; leaves either banana-like, palm-like, bamboo-like, sagittate, or with closely parallel venation from the base, always with a sheathing base clasping the stem; stems either hollow (bamboos), pithy with scattered woody fibers, or composed of compressed leaf sheaths, never forming true wood from a cambial layer 2
1. Shrubby or tree dicots or gymnosperms; leaves generally with reticulate venation or sometimes appearing parallel-veined, usually without a sheathing base; stems mostly solid with true wood derived from a cambial layer 7

2(1). Stems jointed, usually hollow except at the prominent nodes; ligule usually present at the junction of the leaf blade and the sheath; flowers highly modified and reduced, borne in spikelets .... (bamboos) **Poaceae**

2. Stems not jointed, solid throughout; ligules absent; flowers either large with whitish petals or much reduced but then aggregated onto a thickened spadix subtended by a spathe-like bract 3

3(2). Leaves banana-like; individual flowers conspicuous and zygomorphic; inflorescence terminal, racemose 4

3. Leaves not banana-like; individual flowers either inconspicuous and aggregated onto a dense axillary spadix, or solitary, actinomorphic, and long-tubular 5

4(3). Leaves spirally arranged; inflorescence erect or usually bending over, borne below the leaf crown; fruit fleshy, indehiscent; seeds without arils, but usually absent because fruits parthenocarpic; plants cultivated and locally persisting .... **Musaceae**

4. Leaves arranged in a single plane forming a broad fan-like cluster; inflorescence erect, emergent above the leaf crown; fruit capsular and seeds arillate; plants native .... **Strelitziaceae**

5(3). Flowers with a long, white, tubular perianth; ovary inferior; stem often dichotomously branched, the upper parts covered with persistent leaf bases; leaves strap-shaped .... **Velloziaceae**

5. Flowers with perianth parts strongly reduced, usually scale-like if present, grouped onto thickened axes (spadix); ovary superior; stems usually unbranched, the leaf bases persistent or not; leaves usually variously divided or sagittate 6

6(5). Leaves sagittate to triangular-ovate, not divided, thin-fleshy; higher order venation reticulate; plants barely woody, usually growing along river banks or swampy areas .... (*Montrichardia*) **Araceae**

6. Leaves usually variously divided, fibrous; higher order venation finely parallel; plants varying from barely woody with short underground stems to ones with thick trunks, growing in a variety of habitats from river banks and swamps to tepui summits .... **Arecaceae**

7(1). Leaves palmately veined or pliveined 8

7. Leaves pinnately veined (or rarely with a single median vascular bundle and no lateral veins in Podocarpaceae).............. 47

8(7). Plants hemiparasitic; roots absent, replaced by specialized organs (haustoria) which penetrate the host stems 9

8. Plants not hemiparasitic; roots present 10

9(8). Leaf apex obtuse; blade oblanceolate, lanceolate to ovate, or occasionally orbicular .... (*Antidaphne*) **Eremolepiadaceae**

9. Leaf apex apiculate; blade ovate to orbicular; new growth covered with soft scales .... (*Cladocolea intermedia*) **Loranthaceae**

10(8). Stipules conical and enveloping the terminal bud; stipular scar obvious and completely encircling the stem; petiole attached either basally or peltately off-center on the blade; terminal branchlets with watery latex turning dark brown upon exposure to air; plants monoecious, with tiny apetalous flowers usually clustered in spikes or heads; fruit a 1-carpellate nutlet or drupelet; plants often with stilt roots .... **Cecropiaceae**

10. Stipules not conical nor enveloping the terminal bud, the stipular scar usually not encircling the stem, or stipules absent; if leaves peltate then without latex; without the above combination of flowers and fruits . . 11

11(10). Leaves pouch-like, with 3 weak veins from the base, strap-shaped at the base but then becoming revolute until the margins close to form a downward-facing pouch in the upper third of the blade; restricted to Pico da Neblina (Serranía de la Neblina), at ca. 2800 m elevation . . . . **Saccifoliaceae**

11. Leaves not as above 12

12(11). Flowers borne in heads subtended by bracts; ovary inferior; fruit an achene usually crowned with the persistent calyx modified into a pappus of bristles, awns, or scales; corolla tubular, at least basally so; stamens attached to the corolla, the anthers connate and forming a tube; stigmas with 2 branches . . . . **Asteraceae**

12. Plants without the above combination of characters 13

13(12). Leaf margins serrate or dentate (leaves may also be lobed) 14

13. Leaf margins entire (leaves may also be lobed) 24

14(13). Leaf bases strongly unequal or, if weakly unequal then either upper surface of leaf scabrous (sandpaper-like) or twigs with downward-curved stipular spines 15

14. Leaf bases equal; leaves not scabrous; stipular spines, if present, not downward curved 17

15(14). Leaves glossy; inflorescences irregularly umbellate; stipules on young twigs conspicuous, linear; trees of secondary growth forest or light gaps . . . . (*Goupia*) **Celastraceae**

15. Leaves dull; inflorescences cymose, in axillary clusters of 2 or 3 flowers, or the flowers solitary; stipules not linear 16

16(15). Flowers conspicuous; petals 5, white; leaves densely white or pale tomentose on lower surface; fruit a round, red berry, with persistent stigmas; seeds numerous, minute; branches densely glandular-hairy . . . . (*Muntingia*) **Elaeocarpaceae**

16. Flowers tiny and inconspicuous, greenish and without petals; leaves scabrous on upper surface or, if not, then with downward-curving stipular thorns present; fruit a very small, yellow, orange, or red, 1-seeded drupe . . . . (*Celtis, Trema micrantha*) **Ulmaceae**

17(14). Stamen filaments connate and forming a tube surrounding the ovary and style; stamens usually numerous but if 5, 10, or 15 per flower, then the petals hooded and connected to the margin of the staminal tube 18

17. Stamen filaments free; stamens few to many; petals never hooded 19

18(17). Anthers 2-locular; pollen smooth, individual grains not readily visible with 10× magnification; fruit of 2 or 5 mericarps, a globose, bluntly spiny capsule, or a corkscrew-shaped capsule borne on a long gynophore; petals hooded and connected to the margin of the staminal tube . . . . **Sterculiaceae**

18. Anthers 1-locular; pollen spiny, large, individual grains generally visible with 10× magnification; fruit a loculicidal capsule or often separating into mericarps; petals not hooded, free from the staminal tube. . . . . **Malvaceae**

19(17). Flowers unisexual 20
19. Flowers bisexual. 22
20(19). Perianth with both calyx and corolla present . . . . (*Vasivaea*) **Tiliaceae**
20. Perianth usually strongly reduced, with petals and/or sepals often absent 21
21(20). Fruit an achene (sometimes surrounded by an enlarged succulent perianth and then resembling a berry); stamens as many as the tepals and opposite them . . . . **Urticaceae**
21. Fruit usually a 3-locular, explosively dehiscent capsule or occasionally drupe-like or berry-like; stamens various . . . . **Euphorbiaceae**
22(19). Petals connate and forming an urceolate to campanulate corolla; fruit either a berry or a berry-like 5-valved loculicidal capsule surrounded by a fleshy calyx; leaves often coriaceous . . . . **Ericaceae**
22. Petals, if present, free; fruits various; leaves membranous or chartaceous 23
23(22). Petals > 3 cm long, yellow; sepals imbricate; anthers dehiscing by a terminal pore; seeds woolly; latex yellow. . . . . (*Cochlospermum vitifolium*) **Bixaceae**
23. Petals < 3 cm long, variously colored; sepals valvate; anthers generally dehiscing by longitudinal slits; seeds not wooly; latex absent . . . . **Tiliaceae**
24(13). Leaves lobed 25
24. Leaves not lobed 31
25(24). Leaves bifid with a central sinus, resembling a cow's hoof . . . . (*Bauhinia*) **Caesalpinaceae**
25. Leaves not bifid, lobed on the sides 26
26(25). Stinging spines present on stems and leaves; leaves laciniately lobed . . . . (*Urera laciniata*) **Urticaceae**
26. Stinging spines not present 27
27(26). Petals connate at base; corolla tubular; treelets bearing cauliflorous, hollow fleshy fruits with parietal placentation . . . . **Caricaceae**
27. Petals free; placentation axile 28
28(27). Flowers unisexual; perianth usually strongly reduced; petals and/or sepals often absent; plants often with milky latex or abundant clear or colored sap and/or a pair of leaf glands at the junction of petiole and blade; fruit a (2)3(–20)-locular capsule, explosively dehiscent and falling apart, leaving a central column, or rarely drupaceous or berry-like; seeds 1 or 2 per locule . . . . **Euphorbiaceae**
28. Flowers bisexual; petals usually present; plants without latex or glands at the base of the blade; fruit and seeds various 29
29(28). Shrubs or subshrubs; stamen filaments connate and forming a tube around the style; pollen spiny, large, individual grains generally visible using 10× magnification . . . . **Malvaceae**
29. Small to very large trees; stamen filaments either free or, if connate, then the pollen not spiny 30
30(29). Stamens > 10 per flower, the filaments connate and forming a tube surrounding the ovary and style; petals present; fruit either large and indehiscent with 5 broad wings, or a narrow 5-valved capsule with seeds embedded in woolly kapok . . . . . (*Cavanillesia, Ochroma*) **Bombacaceae**

30. Stamens 5 per flower, the filaments free; petals absent; fruit of 5 free follicles, interior walls lined with hispid, irritating hairs .... (*Sterculia*) **Sterculiaceae**

31(24). Leaf nodes swollen; petioles sheathing at base and often with wing-like margins; leaf bases often unequal; inflorescences dense, rarely lax spikes borne opposite the leaves; flowers numerous, minute, subtended by peltate scales; perianth absent; fruit a small 1-seeded drupe .... (*Piper*) **Piperaceae**

31. Leaf nodes not swollen; plants without the above combination of characters 32

32(31). Flowers unisexual; perianth usually strongly reduced; petals and/or sepals often absent; plants often with milky latex or abundant clear or colored sap and/or a pair of leaf glands at the junction of petiole and blade; fruit a (2)3(–20)-locular capsule, explosively dehiscent and falling apart, leaving a central column, or rarely drupaceous or berry-like; seeds 1 or 2 per locule .... **Euphorbiaceae**

32. Plants without the above combination of characters 33

33(32). Petals or tepals connate 34

33. Petals or tepals free or absent 36

34(33). Stipules present; perianth of 1 whorl; fruit an achene.................. .... (*Pouzolzia*) **Urticaceae**

34. Stipules absent; perianth with both calyx and corolla present; fruit a capsule, berry, or drupe 35

35(34). Stems with milky latex; corolla thin, funnelform; fruit capsular; seeds covered with long silky hairs; stems hollow, herbaceous apically; anthers dehiscing by longitudinal slits .... (*Ipomoea carnea*) **Convolvulaceae**

35. Stems without milky latex; corolla generally thickened, tubular or urceolate, never broadly funnelform; fruit fleshy or capsular; seeds without hairs; anthers dehiscing by terminal pores .... **Ericaceae**

36(33). Petioles not pulvinate; stipules absent............................ 37

36. Petioles with an apical pulvinus and/or leaves with stipules 38

37(36). Fruit completely enclosed by a greatly enlarged cupule, dry-drupaceous; anthers dehiscing by longitudinal slits; ovary inferior; inflorescences (corymbs) composed of involucrate clusters of 1 pistillate and 2 staminate flowers .... (*Hernandia*) **Hernandiaceae**

37. Fruit subtended by either a cupule (never completely enclosing the fruit) or an enlarged receptacle, 1-seeded and berry-like; anthers dehiscing by 2 or 4 flap-like valves; ovary superior; inflorescences (panicles) composed of all similar flowers .... **Lauraceae**

38(36). Ovary inferior; leaf bases sheathing the stem and apical pulvinus evident; inflorescences terminal, panicles of globose or ellipsoid heads, the heads nearly sessile, subtended by 2 or more bracteoles; calyx forming a persistent cupule subtending the subglobose fruit .... (*Oreopanax*) **Araliaceae**

38. Ovary superior; leaf bases nonsheathing; inflorescences and flowers not as above 39

39(38). Petioles without an apical pulvinus 40

39. Petioles with an apical pulvinus 42

40(39). Flowers conspicuous, showy, borne singly from leaf axils, yellow with a deep red center, subtended by ca. 9 subulate or lanceolate bracts; stamen filaments connate and forming a tube around the ovary and style; calyx conspicuously dotted with glands .... (*Cienfuegosia*) **Malvaceae**

40. Flowers often small and inconspicuous or sometimes conspicuous, borne in axillary cymes or glomerules, variously colored, not subtended by bracts as above; stamen filaments usually not connate and forming a tube around the style, but if so, then stamens 5 or 15 and connected to small hooded petals.. 41

41(40). Flowers bisexual, conspicuous; petals with a short spathulate appendage; fruit either a large cauliflorous, ellipsoid, ribbed pod with fibrous exocarp and seeds embedded in pulp or, a usually spiny 5-locular capsule .... (*Byttneria*, *Theobroma*) **Sterculiaceae**

41. Flowers unisexual, inconspicuous; petals not appendaged; fruit a small berry .... (*Trema integerrima*) **Ulmaceae**

42(39). Petals absent; leaves and young stems completely glabrous 43

42. Petals present (or absent in *Triumfetta lappula* and *Heliocarpus americanus* of the Tiliaceae, which have indehiscent fruits with numerous spiny or bristly appendages); leaves and/or young stems usually with some hairs or scales.. 44

43(42). Inflorescence and flowers ferruginous-villous; stipules early deciduous; petioles pulvinate at apex; flowers bisexual; stamens numerous; sepals 5; nectary disk prominent, several-lobed; large trees .... **Peridiscaceae**

43. Inflorescence and flowers glabrous; stipules absent; petioles pulvinate at apex and base; flowers unisexual; stamens 6; sepals 6 in 2 series; nectary disk absent; weak shrubs .... (*Abuta grandifolia*) **Menispermaceae**

44(42). Sepals valvate.. 45

44. Sepals imbricate or united to form a cup 46

45(44). Stamens numerous, the filaments free; fruits capsular or dry and indehiscent; seeds dry .... **Tiliaceae**

45. Stamens in 2 or 3 groups of 5, the filaments connate; fruits berry-like or drupaceous, fleshy to woody; seeds surrounded by a fleshy pulp....... .... (*Theobroma*) **Sterculiaceae**

46(44). Sepals imbricate; inflorescence terminal, paniculate; petioles pulvinate at apex; stamen filaments free; anthers dehiscing by apical pores; fruits capsular, dehiscent, covered with short spines or tubercles; seeds with reddish or orange testa .... (*Bixa*) **Bixaceae**

46. Sepals united to form a cup; inflorescence axillary, the flowers solitary or in fascicles; petioles pulvinate at apex and at base; stamen filaments connate; anthers dehiscing longitudinally; fruits drupaceous, without spines or tubercles; seed testa not colored ..... (*Matisia*) **Bombacaceae**

47(7). Petioles pulvinate at base and/or apex 48

47. Petioles not pulvinate 63

48(47). Leaf margin dentate, serrate, or crenate 49

48. Leaf margin entire 53

49(48). Leaves with glands at the base of the blade near the junction with the petiole; flowers < 3 mm long, born on elongate spikes (usually > 10 cm long) .... (*Adenophaedra*, *Pausandra*) **Euphorbiaceae**

49. Leaves without glands at the base of the blade; flowers mostly > 3 mm long,

borne on racemose-paniculate or less elongate inflorescences....... 50

50(49). Petals connate and forming a tubular, 2–3 cm long corolla with spreading lobes; fruit a 1-seeded drupe subtended or mostly enclosed in lower half by the enlarged calyx .... (*Brachynema*) **Olacaceae**

50. Petals free or absent; fruit capsular 51

51(50). Stamens ≤ 10; fruits smooth, somewhat asymmetrical with the stigmatic remains off center; stem and leaves with a turpentine-like smell...... .... (*Protium unifoliolatum*) **Burseraceae**

51. Stamens > 20; fruits ornamented with spines, numerous elongate tubercles, conical or winged-papery projections, or rarely smooth (*Sloanea floribunda*), symmetrical; leaves and stems without a turpentine-like smell... 52

52(51). Petals absent; seeds 1 or 2; fruit valves often widely spreading.......... .... (*Sloanea*) **Elaeocarpaceae**

52. Petals present; seeds (2–)4–many; fruit valves scarcely spreading....... .... (*Carpotroche, Lindackeria, Mayna*) **Flacourtiaceae**

53(48). Flowers unisexual; fruit a 3-locular capsule, explosively dehiscent and falling apart, leaving a central column .... **Euphorbiaceae**

53. Flowers bisexual; fruits various but not 3-locular and falling apart.... 54

54(53). Petals 1 or absent... 55

54. Petals 4–9 (occasionally a few reduced in size in Sabiaceae) 56

55(54). Sepals completely fused in bud, not distinguishable; stamens dimorphic; petals 1 or absent; ovary and fruit usually elevated on a stipe; fruit a 1–many-seeded, oval, moniliform, or elliptic, thin-walled to thick-walled, coriaceous to woody legume .... (*Swartzia*) **Fabaceae**

55. Sepals valvate or imbricate in bud, distinguishable; stamens monomorphic; petals absent; ovary and fruit sessile; fruit capsular, with (3)4(–6) valves, ornamented with spines or tubercles .... (*Sloanea*) **Elaeocarpaceae**

56(54). Fertile stamens ≤ 5 (additional staminodes may be present).......... 57

56. Fertile stamens > 5 58

57(56). Leaves with glandular or pellucid punctations (use 10× magnification), often aromatic with a citrus-like smell; flowers actinomorphic; petals equal or subequal in size and shape; anthers dehiscing by longitudinal slits; fruit a 5-locular drupe, or a capsule of 1–5 mericarps with ventral, loculicidal dehiscence and often with separating, elastic, bony endocarp .... (*Adiscanthus, Hortia*) **Rutaceae**

57. Leaves without glandular punctations, without a citrus-like smell; petals with the 3 outermost ones larger and suborbicular, the inner 2 reduced, thin; anthers dehiscing by transverse slits; fruit a drupe............. .... (*Meliosma*) **Sabiaceae**

58(56). Carpels free, 3; fruit a flattened drupe .... (*Simaba*) **Simaroubaceae**

58. Carpels connate or solitary.................................... 59

59(58). Petals connate for most their length, with some stamens attached to the corolla tube; leaves with 10–20 strongly parallel secondary veins on each side of the midvein and very finely parallel tertiary veins ± perpendicular to the secondaries; fruit a 1-seeded ellipsoid drupe .... (*Minquartia*) **Olacaceae**

59. Petals free and the stamens not attached; fruit a dehiscent or indehiscent capsule, legume, or berry (sometimes woody or coriaceous); leaves with fewer and less obviously parallel secondary and tertiary veins...... 60

60(59). Fruit an echinate or tuberculate capsule; petals 6–9 .... (*Lindackeria*) **Flacourtiaceae**

60. Fruit smooth, or roughened but neither echinate nor tuberculate; petals 4 or 5 61

61(60). Fruit a 1-seeded, flattened, ± circular to semilunate legume; plants glabrous or with simple hairs; flowers either zygomorphic with a keel and lateral standard petals (*Dalbergia*) or with the sepals connate and forming a narrow pubescent tube, petals strap-shaped and ca. 2 mm wide, yellow, twisted, and much longer than the calyx (*Etaballia*)..... .... **Fabaceae**

61. Fruit a berry or a dehiscent or indehiscent capsule; plants glabrous or stellate-pubescent or lepidote-pubescent; flowers not as above 62

62(61). Petals 4; stamen filaments free; ovary and fruit borne on a stipe; fruit a many-seeded, torulose or rounded, coriaceous berry .... (*Capparis, Morisonia, Steriphoma*) **Capparaceae**

62. Petals 5; stamen filaments connate; ovary and fruit sessile; fruit a 3-valved capsule or hard-shelled, 1–4-seeded, indehiscent and capsule-like .... (*Catostemma, Scleronema*) **Bombacaceae**

63(47). Stipules (or bud scales in Podocarpaceae) present, especially on young leaves near the stem apex, often conspicuous, generally > 2 mm long, sometimes enveloping the terminal bud or surrounding the stem, sometimes quickly deciduous but then stipular scars usually evident at the leaf nodes 64

63. Stipules absent or small and inconspicuous (< 1 mm long), without conspicuous stipular scars 100

64(63). Leaf nodes swollen; petioles sheathing at base and often with wing-like margins; leaf bases often unequal; inflorescences dense, rarely lax spikes borne opposite the leaves; flowers numerous, minute, subtended by peltate scales; perianth absent; fruit a small 1-seeded drupe .... (*Piper*) **Piperaceae**

64. Leaf nodes not swollen; plants without the above combination of characters 65

65(64). Leaves with only the central vein differentiated, firm-coriaceous, entire, generally narrowly elongate to lanceolate; plants without true flowers; ovules borne in a highly modified cone covered by a coriaceous bract that enlarges and becomes a fleshy aril when the seed matures; pollen in small cones .... **Podocarpaceae**

65. Leaves with the secondary or higher order venation usually visible, of various textures and thicknesses, entire to lobed or with various kinds of serrations; plants with true flowers; ovules and seeds enclosed in an ovary; pollen in anthers 66

66(65). Terminal buds or leaf nodes enveloped by a conical or cylindrical, usually conspicuous stipule or scale, leaving a circular scar at the leaf node when it falls off (except *Euceraea* where scars are solitary at the leaf bases) 67

66. Terminal buds or leaf nodes not enveloped by a conspicuous, conical or cylindrical stipule; leaf nodes without a circular scar 74

67(66). Stipules forming an ochrea (a cylindrical sheath surrounding the stem above the leaf nodes, which upon abscision leaves a circular scar at the node) . . . . **Polygonaceae**

67. Stipules conical, enveloping the terminal bud 68

68(67). Leaves with margins crenate to serrulate, often pellucid-punctate; inflorescence a branched panicle of spikes with tiny flowers; fruit a 1- or 2-seeded berry with arillate seeds . . . . (*Euceraea*) **Flacourtiaceae**

68. Leaves with margins entire, not pellucid-punctate; inflorescence not as above; fruit an aggregation of follicles or berries, a capsule, or schizocarp, or fleshy with nonarillate seeds. 69

69(68). Flowers bisexual, showy; petals whitish; stamens numerous; carpels free or aggregated into a cone-like structure. 70

69. Flowers unisexual (bisexual in Lacistemataceae), usually strongly reduced, petals absent or inconspicuous; flowers aggregated into dense inflorescences with minute flowers or, if flowers larger and individually visible without magnification, the pistillate flowers with 3 or fewer connate carpels 71

70(69). Lower surface of leaves glabrous and silvery to greenish white; inflorescences axillary, pseudoumbellate, usually subtended by imbricate bracts; fruiting pedicels maroon-red; fruit of 3–12 free, purplish black, berry-like carpels in 1 whorl . . . . **Winteraceae**

70. Lower surface of leaves pubescent along the midrib; flowers solitary, terminal or subterminal; fruit a cone-like aggregation of separate follicles . . . . **Magnoliaceae**

71(69). Inflorescence lax, the individual flowers readily distinguishable; fruit a 3-locular capsule, explosively dehiscent and falling apart, leaving a central column . . . . (*Micrandra*) **Euphorbiaceae**

71. Inflorescence dense, flowers reduced; fruit a 3-valved capsule or a minute achene or drupe that is sometimes aggregated into a fleshy syncarp or fig-like receptacle. 72

72(71). Flowers bisexual, borne in short, axillary, densely bracteate spikes < 3 cm long, often several emerging at different angles from the same node; flowers with 1 stamen, no petals, and a concave bract encircling each flower; fruit a capsule with 1 arillate seed . . . . **Lacistemataceae**

72. Flowers unisexual, often in single-sex inflorescences or in specialized fig-like receptacles; fruit never capsular with arillate seeds 73

73(72). Inflorescences mostly in pairs in the leaf axils, branched or sometimes unbranched, with sessile flowers on globose or subglobose heads; stigma 1; hemiepiphytes or terrestrial, with prominent aerial or stilt roots; stems sometimes exuding clear watery sap that dries dark-colored. . . . . (*Coussapoa*) **Cecropiaceae**

73. Inflorescences various, if paired, then flowers borne within a closed receptacle; stigmas 1 or 2; stilt roots or aerial roots present or not; stems with white or colored latex . . . . **Moraceae**

74(66). Leaf margins dentate, serrate or crenate 75

74. Leaf margins entire 84

75(74). Flowers unisexual; perianth usually strongly reduced; petals and/or sepals often absent 76

75. Flowers bisexual, floral parts not reduced or modified as above (petals absent in some Flacourtiaceae) 77

76(75). Ovary (2)3(–20)-locular; ovules 1 or 2 per locule; placentation axile; fruit usually capsular and separating into 3 cocci or rarely drupaceous or berry-like; stamens 2–many . . . . **Euphorbiaceae**

76. Ovary 1-locular, ovule 1; placentation apical; fruit an achene or more or less drupaceous or forming a 1–many-seeded drupaceous syncarp; stamens 1–4 . . . . **Moraceae**

77(75). Anthers dehiscing by terminal pores; ovary either 2–5-syncarpellate with an apical style and fruit usually a spindle-shaped capsule, or ovary strongly lobed with the style gynobasic (borne at the base of the ovary) and fruit of separate drupes on an enlarged receptacle . . . . . **Ochnaceae**

77. Anthers dehiscing by longitudinal slits; ovary with apical style, fruit of fused carpels or of a single carpel 78

78(77). Stamen filaments connate and forming a tube surrounding the style . . 79

78. Stamen filaments free 81

79(78). Fruit a 1–few-seeded drupe; plants glabrous or with simple hairs (use 10× magnification) . . . . **Hugoniaceae**

79. Fruit a capsule; plants usually pubescent with simple and/or stellate hairs 80

80(79). Fruit a schizocarp of 5–14 segments; anthers 1-locular; styles 5–14; stigmas capitate; pollen spiny, large, individual grains generally visible using 10× magnification . . . . (*Pavonia, Sida*) **Malvaceae**

80. Fruit a 2- or 5-valved capsule with 1 or 5 seeds; anthers 2-locular; styles 1 or 5; stigmas often clavate or fimbriate; pollen not spiny, small (individual grains not readily distinguishable using 10× magnification). . . . . (*Melochia, Waltheria*) **Sterculiaceae**

81(78). Calyx tube campanulate or totally enclosing the bud and then splitting in 2; fruit a stipitate legume . . . . . . (*Lecointea, Zollernia*) **Caesalpiniaceae**

81. Calyx and fruit not as above 82

82(81). Stamens as many as and opposite the petals; ovary 2- or 3-locular, 1 basal ovule per locule . . . . **Rhamnaceae**

82. Stamens as many as the petals and alternate with them or > 8; ovary usually 1-locular, occasionally more; ovules several to many on a parietal or basal placenta 83

83(82). Stamens 8 or more; flowers actinomorphic; leaf blades often pellucid-punctate, with stellate hairs, or blade base or petiole apex bearing glands; petals often absent . . . . **Flacourtiaceae**

83. Stamens 5; flowers strongly zygomorphic to actinomorphic; leaves not punctate, hairs if present, simple, blade or petiole glands lacking; petals present . . . . **Violaceae**

84(74). Stipules intrapetiolar; triangular or narrowly so, often longitudinally striate or ribbed, sometimes densely clustered as cataphylls on stem apices or on leafless side shoots; sometimes the lower surface of the leaf with 2 lines paralleling the midvein and defining a lighter colored panel on either side of the midvein; fruit a reddish, narrowly ellipsoid drupe . . . . **Erythroxylaceae**

84. Plants without the above combination of characters 85

85(84). Stipules appearing as ± fused ligular projections from the sheathing petiole; ovary inferior; inflorescences terminal or subterminal, umbellate . . . . (*Schefflera*) **Araliaceae**

85. Stipules not as above, petiole not sheathing the stem; ovary superior; inflorescences not terminal and umbellate 86

86(85). Perianth of 1 whorl or absent; flowers unisexual or bisexual 87

86. Perianth of 2 whorls; flowers bisexual 92

87(86). Flowers unisexual 88

87. Flowers bisexual 89

88(87). Ovary 3(–20)-locular; ovules 1 or 2 per locule; placentation axile; fruit usually capsular and separating into 3 cocci or rarely drupaceous or berry-like; stamens 2–many . . . . **Euphorbiaceae**

88. Ovary 1-locular; ovule 1; placentation apical; fruit an achene or more or less drupaceous or forming a 1–many-seeded drupaceous syncarp; stamens 1–4 . . . . **Moraceae**

89(87). Inflorescences dense, short axillary spikes, often several emerging at different angles from the same node; flowers minute, each encircled by a concave bract; stamen 1; fruit a 3-valved capsule with 1 arillate seed . . . . **Lacistemataceae**

89. Inflorescences racemes, panicles, cymes, fascicles, or flowers solitary; stamens > 1; fruit fleshy or, if capsular, then with > 1 seed. 90

90(89). Staminal filaments thickened and expanded near the base (flat when dried); styles 2, short, divergent . . . . . . (*Ampelocera edentula*) **Ulmaceae**

90. Staminal filaments filiform; style 1 (sometimes divided apically) or absent 91

91(90). Fruit a dry or fleshy drupe; flowers markedly perigynous, zygomorphic or actinomorphic . . . . **Chrysobalanaceae**

91. Fruit a capsule or berry; flowers hypogynous or weakly perigynous, actinomorphic . . . . **Flacourtiaceae**

92(86). Petals connate, sometimes only basally so; stamens adnate to the corolla 93

92. Petals free; stamens free from the corolla 94

93(92). Flowers fasciculate in leaf axils; fruit a 1–several- seeded, often irregularly shaped berry; seed smooth and shiny with a broad scar covering much of the circumference; leaves and stems with milky latex; hairs, if present, 2-branched . . . . (*Chromolucuma, Ecclinusa*) **Sapotaceae**

93. Flowers borne in strongly dichotomously branched inflorescences, often emerging from the petiole or from the midrib of the leaf blade, sometimes axillary; fruit a dry drupe with 1–3 locules and 1 seed per locule; seed without a large scar; leaves and stems without milky latex; hairs, if present, not 2-branched . . . . **Dichapetalaceae**

94(92). Anthers dehiscing by 1 or 2 terminal pores; fruit either of several separate 1-seeded drupes on a common fleshy receptacle or a spindle-shaped capsule . . . . **Ochnaceae**

94. Anthers dehiscing by longitudinal slits; fruit capsular or drupaceous, never of separate drupes on a common receptacle 95

95(94). Petals 3; fertile stamens 4 with additional staminodes; fruit a finely pubescent 3-locular capsule with 3 basally winged seeds . . . . **Euphroniaceae**

95. Petals ≥ 4; fertile stamens (3)4–many; fruit a dry or fleshy drupe or dry, indehiscent and winged 96
96(95). Flowers hypogynous 97
96. Flowers perigynous 98
97(96). Stamens 40–50, the filaments free; fruit dry, indehiscent, with accrescent, wing-like, persistent sepals .... **Monotaceae**
97. Stamens 10–15, the filaments connate and forming a tube; fruit a 1- or few-seeded drupe .... **Hugoniaceae**
98(96). Ovary stipitate; corona of 1 or more rows of thread-like filaments or scales present .... **Passifloraceae**
98. Ovary sessile; corona absent 99
99(98). Leaves with a pair of glands near the base of the blade or near the midvein in the lower half of the blade; style terminal. . (*Prunus*) **Rosaceae**
99. Leaves without glands on the leaf blade, but glands sometimes present on the petiole; style emerging from the base of the ovary. .... **Chrysobalanaceae**
100(63). Petals connate, sometimes only basally so; calyx and corolla present (corolla absent in Proteaceae and Thymeleaceae but the calyx forming a petaloid tube; calyx modified into a pappus or absent in Asteraceae) 101
100. Petals or tepals free or absent 130
101(100). Flowers borne in heads subtended by bracts; ovary inferior; fruit an achene usually crowned with the persistent calyx modified into a pappus composed of bristles, awns, or scales; corolla tubular, at least basally so; stamens attached to the corolla, the anthers connate and forming a tube; stigmas with 2 branches .... **Asteraceae**
101. Plants without the above combination of characters.............. 102
102(101). Flowers zygomorphic 103
102. Flowers actinomorphic....................................... 107
103(102). Leaf margins entire.. 104
103. Leaf margins serrate, dentate, or sinuate....................... 105
104(103). Ovary inferior or semi-inferior; corollas red, orange, violet, or purple; inflorescence bracts greenish; stamen filaments connate; stems and leaves with copious milky latex . . . (*Siphocampylus*) **Campanulaceae**
104. Ovary superior; corollas yellowish or cream-colored; inflorescence bracts reddish, conspicuous; stamen filaments free; stems and leaves without latex .... (*Amasonia arborea*) **Verbenaceae**
105(103). Leaves glandular-punctate and/or pellucid-punctate; inflorescence usually a congested unilateral spike or raceme; fruit of 1–5 mericarps with 1 seed per carpel .... (*Raveniopsis*) **Rutaceae**
105. Leaves not glandular-punctate nor pellucid-punctate; flowers emerging singly or in fascicles directly off the trunk, solitary in leaf axils, or clustered and terminal-appearing; fruit a hard-shelled or coriaceous berry 106
106(105). Leaves borne in alternate fascicles; corolla broadly campanulate, off-white or tannish, usually with maroon penciling, especially on lobes and on the tube inside; flowers cauliflorous; fruit large, round, berry-like but with a hard, gourd-like shell ..... (*Crescentia*) **Bignoniaceae**

106. Leaves solitary; corollas salverform, white, blue, or purple; flowers solitary or clustered and terminal-appearing; fruit a coriaceous berry . . . . (*Brunfelsia*) **Solanaceae**

107(102). Ovary inferior or semi-inferior 108

107. Ovary superior 112

108(107). Corolla forming a radiate corona-like structure with ca. 25 veins, pleated in bud and unfolding like an umbrella; fruit a semi-inferior, oblong-pyramidal, tardily dehiscent capsule with a broad circular rim formed by the persistent calyx, with 1 cone-shaped seed. . . . . (*Asteranthos*) **Lecythidaceae**

108. Corolla tubular or urceolate, with < 10 lobes; fruit inferior or semi-inferior, fleshy 109

109(108). Anthers dehiscing by terminal pores; fruit generally globose, with numerous small seeds . . . . **Ericaceae**

109. Anthers dehiscing by longitudinal slits; fruit usually oblong, with 1 or 2 seeds 110

110(109). Stamens either 4 or 5 and as many as the corolla lobes, or 3 fertile ones alternating with 6 staminodes . . . . **Olacaceae**

110. Stamens 8 or 12–many 111

111(110). Anthers linear; corolla throat bearing an 8-toothed corona; stamens 8 in 1 series; corolla lobes 4 . . . . **Lissocarpaceae**

111. Anthers broadly ovate or rotund; corolla without a corona; stamens 12–many in more than 1 series or grouped into fascicles; corolla lobes 5, rarely 3 . . . . **Symplocaceae**

112(107). Leaves and stems with stellate or lepidote hairs; inflorescences racemes 113

112. Leaves and stems usually without stellate or lepidote hairs, but if such hairs present, then either the anthers dehiscing by terminal pores (*Solanum*, Solanaceae) or the flowers in scorpioid cymes or head-like inflorescences (*Cordia*, Boraginaceae); inflorescences various . . . 114

113(112). Stamens 5; fruits drupaceous . . . . (*Dendrobangia*) **Icacinaceae**

113. Stamens 10; fruits capsular . . . . **Styracaceae**

114(112). Perianth of 1 whorl (a petaloid calyx tube) 115

114. Perianth of 2 whorls (calyx and corolla). 116

115(114). Flowers borne singly on racemes or clustered in heads; bark thick, homogenous, tearing in long strips . . . . **Thymeleaceae**

115. Flowers borne in pairs on racemes; bark not tearing in long strips . . . . **Proteaceae**

116(114). Leaves, flowers, and fruits with linear or punctate glands; petals only basally connate; stamens opposite the petals and as many as them (often 4); fruit a 1-seeded berry . . . . **Myrsinaceae**

116. Plants not as above 117

117(116). Flowers small (< 4 mm wide), borne in fascicles on the stem or in the leaf axils; corolla lobes 4–18; fruit a berry or drupe, usually several-seeded, with 1 seed per locule; seed usually laterally compressed, shiny with a large scar covering much of the circumference; stem and leaves with whitish or colored latex; indument, when present, usually sericeous and of 2-branched hairs . . . . **Sapotaceae**

117. Plants not as above 118

118(117). Plants with abundant milky latex (or colored, watery, or inconspicuous in *Aspidosperma*); corolla contorted, 5-lobed; stamens 5; stigma with a thickened head and a ring of hairs below; ovary 2-carpellate, carpels usually free and united above by the style; fruit of 1 or pair of berries, follicles, or an angled capsule . . . . **Apocynaceae**

118. Plants without latex; flowers various; stamens ≥ 4; stigma not as above; ovary with carpels connate in both flower and fruit, 1-, 3- or more-locular; fruits various but not paired on the same floral receptacle 119

119(118). Anthers dehiscing by terminal pores 120

119. Anthers dehiscing by longitudinal slits 121

120(119). Corollas urceolate; stamens 10; fruit a capsule or a berry . . . . **Ericaceae**

120. Corollas rotate or campanulate; stamens 5; fruit a berry . . . . (*Cyphomandra*, *Solanum*) **Solanaceae**

121(119). Sepals overlapping, free or basally connate 122

121. Sepals not overlapping, valvate or connate for varying lengths 125

122(121). Stamens > 12 . . . . **Theaceae**

122. Stamens ≤ 10 123

123(122). Stamen filaments connate; staminodes 5, alternate with the stamens, petaloid or gibbous; ovary 1-locular, with free-central placentation . . . . **Theophrastaceae**

123. Stamen filaments free; staminodes absent; ovary 2- or more locular, the ovules pendent from the apex 124

124(123). Petals connate only basally . . . . **Aquifoliaceae**

124. Petals connate for much of their length and forming a tube; inflorescences terminal or axillary cymes, the branches elongate to condensed, often scorpioid, or rarely the flowers solitary and axillary . . . . (*Heliotropium*, *Lepidocordia*, *Tournefortia*) **Boraginaceae**

125(121). Calyx not inflated nor collar-like in flower or fruit; stamens ≤ 5 126

125. Calyx accrescent, inflated or collar-like in flower and/or fruit; stamens 5 or ≥ 10 127

126(125). Inflorescences cymose, either open, scorpioid, or condensed into short spikes or heads; ovules usually 4; stigmas 1–4; fruits drupaceous or dry and enclosing 1 or 4 pyrenes. . . . . (*Bourreria*, *Cordia*, *Tournefortia*) **Boraginaceae**

126. Inflorescences not as above, the flowers either solitary and axillary with large flowers > 10 cm long and coriaceous berries (*Solandra*), or in terminal or axillary cymose panicles with fleshy berries and smaller flowers (*Cestrum*); ovules numerous; style and stigma 1 . . . . . . **Solanaceae**

127(125). Weak shrub < 2 m tall; flowers purplish, campanulate; fruit enclosed by an apically toothed bladder-like calyx; fruit a many-seeded berry. . . . . (*Deprea*) **Solanaceae**

127. Plants not as above 128

128(127). Fruits several-seeded; calyx forming a lobed cupule around the fruit base; flowers unisexual, plants dioecious; stamens > 10 . . **Ebenaceae**

128. Fruits 1-seeded; calyx forming either a continuous collar around the fruit or enclosing it for most of its length; flowers bisexual; stamens 5 or 10 129

129(128). Stigmas 4; petals overlapping in bud; vegetative parts of plants with some simple, stellate, or 2-branched trichomes; stamens 5 . . . . (*Cordia*) **Boraginaceae**
129. Stigmas 1, 2–5 lobed; petals valvate in bud; vegetative parts of plants mostly glabrous; stamens 5 or 10 . . . . **Olacaceae**
130(100). Ovary inferior 131
130. Ovary superior or half-inferior 136
131(130). Trees > 3 m tall 132
131. Shrubs < 3 m tall 134
132(131). Flowers > 15 mm across; stamens numerous, connate basally and forming a staminal ring, the ring either actinomorphic or extended on one side into a strap-like structure which arches to form a hood over the pistil; fruits large, either indehiscent and cauliflorous, and globose with a hard, thin shell (*Couroupita*), or berry-like (> 25 mm diameter, *Gustavia*), or dehiscent, woody, and releasing seeds via a round lid (several genera) . . . . **Lecythidaceae**
132. Flowers < 10 mm across; stamens < 20, free; fruit either small and berry-like (< 10 mm diameter), an oblong drupe, or 2-, 4-, or 5-winged 133
133(132). Inflorescence a once-compound terminal umbel; petals present; stigma 5–9 lobed; fruit ellipsoid to subglobose, berry-like, crowned by the persistent calyx and style; leaves usually scattered along the stems. . . . . (*Dendropanax*) **Araliaceae**
133. Inflorescences of terminal or axillary spikes or racemes; petals absent; stigma simple; fruit an oblong drupe or dry and with 2, 4, or 5 wings; leaves often clustered at tips of branches . . . . (*Buchenavia*, *Terminalia*) **Combretaceae**
134(131). Leaves generally thin-membranous; petals yellow, generally falling off quickly; stamens as many as or twice the number of petals or sepals; fruit a capsule, readily or barely dehiscent; mostly lowland plants or occasionally upland plants to 1200 m elevation, of moist or wet areas . . . . **Onagraceae**
134. Leaves coriaceous; petals white or pink, stamens more than twice the number of petals or sepals; fruit fleshy, a berry or a pome; highland plants above 1500 m elevation 135
135(134). Leaf margins entire; new growth and lower surface of leaves with dense, whitish pubescence; flowers solitary, axillary . . . . (*Calycolpus alternifolius*) **Myrtaceae**
135. Leaf margins crenate to serrate, leaves glabrous; spines present at branch tips; flowers in corymbose cymes 2–5 cm long. . . . . (*Hesperomeles*) **Rosaceae**
136(130). Leaf margins serrate, dentate, crenulate, or spinose 137
136. Leaf margins entire 154
137(136). Fruits dehiscent 138
137. Fruits indehiscent 146
138(137). Flowers unisexual; fruit usually a 3-locular capsule, explosively dehiscent and falling apart, leaving a central column . . . . . **Euphorbiaceae**
138. Flowers bisexual; fruit not as above 139
139(138). Leaves with round or elongate nectariferous glands borne on the petiole,

rachis on lower surface, or at the petiole/blade junction (these occasionally small and best seen with 10× magnification) 140
139. Leaves lacking nectaries 141
140(139). Petiole or petiole/blade junction with a pair of round, conspicuous nectariferous glands; stems without thorns; petals yellow or orange; ovary 1-locular with parietal placentation; styles 3 .... (*Turnera*) **Turneraceae**
140. Petiole or rachis on lower surface with 1–5 round or elongate nectariferous glands, these either smooth or with a longitudinal opening; stems usually with scattered recurved thorns; petals purple, greenish yellow, or white; ovary 5-locular with axile placentation; style 1 .... (*Byttneria*) **Sterculiaceae**
141(139). Leaves with stellate hairs (use 10× magnification) 142
141. Leaves glabrous or with simple hairs 143
142(141). Anthers dehiscing by terminal pores; stamens 10(–12), in 2 whorls; petals 5; fruit a 3-lobed loculicidal capsule; seeds numerous, not arillate .... **Clethraceae**
142. Anthers dehiscing by longitudinal slits; stamens numerous; petals 3 or 4; fruit a globose, pilose capsule, dehiscent along the ventral and lateral sutures into 4 valves, red within; seeds 2–4, covered by a striate, ± entire white aril .... (*Curatella*) **Dilleniaceae**
143(141). Stamens 4 or 5 144
143. Stamens 6–many.. 145
144(143). Fruit a prismatically cylindric, elongate capsule, pointed at apex, 5-valved; seeds obliquely oblong and curved at the basal end, with a conspicuous, pale, thin wing attached at the distal end; petals white to pink; style simple, filiform, the stigma capitate; sinus of each leaf margin crenation with a caducous glandular tooth............... .... (*Ochthocosmus*) **Ixonanthaceae**
144. Fruit obovoid, 2-valved; seeds white-arillate; petals greenish; style absent, the stigma sessile, 2–4-lobed; leaves without glandular teeth.. .... (*Maytenus*) **Celastraceae**
145(143). Leaves often clustered in whorls at stem tips, coriaceous to nearly fleshy, epunctate; calyx often closely subtended by 2–several bracteoles; terminal bud usually large and conspicuous; petals conspicuous, yellow, pink, or white; stamens many; fruit a dry capsule; mostly high-altitude and/or moist-forest species .. (*Bonnetia, Gordonia*) **Theaceae**
145. Leaves usually distributed evenly along the stem, often distichous, chartaceous or membranaceous, often linear-punctate or pellucid-punctate; calyx not subtended by bracteoles; terminal bud small and inconspicuous; petals ± conspicuous, inconspicuous, or absent; stamens 6–many; fruit often a fleshy capsule; mostly low-altitude and dry-forest species .... **Flacourtiaceae**
146(137). Perianth of 1 whorl; flowers unisexual (or occasionally some bisexual flowers present)... 147
146. Perianth of 2 whorls (calyx and corolla); flowers bisexual 149
147(146). Leaves with yellowish gland dots on lower surface, aromatic; inflorescences axillary spikes or aments; fruit a 1- seeded tuberculate drupe

with waxy glands .... **Myricaceae**

147. Leaves without yellowish gland dots or sometimes with dark punctations on lower surface, not aromatic; inflorescences various; fruit a berry or smooth drupe 148

148(147). Leaves usually spine-tipped or with a few spinose teeth towards the apex; inflorescences spicate, capitate, or 1-flowered; plants with milky latex; stamens 1–3; fruit drupaceous; trunk and branches unarmed; nectary disk absent .... (*Clarisia ilicifolia*) **Moraceae**

148. Leaves glandular-crenate, -serrate, or -dentate; inflorescences fascicles or short racemes; plants lacking milky latex; stamens 15–30; fruit a coriaceous berry; trunk or branches with spines; nectary disk 8–12-lobed .... (*Xylosma*) **Flacourtiaceae**

149(146). Stamens 5, opposite the petals .... (*Rhamnus*) **Rhamnaceae**

149. Stamens > 5, opposite and/or alternate the petals................ 150

150(149). Inflorescences axillary fascicles or flowers solitary or paired; fruit a small to large coriaceous berry with a conical or pointed apex formed by the persistent style .... (*Freziera, Ternstroemia*) **Theaceae**

150. Inflorescences panicles or racemes; fruit not as above 151

151(150). Carpels free, several; fruit of several metallic-colored drupes borne on a common fleshy receptacle; petals bright yellow; anthers dehiscing by terminal pores .... (*Ouratea*) **Ochnaceae**

151. Carpels connate or solitary; fruit a solitary drupe or berry; petals white, greenish, greenish yellow, pink, or red; anthers dehiscing by longitudinal slits 152

152(151). Flowers zygomorphic; sepals connate and totally enclosing the flower in bud, then apical portions splitting in 2 and deciduous, the lower portion persisting; petals pink; ovary elevated above the receptacle by a stipe, forming a flattened, drupaceous legume................... .... (*Zollernia*) **Caesalpiniaceae**

152. Flowers actinomorphic; sepals not as above; petals white, greenish, or yellowish green; fruits sessile, round in cross-section, drupes or berries.. 153

153(152). Flowers 5-merous; stamen filaments connate to varying degrees, from basally united to fused into a short tube, filaments flattened (filiform in *Vantanea*), often of alternating lengths, sometimes trifurcate with 3 anthers; leaves not glandular-serrate nor glandular-crenate, coriaceous and lacking tomentum, without glands at base of lamina or petiole apex; seeds 1–4 .... **Humiriaceae**

153. Flowers 3(4)-merous; stamen filaments free, filiform, more or less of equal length, never trifurcate; leaves glandular-serrate or glandular-crenate, with tomentum present at least on the lower surface; petiole apex or base of lamina often bearing 1(2) glands; seeds 10–30 .... (*Banara*) **Flacourtiaceae**

154(136). Flowers unisexual 155

154. Flowers bisexual.. 161

155(154). Fruits indehiscent.. 156

155. Fruits dehiscent.. 158

156(155). Ovary 2- or more-locular .... **Euphorbiaceae**

156. Ovary 1-locular 157
157(156). Pistillate flowers with 2 styles, usually persistent in fruit; staminate flowers with 2–7 glabrous tepals and 1–3 stamens; plants with milky latex .... (*Clarisia*) **Moraceae**
157. Pistillate flowers with style absent, the stigma sessile; staminate flowers with 4(5) villous petals and 4(5) stamens opposite the petals; plants without milky latex .... **Opiliaceae**
158(155). Plants often with milky latex or abundant clear or colored sap and/or a pair of leaf glands at the junction of petiole and blade; fruit a (2)3(–20)-locular capsule, explosively dehiscent and falling apart, leaving a central column; seeds 1 or 2 per locule ..... **Euphorbiaceae**
158. Plants without the above combination of characters.............. 159
159(158). Leaves distichous; sap thin, reddish; plants usually with stellate or 2-branched hairs; fruit a 2-valved, coriaceous capsule; seeds arillate; flowers mostly 3-lobed, pistillate ones with a single carpel with 1 basal ovule, staminate ones with 2–many anthers and connate filaments that form a tube .... **Myristicaceae**
159. Leaves polystichous; sap or latex white, yellow, or not colored; plants glabrous or with simple hairs; fruits 4- or 5-valved or winged; seeds arillate or not; flowers not as above 160
160(159). Fruit a thin-walled capsule with 3 papery, rounded wings ca. 1 cm across; petals absent; stamens 10; leaves strongly resinous, appearing covered by a coat of varnish; plants lacking latex .... (*Dodonaea*) **Sapindaceae**
160. Fruit a 4- or 5-valved, fleshy capsule; flowers with 2 or more series of decussate sepals and petals; stamens > 10; leaves not as above; plants with white or yellow latex .... (*Tovomita*) **Clusiaceae**
161(154). Fruits indehiscent, generally a fleshy or coriaceous drupe or berry, or sometimes dry 162
161. Fruits dehiscent, generally a dry or sometimes ± fleshy capsule 181
162(161). Anthers dehiscing by valves or flaps; flowers mostly 3-merous with 6 tepals; stamens (3–)9(–12); fruit a 1-seeded berry, usually subtended by a cupule or fruit seated on a woody, slightly swollen pedicel; inflorescences paniculate; leaves or bark often aromatic .. **Lauraceae**
162. Anthers dehiscing by longitudinal slits or pores; flowers not as above; fruits not as above (except in Olacaceae) 163
163(162). Flowers mostly 3-merous; tepals or petals free; stamens ≥ 10, usually numerous; carpels small, many, and free; fruits either a cluster of monocarps or an aggregate fruit with the stigmatic remains of each carpel still visible; bark generally peeling off in long strips, often aromatic; leaves distichous (but polystichous in *Tetrameranthus*) .... **Annonaceae**
163. Plants without the above combination of characters; leaves polystichous (or distichous in many Flacourtiaceae)....................... 164
164(163). Sepals and petals 4; ovary stipitate; stamens 8 or many; fruit a round or elongate berry; placentation parietal; seeds usually arillate 165
164. Sepals and petals usually more or fewer than 4, or calyx and/or corolla

absent; ovary sessile; stamens few to many; fruit a drupe, berry, or capsule; seeds arillate or not. 166

165(164). Leaves > 15 cm long, obovate with an acuminate tip; petiole much thickened at base; stamens 8; flowers with coronal filaments; styles 4, diverging, each with a globose stigma; fruit globular with a pointed apex, somewhat hollow inside . . . . (*Dilkea*) **Passifloraceae**

165. Leaves generally smaller or not obovate with an acuminate tip; petiole not markedly thicker at the base; stamens usually numerous; flowers without coronal filaments; style and stigma 1; fruit rounded to elongate, not beaked at the apex, generally solid throughout with a pulpy matrix surrounding the seeds, fruit torulose, sometimes dehiscing tardily and irregularly to expose the seeds . . (*Capparis*) **Capparaceae**

166(164). Leaves somewhat fleshy, the petiole base surrounded by several sharp, unbranched spines; flowers with numerous colored tepals, numerous stamens, and a fleshy ovary with reduced leaves emerging from it . . . . (*Pereskia guamacho*) **Cactaceae**

166. Plants without the above combination of characters. 167

167(166). Fruit apocarpous, of several metallic-colored drupes borne on a common fleshy receptacle; petals bright yellow; anthers dehiscing by terminal pores . . . . (*Ouratea*) **Ochnaceae**

167. Fruit syncarpous; petals of various colors, usually not bright yellow; anthers dehiscing by longitudinal slits or, if poricidal, then the petals not yellow 168

168(167). Calyx and corolla zygomorphic; anthers dehiscing by terminal pores; outer 3 sepals differentiated from inner 2 and usually petaloid. . 169

168. Calyx and corolla actinomorphic; anthers dehiscing by longitudinal slits (except by terminal pores in *Purdiaea*, Cyrillaceae); sepals not as above. 170

169(168). Fruit a dry, 1-seeded bur; stamens 4; leaves < 3 cm long; shrubs < 2 m tall . . . . **Krameriaceae**

169. Fruit a 1-seeded blue or blackish drupe; stamens 8; leaves > 3 cm long; shrubs > 3 m tall . . . . (*Monnina*) **Polygalaceae**

170(168). Flowers with a conspicuous corona of 1 or more rows of thread-like filaments or scales; ovary stipitate; leaves with nectary-glands present on the petioles and/or leaf blade; fruit thin-walled, brittle; seeds with a juicy aril . . . . **Passifloraceae**

170. Flowers lacking a corona; ovary sessile; leaves without nectary-glands; fruits not as above; seed arils if present, not juicy 171

171(170). Fruit large, to 6 cm wide, thick-coriaceous, tan-mottled, with numerous seeds embedded in a pulpy matrix; stamen filaments connate . . . . (*Leonia*) **Violaceae**

171. Fruit smaller than above or asymmetrical and 1-seeded; stamen filaments free or connate. 172

172(171). Fruit with a hard endocarp, 1–3-seeded; flowers with 5 petals and 5 alternating stamens; plants with lower surface of leaves with smooth velvety pubescence (*Emmotum*), with glabrous leaves and the fruit an asymmetrical bicolored drupe (glossy, black, and ridged on one side, the other side white and concave, *Discophora*), or with flowers inconspicuous on short, narrow panicles < 5 cm long (*Poraqueiba*) .... **Icacinaceae**

172. Fruit with or without a hard endocarp, 1–many-seeded; stamens 1–many; lower surface of leaves not velvety; otherwise not as above 173

173(172). Leaves glandular-punctate and/or pellucid-punctate (use 10× magnification), often with a citrus-like smell; inflorescence terminal, a robust, many times dichotomous, many-flowered, flat-topped or round-topped dichasium; fruit a 5-locular, ± globose drupe........ (*Hortia*) **Rutaceae**

173. Leaves without glandular or pellucid punctations, but sometimes with dark dots on the lower surface of the leaf, not aromatic; inflorescence axillary, not as above; fruit a drupe, berry, or dry, indehiscent and capsule-like 174

174(173). Stamens usually ≥ 10, often of different lengths, the filaments connate at the base and sometimes forming a short tube or else free 175

174. Stamens ≤ 10, subequal in length, the filaments free 177

175(174). Flowers numerous in axillary panicles; fruit a 1-seeded or 2-seeded oblong to subglobose drupe .... **Humiriaceae**

175. Flowers solitary, in pairs, or fasciculate in the leaf axils; fruit a fleshy or ± coriaceous berry, but not drupaceous........ 176

176(175). Fruit usually beaked or pointed at the apex, with a persistent style and prominent, persistent sepals; seeds > 2, without arils; carpels 2 or 3 .... (*Ternstroemia*) **Theaceae**

176. Fruit rounded at the apex, lacking a persistent style; seeds 1 or 2, arillate; carpel 1 .... (*Doliocarpus*) **Dilleniaceae**

177(174). Leaves with a pair of glands near the basal edge of the lamina; flowers with a hypanthium and whitish petals, in axillary racemes; fruit a drupe with a faint line or constriction encircling it........ (*Prunus*) **Rosaceae**

177. Leaves without glands as above; inflorescence fasciculate, racemose, paniculate, or flowers solitary; if flowers with a hypanthium then not borne in simple racemes; fruit not as above 178

178(177). Leaves either with stellate hairs and peltate scales or the fruit a large, slightly asymmetrical drupe (*Mangifera*) or a kidney-shaped drupe on a fleshy receptacle (*Anacardium*) .... **Anacardiaceae**

178. Leaves without stellate hairs and peltate scales; fruit not as above.... 179

179(178). Stamen filaments basally connate and flattened; fruit a 5-locular and 5-seeded, globose-oblong berry; inflorescence 5–15 cm long, axillary, an umbellate to corymbose raceme with a whorl of foliaceous bracts, each flower subtended by a pair of persistent or deciduous bracteoles; flowers 5-merous .... **Tetrameristaceae**

179. Stamen filaments free; fruit either a 1-seeded drupe with an accrescent calyx or dry and 2–5-locular; inflorescence usually shorter than above or, if ≥ 5 cm long, of spicate racemes and without bracts and bracteoles as above; flowers 4-, 5-, or more-merous 180

180(179). Fruit dry and capsule-like but indehiscent, several-seeded; flowers numerous in straight, axillary racemes (4–)5–12 cm long, either with > 20 small flowers per raceme and each with 5 stamens (*Cyrilla*), or flowers few, lavender, and subtended by conspicuous whitish bracts, with 10 poricidal stamens (*Purdiaea*) .... **Cyrillaceae**

180. Fruit fleshy, 1-seeded, with an accrescent calyx or disk either enclosing much or most of the drupe and leaving a depressed nipple-like apical portion, or the calyx flaring into a collar-like structure around the base of the drupe; flowers fasciculate or in short corymbose racemes or panicles < 5 cm long .... **Olacaceae**

181(161). Inflorescences lacking an obvious axis (fasciculate, glomerate, or flowers solitary) 182

181. Inflorescence with an evident axis (racemose, spicate, paniculate, cymose, or corymbose) 186

182(181). Petals absent .... (*Casearia, Laetia*) **Flacourtiaceae**

182. Petals present 183

183(182). Stamens 4 or 5, inserted at the margin of a prominent nectary disk; fruits 2-valved, coriaceous, obovate capsules, usually orange or reddish and containing 1–3 white-arillate seeds .... (*Maytenus*) **Celastraceae**

183. Stamens > 5, not borne on a nectary disk; fruits not as above...... 184

184(183). Flowers mostly 3-merous; carpels many, free; fruit a cluster of dehiscent monocarps with the stigmatic remains of each carpel still visible; bark often aromatic; leaves distichous .... **Annonaceae**

184. Flowers 5-merous; carpels 3 or 4, connate; fruit a solitary capsule; bark not aromatic; leaves polystichous 185

185(184). Sepals connate and forming a short, cylindrical tube; ovary 1-locular with parietal placentation; seeds arillate; petioles sometimes bearing paired nectary glands; leaves membranous or chartaceous .... (*Turnera*) **Turneraceae**

185. Sepals free, imbricate; ovary 3(4)-locular with axile placentation; seeds not arillate; petioles without nectary glands; leaves coriaceous .... (*Bonnetia*) **Theaceae**

186(181). Fruit a thin-walled capsule with 3 papery, rounded wings ca. 1 cm across; petals absent; leaves strongly resinous, appearing covered by a coat of varnish when dry .... (*Dodonaea*) **Sapindaceae**

186. Fruit not winged; petals present; leaves not resinous............. 187

187(186). Stems with recurved thorns; petals with a 2-lobed hood and inflexed at the apex, adnate to a short staminal tube; fruits usually spiny, 5-parted schizocarps .... (*Byttneria*) **Sterculiaceae**

187. Stems without thorns; flowers and fruits not as above............ 188

188(187). Leaves usually noticeably glandular-punctate and/or pellucid-punctate (use 10× magnification), often with a citrus-like odor, the blades commonly oblanceolate with a long-decurrent base (*Adiscanthus, Angostura*), but subsessile, obovate, and emarginate to rounded in *Rutaneblina*; fertile stamens 2 or 5; fruit of 1–5 mericarps with ventral, loculicidal dehiscence and often with separating, elastic, bony endocarp .... **Rutaceae**

188. Leaves not punctate, lacking a citrus-like odor; blades usually not

shaped as above; stamens few to many; fruits not as above 189

189(188). Corollas zygomorphic; fertile stamens ≤ 5 or stamens 8........... 190

189. Corollas actinomorphic; fertile stamens ≥ 5 192

190(189). Stamens 8; fruit a 2-locular, fleshy, oblanceolate, flattened capsule with an apical notch .... (*Bredemeyera*) **Polygalaceae**

190. Stamens 5 and all fertile, or 4 fertile ones with 2–5 staminodia; fruit a 1- or 3-locular capsule 191

191(190). Leaves with whitish pubescence on lower surface, generally mucronate or narrowly acute; petals 3; fertile stamens 4 with 2–5 staminodia; fruit a finely pubescent, 3-locular capsule, each locule with a single winged seed .... **Euphroniaceae**

191. Leaves mostly glabrous, not mucronate; petals and stamens 5; fruit a 3-lobed, 1-locular capsule with several rounded seeds borne on parietal placentas .... (*Amphirrhox*) **Violaceae**

192(189). Anthers dehiscing by terminal pores; trichomes frequently gland-tipped; style well exserted, bent to one side; petals 5–9... (*Befaria*) **Ericaceae**

192. Anthers dehiscing by longitudinal slits; trichomes not gland-tipped; style not exserted beyond the perianth nor bent to one side 193

193(192). Stamens numerous, > 20; petals showy, pink, white, or yellow..... 194

193. Stamens ≤ 20; petals white, greenish white, or yellow 195

194(193). Anthers with a conspicuous cupular gland at apex; seeds either very numerous and linear with prominent wings or just 1–few and flattened; inflorescences many-flowered terminal or axillary panicles or axillary racemes .... (*Caraipa, Mahurea*) **Clusiaceae**

194. Anthers without a cupular gland; seeds numerous and linear, slightly winged or not; inflorescences 3–many-flowered cymes or panicles... .... (*Archytaea, Bonnetia*) **Theaceae**

195(193). Inflorescences terminal to subterminal cymes; petals dull yellow; styles 3; fruit a bony, loculicidal, finely sericeous capsule.. **Tepuianthaceae**

195. Inflorescences axillary racemes or panicles; petals whitish or greenish white; style 1; fruit a 2- or 5-locular, glabrous capsule .... **Ixonanthaceae**

Literature Cited

Aero-Service Corporation. 1972. *Levantamiento Radar de Venezuela Sur.* 3 volumes, maps 1:250,000. Caracas: MOP-CODESUR.

Aguerrevere, S. E., V. M. López, C. Delgado, and C. A. Freeman. 1939. Exploración de la Gran Sabana. *Revista de Fomento* 3(19): 501–729.

Ahti, T. 1992. La flora: Plantas inferiores. Pages 133–138 in *El Macizo del Chimantá, Escudo de Guayana, Venezuela: Un Ensayo Ecológico Tepuyano,* edited by O. Huber. Caracas: Oscar Todtmann.

Anderson, A. B. 1981. White-sand vegetation of Brazilian Amazonia. *Biotropica* 13: 188–210.

André, E. 1904. *A Naturalist in the Guianas.* London: Smith, Elder.

Anduze, P. 1960. *Shailili-Ko: Descubrimiento de las Fuentes del Orinoco.* Caracas: Talleres Gráficos Ilustraciones S.A.

Armellada, C. de, and M. Gutiérrez S. 1981. *Diccionario Pemón. Pemón-Castellano, Castellano-Pemón.* Caracas: CORPOVEN (Filial de Petróleos de Venezuela).

Arnal, P. 1943. *Exploraciones Botánicas en Venezuela.* Caracas: Instituto Pedagógico Nacional.

Aymard, G., B. Stergios, and N. Cuello. 1989. Informe preliminar sobre la vegetación del área de interfluvio Orinoco-Atabapo (3°10' N, 67°17' W), Departamento Atabapo, Territorio Federal Amazonas, Venezuela. *Informe Técnico del Vice-Rectorado de Producción Agrícola* (UNELLEZ, Guanare, Venezuela) 9(15): 170–219.

Baralt, R. M. 1841. *Resúmen de la Historia de Venezuela.* Paris: H. Fournier.

Beard, J. S. 1944. Climax vegetation in tropical America. *Ecology* 25: 127–158.

———. 1955. The classification of tropical American vegetation-types. *Ecology* 36: 89–100.

Bernardi, A. L. 1956. *Contribución a la Flora de la Guayana Venezolana.* Mérida, Venezuela: Publicaciones de la Dirección Cultural de la Universidad de los Andes.

———. 1957. *Estudio Botánico-forestal de las Selvas Pluviales del Río Apacará, región de Urimán, Estado Bolívar.* Mérida, Venezuela: Publicaciones de la Dirección Cultural de la Universidad de los Andes.

Berry, P. E. 1976. *Estudio Bibliográfico y Taxonómico Preliminar Sobre Palma "Seje."* Caracas: CODESUR.

Blancaneaux, P., and M. Pouyllau. 1977. Formes d'altération pseudokarstiques en relation avec la géomorphologie des granites précambriens du type Rapakivi dans le territoire fédéral de l'Amazone, Vénézuéla. *Cahiers ORSTOM, Série Pédologie* 15: 131–142.

Boadas, A. R. 1983. Geografía del Amazonas Venezolano. Volume 15 in *Colección Geografía de Venezuela Nueva,* edited by P. Cunill Grau. Caracas: Editorial Ariel-Seix Barral Venezolana.

Boggan, J., V. Funk, C. Kelloff, M. Hoff, G. Cremers, and C. Feuillet. 1992. *Checklist of the Plants of the Guianas (Guyana, Surinam, French Guiana).* Washington, D.C.: Smithsonian Institution.

Bongers, F., D. Engelen, and H. Klinge. 1985. Phytomass structure of natural plant communities on spodosols in southern Venezuela: the Bana woodland. *Vegetatio* 63: 13–34.

Boom, B. M. 1990a. Flora and vegetation of the Guayana-Llanos ecotone in Estado Bolívar, Venezuela. *Memoirs of the New York Botanical Garden* 64: 254–278.

———. 1990b. Useful plants of the Panare Indians of the Venezuelan Guayana. *Advances in Economic Botany* 8: 57–76.

Boom, B. M., and S. Moestl. 1990. Ethnobotanical notes of José M. Cruxent from the Franco-Venezuelan expedition to the headwaters of the Orinoco River, 1951–1952. *Economic Botany* 44: 416–419.

Brako, L., and J. L. Zarucchi. 1993. Catalogue of the flowering plants and gymnosperms of Peru. *Monographs in Systematic Botany from the Missouri Botanical Garden* 45: i–xl, 1–1286.

Brewer-Carías, C. 1976. Las simas de Sarisariñama. *Boletín de la Sociedad Venezolana de Ciencias Naturales* 22(132/133): 549–624.

———. 1978. *La Vegetación del Mundo Perdido.* Caracas: Fundación Eugenio Mendoza.

———, editor. 1988. *Cerro de la Neblina: Resultados de la Expedición 1983–1987.* Caracas: FUDECI.

Briceño, H., C. Schubert, and J. Paolini. 1990. Table-mountain geology and surficial geochemistry: Chimantá Massif, Venezuelan Guayana Shield. *Journal of South American Earth Sciences* 3: 179–194.

Brown, N. E. [and collaborators]. 1901. Report on two botanical collections made by Messrs. F. V. McConnell and J. J. Quelch at Mount Roraima in British Guiana. *Transactions of the Linnean Society of London* 6: 1–107.

Brünig, E. F., J. Heuveldop, J. Smith, and D. Alder. 1978. Structure and functions of a rainforest in the International Amazon Ecosystem Project: floristic stratification and variation of some features of stand structure and precipitation. Pages 125–144 in *Glimpses of Ecology,* edited by J. B. Singh and B. Gopal. Jaipur, India: International Scientific Publications.

Buck, W. R. 1990. Indices to *The Botany of the Guayana Highland. Memoirs of the New York Botanical Garden* 64: 45–122.

Butt-Colson, A. 1985. Routes of knowledge: an aspect of regional integration in the circum-Roraima area of the Guiana Highlands. *Antropológica* 63/64: 103–149.

Cabrera, A. L., and A. Willink. 1973. *Biogeografía de América Latina.* Serie de Biología, Monografía 13. Washington, D.C.: Programa Regional de Desarrollo Científico y Tecnológico, Secretaría General de la Organización de los Estados Americanos.

Canales, H. 1985. *La Cobertura Vegetal y el Potencial Forestal del T.F.D.A.* [Territorio Federal Delta Amacuro] *(Sector Norte del Río Orinoco).* Serie Informes Técnicos, Zona 12/IT/270, 3 maps 1:250,000. Maturín, Venezuela: MARNR.

Canales, H., and A. Catalán. 1981. *Evaluación de los Efectos de un Aprovechamiento Forestal en el Bosque de Transición-alto-medio-denso (Limón de Parhueña, Territorio Federal Amazonas).* Serie Informes Científicos, DGSIIA/IC/06. Caracas: MARNR.

Catalán, A. 1980. *Inventario de los Recursos Forestales de la Reserva Forestal del Sipapo, Territorio Federal Amazonas.* Serie Informes Científicos, Zona 10/IC/1980, 2 volumes. Puerto Ayacucho: MARNR.

CBR (Consejo de Bienestar Rural). 1961. *Reconocimiento Agropecuario Forestal del Oriente de la Guayana Venezolana.* Caracas: CBR.

Chaffanjon, J. 1889. *L'Orénoque et le Caura.* Paris: Librairie Hachette et Cie. (*El Orinoco y el Caura.* Spanish edition, edited by M. A. Perera, 1986, Fundación Cultural Orinoco, Caracas.)

Chesney, L. 1979. *Inventario Forestal de la Cuenca del Río Manapiare.* Caracas: Informe presentado a la DGSIIA, MARNR.

Codazzi, A. 1841. *Resúmen de la Geografía de Venezuela.* Paris: H. Fournier.

CODESUR. 1973. *La Conquista del Sur: Atlas del Territorio Federal Amazonas y del Distrito Cedeño del Estado Bolívar.* Caracas: MOP.

Colonnello, G. 1984. Contribución al conocimiento del microclima y medio físico-biótico de la cima del tepuy Marahuaca. *Memoria de la Sociedad de Ciencias Naturales La Salle* 44(122): 9–35.

———. 1986. El Abismo, una región inexplorada en el confín sur-este de Venezuela. *Natura* 78: 22–28.

Colonnello, G., S. Castroviejo, and G. López. 1986. Comunidades vegetales asociadas al Río Orinoco en el Sur de Monagas y Anzoátegui (Venezuela). *Memoria de la Sociedad de Ciencias Naturales La Salle* 46(125/126): 127–165.

Colonnello, G., L. Sánchez, and E. Vásquez. 1988. Investigaciones hidrobiológicas en la planicie de inundación del Bajo Orinoco, Venezuela. *Pantepui* 4: 3–10.

Conaway, M. E. 1984. Still Guahibo, still moving: a study of circular migration and marginality in Venezuela. *Relaciones Antropológicas; Occasional Publications on South American Anthropology* 1: 107 + 224 pages.

COPLANARH. 1969. *Inventario Nacional de Aguas Superficiales.* Volume 1. Caracas: COPLANARH.

———. 1972. *Plan Nacional de Aprovechamiento de los Recursos Hidráulicos.* Volume 1. Caracas: COPLANARH.

Coppens, W., editor. 1980, 1983, 1988. *Los Aborígenes de Venezuela.* 3 volumes. Caracas: Fundación La Salle de Ciencias Naturales, Instituto Caribe de Antropología y Sociología.

Corpovoz. 1984. El fascinante mundo del Marahuaka. *Corpovoz* (April–May 1984): 44 pages (unnumbered).

Cronquist, A. 1981. *An Integrated System of Classification of Flowering Plants.* New York: Columbia University Press.

Cuervo, A. 1893. *Colección de Documentos Inéditos Sobre la Geografía y la Historia de Colombia.* Bogotá: Imprenta Zalamea.

Cuevas, E. 1987. Perfil nutricional de la vegetación de turberas en el Macizo del Chimantá, Estado Bolívar, Venezuela: Resultados preliminares. *Acta Científica Venezolana* 38: 366–375.

———. 1992. Relaciones nutricionales de la vegetación de turberas alto-tepuyanas. Pages 203–218 in *El Macizo del Chimantá, Escudo de Guayana, Venezuela: Un Ensayo Ecológico Tepuyano,* edited by O. Huber. Caracas: Oscar Todtmann.

CVG-EDELCA. 1985. *Guri: Eje de Desarrollo del Río Caroní, 26.000 MW.* Caracas: CVG-EDELCA.

CVG-TECMIN. 1987. *Proyecto Inventario de los Recursos Naturales de la Región Guayana.* Hojas NB-20-4, NB-20-8, NB-20-12, NB-20-16, 8 volumes, maps 1:500,000. Ciudad Bolívar: CVG-TECMIN.

———. 1989. *Proyecto Inventario de los Recursos Naturales de la Región Guayana.* Hojas NB-20-3, NB-20-7, NB-20-11, NB-20-15, NA-20-3, 8 volumes, maps 1:500,000. Ciudad Bolívar: CVG-TECMIN.

———. 1991a. *Proyecto Inventario de los Recursos Naturales de la Región Guayana.* Informe de Avance NC-20-11, NC-20-12, 3 volumes, maps 1:500,000. Ciudad Bolívar: CVG-TECMIN.

———. 1991b. *Proyecto Inventario de los Recursos Naturales de la Región Guayana.* Informe de Avance NC-20-15, 3 volumes, maps 1:500,000. Ciudad Bolívar: CVG-TECMIN.

———. 1991c. *Proyecto Inventario de los Recursos Naturales de la Región Guayana.* Informe de Avance NC-20-16, 3 volumes, maps 1:500,000. Ciudad Bolívar: CVG-TECMIN.

———. 1991d. *Proyecto Inventario de los Recursos Naturales de la Región Guayana.* Informe de Avance NC-20-14 and NB-20-2, 3 volumes, maps 1:500,000. Ciudad Bolívar: CVG-TECMIN.

———. 1991e. *Proyecto Inventario de los Recursos Naturales de la Región Guayana.* Informe de Avance NB-20-6, 3 volumes, maps 1:500,000. Ciudad Bolívar: CVG-TECMIN.

Dalton, L. V. 1912. On the geology of Venezuela. *Geological Magazine; or, Monthly Journal of Geology (London)* 9: 203–210.

Danielo, A. 1976. Végétation et sols dans le Delta de l'Orénoque. *Annales de Géographie* 85(471): 555–577.

Delascio, F. 1985. *Aspectos Biológicos del Delta del Orinoco.* Caracas: INPARQUES, Dirección de Investigaciones Biológicas, División de Vegetación.

Delascio, F., and J. A. Steyermark. 1989. Notas sobre la vegetación del Marahuaka. *Acta Terramaris* 1: 15–20.

Dezzeo, N. 1990. Bodeneigenschaften und Nährstoffvorratsentwicklung in autochthon degradierenden Wäldern SO-Venezuelas. *Göttinger Beiträge zur Land- und Forstwirtschaft in den Tropen und Subtropen* 53: 1–104.

Donselaar, J. van. 1968. Phytogeographic notes on the savanna flora of southern Surinam (South America). *Acta Botanica Neerlandica* 17: 393–404.

Doyle, A. C. 1912. *The Lost World.* New York: Hodder and Stoughton.

Ducke, A., and G. A. Black. 1953. Phytogeographical notes on the Brazilian Amazon. *Anais da Academia Brasileira de Ciências* 25: 1–46.

Duivenvoorden, J. F., and A. M. Cleef. 1994. Amazonian savanna vegetation on the sandstone plateau near Araracuara, Colombia. *Phytocoenologia* 24: 197–232.

Eden, M. J. 1968. Geographers on the Orinoco. *Geographical Magazine (London)* 41(2): 107–109.

———. 1974. Palaeoclimatic influences and the development of savanna in Southern Venezuela. *Journal of Biogeography* 1: 95–109.

Ek, R. C. 1990. Index of Guyana plant collectors. Fascicle 1, 85 pages, in *Flora of the Guianas, Supplementary Series,* edited by A. R. A. Görts-van Rijn. Koenigstein, Germany: Koeltz Scientific Books.

Ernst, A. 1888. Sertulum Aturense, o sea, lista de una pequeña colección de plantas que recogió el Señor Alfredo Jahn en Octubre de 1887 cerca de Atures, alto Orinoco. *Revista Científica (Caracas)* 1: 219–223.

Estrada, J., and J. Fuertes. 1993. Estudios botánicos en la Guayana colombiana. IV. Notas sobre la vegetación y la flora de la Sierra de Chiribiquete. *Revista de la Academia Colombiana de Ciencias* 18(71): 483–498.

Ewel, J. J., and A. Madriz. 1968. *Zonas de Vida de Venezuela.* Memoria explicativa sobre el mapa ecológico. Caracas: Ministerio de Agricultura y Cría.

Fanshawe, D. B. 1952. *The Vegetation of British Guiana: A Preliminary Review.* Institute paper no. 29. Oxford: Imperial Forestry Institute, University of Oxford.

Fearnside, P. M. 1989. Brazil's Balbina dam: environment versus the legacy of the pharaohs in Amazonia. *Environmental Management* 13: 401–423.

Finol, H. 1973. La silvicultura en la Orinoquia venezolana. *Revista Forestal Venezolana* 25: 37–114.

———. 1974. *La Silvicultura en la Orinoquia Venezolana.* Mérida: ULA, Facultad de Ciencias Forestales. Mimeograph.

Fittkau, E. J. 1983. Grundlagen der Oekologie Amazoniens. Versuch einer Zusammenschau. *Spixiana,* supplement 9: 201–218.

Fölster, H. 1986. Forest-savanna dynamics and desertification processes in the Gran Sabana. *Interciencia* 11: 311–316.

Fölster, H., and O. Huber. 1984. *Interrelaciones Suelos-vegetación en el Área de Galipero, Territorio Federal Amazonas, Venezuela.* Serie Informes Técnicos, DGSIIA/IT/144. Caracas: MARNR, DGSIIA, Dirección de Suelos, Vegetación y Fauna.

Friedmann, H. 1948. Birds collected by the National Geographic Society's expeditions to northern Brazil and southern Venezuela. *Proceedings of the United States National Museum* 97(3219): 373–569.

Friel, A. O. 1924. *The River of Seven Stars.* New York: Harper.

FUDECI. 1990. *Memoria.* Caracas: FUDECI.

Fuentes, E. 1980. Los Yanomami y las plantas silvestres. *Antropologica* 54: 3–138.

Gabaldón, M. 1992. *Parques Nacionales de Venezuela.* Parques Nacionales y Conservación Ambiental, Volume 1. Caracas: Instituto Nacionales de Parques.

Gansser, A. 1974. The Roraima problem (South America). *Verhandlungen der Naturforschenden Gesellschaft in Basel* 84: 80–97.

García, P., compiler. 1987. *Proyecto Inventario de los Recursos Naturales de la Región Guayana (PIRNRG). Manual Metodológico (Versión Preliminar).* Ciudad Bolívar: CVG-TECMIN.

George, U. 1988. *Inseln in der Zeit. Venezuela-Expeditionen zu den letzten weissen Flecken der Erde.* Hamburg: GEO.

Gibbs, A. K., and C. N. Barron. 1993. *The Geology of the Guiana Shield.* Oxford Monographs on Geology and Geophysics, No. 22. New York: Oxford University Press.

Gili [Gilif], F. S. 1780–1784. *Saggio di Storia Americana.* Rome.

Givnish, T. J., R. W. McDiarmid, and W. R. Buck. 1986. Fire adaptation in *Neblinaria celiae* (Theaceae), a high-elevation rosette shrub endemic to a wet equatorial tepui. *Oecologia (Berlin)* 70: 481–485.

Gleason, H. A. 1929. A collection of plants from Mt. Duida. *Journal of the New York Botanical Garden* 30(355): 166–168.

Gleason, H. A. [and collaborators]. 1931. Botanical results of the Tyler-Duida Expedition. *Bulletin of the Torrey Botanical Club* 58: 277–516.

Gleason, H. A., and E. P. Killip. 1939. The flora of Mount Auyantepui, Venezuela. *Brittonia* 3: 141–204.

Goetz, P. W., editor-in-chief. 1985. *The New Encyclopaedia Britannica.* 15th edition, volume 20. Chicago: Encyclopaedia Britannica, Inc.

Gómez Picón, R. 1953. *Orinoco, Río de Libertad.* Madrid: Afrodisio Aguado.

Gómez-Pompa, A., T. C. Whitmore, and M. Hadley, editors. 1991. *Rain Forest Regeneration and Management.* Volume 6 in *Man and the Biosphere.* Paris: UNESCO.

Gondelles, R., editor. 1974. *Parque Nacional Canaima. La Gran Sabana, Plan Rector.* Caracas: CONAHOTU (Corporación Nacional de Hoteles y Turismo).

———. 1992. *El Régimen de Áreas Protegidas en Venezuela.* Caracas: Fundación Banco Consolidado.

Gondelles, R., J. García, and J. A. Steyermark. 1983. *Los Parques Nacionales de Venezuela.* 2nd edition. Madrid: INCAFO (Instituto de la Caza Fotográfica).

Good, R. 1969. *The Geography of the Flowering Plants.* 3rd edition. London: Longmans.

Gorzula, S., and O. Huber. 1992. Consideraciones finales. Pages 325–330 in *El Macizo del Chimantá, Escudo de Guayana, Venezuela: Un Ensayo Ecológico Tepuyano,* edited by O. Huber. Caracas: Oscar Todtmann.

Gosh, S. K. 1985. Geology of the Roraima Group and its implications. Pages 33–50 in *Memoria I Simposium Amazónico (Puerto Ayacucho 1981),* edited by M. Muñóz. Boletín de Geología, Publicación Especial no. 10. Caracas: Ministerio de Energía y Minas, Dirección de Geología.

Goulding, M., M. Leal Carvalho, and E. G. Ferreira. 1988. *Rio Negro, Rich Life in Poor Water.* The Hague: SPB Academic Publisher.

Grabert, H. 1976a. Alter und Geschichte der Roraima-Folge aus Guayana (Süd-Amerika). *Münsteraner Forschungsberichte in Geologie und Paläontologie* 38/39: 29–45.

———. 1976b. Die Inselberglandschaft des Roraima in venezolanisch Guayana. *Die Erde* 107: 57–69.

Granville, J. J. de. 1988. Phytogeographical characteristics of the Guianan forests. *Taxon* 37: 578–594.

———. 1991. Remarks on the montane flora and vegetation types of the Guianas. *Willdenowia* 21: 201–213.

Granville, J. J. de, and C. Sastre. 1973. Aperçu sur la végétation des inselbergs du sud-ouest de la Guyane Française. *Compte Rendu des Séances de la Société de Biogéographie* 439: 54–58.

Grubert, M. 1974. Podostemaceen-Studien. Teil I. Zur Ökologie einiger venezolanischer Podostemaceen. *Beiträge zur Biologie der Pflanzen* 50: 321–391. Berlin.

Grupo Científico Chimantá. 1986. Reconocimiento preliminar del macizo del Chimantá, Estado Bolívar (Venezuela). *Acta Científica Venezolana* 37: 25–42.

Gumilla, P. J. 1741. *El Orinoco Ilustrado y Defendido, Historia Natural, Civil y Geographica de Este Gran Río y de Sus Caudalosos Vertientes.* Madrid.

Haffer, J. 1974. *Avian Speciation in Tropical South America.* Publication no. 14. Cambridge, Massachusetts: Nuttal Ornithological Club.

Halliburton, A. 1952. Exploring a living 'Lost World' of 1,000,000 B.C. *Sunday Mirror Magazine,* December 14.

Hammen, T. van der. 1982. Paleoecology of tropical South America. Pages 60–66 in *Bi-*

ological Diversification in the Tropics, edited by G. T. Prance. New York: Columbia University Press.

Harris, D. R. 1968. Venezuela's empty rain forests. *Geographical Magazine (London)* 41: 216–220.

Heinen, D. 1988. Los Warao. Pages 585–689 in *Etnología Contemporánea II,* Monografía no. 35, edited by J. Lizot. Volume 3 in *Los Aborígenes de Venezuela,* general editor W. Coppens. Caracas: Fundación La Salle de Ciencias Naturales, Instituto Caribe de Antropología y Sociología.

Henley, P. 1982. *The Panare: Tradition and Change on the Amazonian Frontier.* New Haven: Yale University Press.

———. 1988. Los E'ñepá (Panare). Pages 215–306 in *Etnología Contemporánea II,* Monografía no. 35, edited by J. Lizot. Volume 3 in *Los Aborígenes de Venezuela,* general editor W. Coppens. Caracas: Fundación La Salle de Ciencias Naturales, Instituto Caribe de Antropología y Sociología.

Hernández, L. 1987. Degradación de los bosques de la Gran Sabana. *Pantepui* 3: 11–25.

———. 1992. Gliederung, Struktur und floristische Zusammensetzung von Wäldern und ihrer Degradations-und Regradationsphasen im Guayana-Hochland, Venezuela. *Göttinger Beiträge zur Land-und Forstwirtschaft in den Tropen und Subtropen* 70: 1–227.

Herrera, R., C. F. Jordan, H. Klinge, and E. Medina. 1978. Amazon ecosystems: their structure and functioning with particular emphasis on nutrients. *Interciencia* 3: 223–232.

Hickman, J. C., ed. 1993. *The Jepson Manual: Higher Plants of California.* Berkeley: University of California Press.

Hitchcock, C. B. 1947. The Orinoco-Ventuari region, Venezuela. *Geographical Review (New York)* 37: 525–566.

Holdridge, L. R. 1979. *Ecología Basada en Zonas de Vida.* Serie Libros y Materiales Educativos, no. 34. San José, Costa Rica: Instituto Interamericano de Ciencias Agrícolas.

Holmgren, P. K., N. H. Holmgren, and L. C. Barnett. 1990. *Index Herbariorum—Part 1: The Herbaria of the World.* 8th edition. Bronx: New York Botanical Garden.

Holst, B. K. 1987. Aparamán-tepui . . . conquered! *Missouri Botanical Garden Bulletin* 75(5): 5–6.

Holst, B. K., and C. A. Todzia. 1990. Léon Croizat's plant collections from the Franco-Venezuelan expedition to the headwaters of the Río Orinoco. *Annals of the Missouri Botanical Garden* 77: 485–516.

Holt, E. G. 1931. In Humboldt's wake: narrative of a National Geographic Society expedition up the Orinoco and through the strange Casiquiare Canal to Amazonian waters. *National Geographic Magazine* 60: 620–644.

———. 1933. A journey by jungle rivers to the home of the cock-of-the-rock. *National Geographic Magazine* 64: 585–630.

Hoock, J. 1971. *Les Savanes Guyanaises: Kourou: Essai de Phytoécologie Numérique.* Mémoires ORSTOM, no. 44. Paris: ORSTOM.

Hoyos, J. 1973. Expedición a la Laguna Asisa (Territorio Amazonas, Venezuela). *Natura* 51: 20–23.

Huber, O. 1980. Die Felsvegetation am oberen Orinoko in Südvenezuela. Pages 200–203 in *Blumenparadiese und Botanische Gärten der Erde,* edited by H. Reisigl.

Innsbruck: Pinguin-Verlag.
———. 1982. *Esbozo de las Formaciones Vegetales del Territorio Federal Amazonas, Venezuela.* Serie Informes Técnicos, DGSIIA/IT/103. Caracas: MARNR.
———. 1985a. Resultados preliminares del inventario botánico-ecológico del bioma sabana en el Territorio Federal Amazonas. Pages 679–696 in *Memoria I Simposium Amazónico (Puerto Ayacucho 1981),* edited by M. I. Muñóz. Boletín de Geología, Publicación Especial no. 10. Caracas: Ministerio de Energía y Minas, Dirección General Sectorial de Minas y Geología, Dirección de Minas.
———. 1985b. Sabanas y formaciones abiertas del Territorio Federal Amazonas. Pages 54–55 in *Atlas de la Vegetación de Venezuela,* coordinated by E. Ara. Caracas: MARNR.
———. 1986. La vegetación de la Cuenca del Caroní. *Interciencia* 11: 301–310.
———. 1987. Consideraciones sobre el concepto de Pantepui. *Pantepui* 2: 2–10.
———. 1988a. Guayana highlands versus Guayana lowlands, a reappraisal. *Taxon* 37: 595–614.
———. 1988b. Vegetación y flora de Pantepui, Región Guayana. *Acta Botânica Brasileira* 1(2), supplement: 41–52.
———. 1989. Shrublands of the Venezuelan Guayana. Pages 271–285 in *Tropical Forests: Botanical Dynamics, Speciation and Diversity,* edited by L. B. Holm-Nielsen, I. C. Nielsen, and H. Balslev. London: Academic Press.
———. 1990. Savannas and related vegetation types of the Guayana Shield region in Venezuela. Pages 57–97 in *Las Sabanas Americanas: Aspectos de su Biogeografía, Ecología y Utilización,* compiled by G. Sarmiento. Mérida: ULA.
———, editor. 1992a. *El Macizo del Chimantá, Escudo de Guayana, Venezuela: Un Ensayo Ecológico Tepuyano.* Caracas: Oscar Todtmann.
———. 1992b. La vegetación. Pages 161–178 in *El Macizo del Chimantá, Escudo de Guayana, Venezuela: Un Ensayo Ecológico Tepuyano,* edited by O. Huber. Caracas: Oscar Todtmann.
———. 1992c. Consideraciones fitogeográficas sobre la flora del Chimantá. Pages 189–202 in *El Macizo del Chimantá, Escudo de Guayana, Venezuela: Un Ensayo Ecológico Tepuyano,* edited by O. Huber. Caracas: Oscar Todtmann.
Huber, O., and C. Alarcón. 1988. *Mapa de Vegetación de Venezuela.* 1:2,000,000. Caracas: MARNR, The Nature Conservancy.
Huber, O., and F. Guánchez. 1988. *Flora y Vegetación del Área de Los Pijiguaos, Distr. Cedeño, Estado Bolívar: Informe Final.* List of species collected, vegetation map 1:250,000. Caracas: Convenio MARNR-BAUXIVEN. Mimeograph.
Huber, O., and J. J. Wurdack. 1984. History of botanical exploration in Territorio Federal Amazonas, Venezuela. *Smithsonian Contributions to Botany* 56: 1–83, 1 map.
Huber, O., J. A. Steyermark, G. T. Prance, and C. Alès. 1984. The vegetation of the Sierra Parima, Venezuela-Brazil: some results of recent exploration. *Brittonia* 36: 104–139.
Hueck, K. 1960. Mapa de vegetación de la República de Venezuela. *Boletín. Instituto Forestal Latino-Americano de Investigación y Capacitación* 7: 3–15, 1 map, 1:2,000,000.
———. 1961. Verbreitung, Ökologie und wirtschaftliche Bedeutung der "Chaparrales" in Venezuela. *Berichte des Geobotanischen Instituts der Eidgenössischen Technischen Hochschule Stiftung Rübel* 32 (1960): 192–203.

Humboldt, A. von. 1816–1831. *Voyages aux Régions Equinoxiales du Nouveau Continent.* 13 volumes. Paris: Librairie Grecque-Latine-Allemande, N. Maze, Librairie, and J. Smith & Gide fils.

———. 1818–1829. *Personal Narrative of Travels to the Equinoctial Regions of the New Continent, During the Years 1799–1804.* Translated by H. M. Williams. London: Henry G. Bohn.

———. 1956. *Viaje a las Regiones Equinocciales del Nuevo Continente.* Volume 4. Translated by L. Alvarado. Caracas: Ediciones del Ministerio de Educación.

Im Thurn, E. F. 1885a. The first ascent of Roraima. *Timehri* 4: 1–48.

———. 1885b. Roraima. *Timehri* 4: 255–267.

———. 1886. Notes on the plants observed during the Roraima expedition of 1884. *Timehri* 5: 147–223.

———. 1887. The botany of the Roraima Expedition of 1884: being notes on the plants observed; with a list of the species collected, and determinations of those that are new. *Transactions of the Linnean Society of London* ser. 2, 2: 249–300.

Jahn, A., Jr. 1909. Beiträge zur Hydrographie des Orinoco und Río Negro. *Zeitschrift der Gesellschaft für Erdkunde zu Berlin* 1909(2): 98–121.

Jam Lander, P. 1958. Expedición al Territorio Amazonas. *Memoria de la Sociedad de Ciencias Naturales La Salle* 18: 77–89.

Janzen, D. H. 1973. Rate of regeneration after a tropical high elevation fire. *Biotropica* 5: 117–122.

———. 1974. Tropical blackwater rivers, animals, and mast fruiting by the Dipterocarpaceae. *Biotropica* 6: 69–103.

Jordan, C. F., editor. 1987. *Amazonian Rain Forests: Ecosystem Disturbance and Recovery.* Ecological Studies 60. New York: Springer-Verlag.

———, editor. 1989. *An Amazonian Rain Forest: The Structure and Function of a Nutrient Stressed Ecosystem and the Impact of Slash-and-Burn Agriculture.* Volume 2 in *Man and the Biosphere.* Paris: UNESCO.

Klinge, H. 1978. Studies on the ecology of Amazon caatinga forest in southern Venezuela. Part 2: Biomass dominance of selected tree species in the Amazon caatinga near San Carlos de Río Negro. *Acta Científica Venezolana* 29: 258–262.

Klinge, H., and E. Medina. 1979. Río Negro caatingas and campinas, Amazonas States of Venezuela and Brazil. Pages 483–488 in *Heathlands and Related Shrublands,* edited by R. L. Specht. Volume 9A in *Ecosystems of the World,* editor-in-chief D. W. Goodall. Amsterdam: Elsevier Scientific.

Klinge, H., E. Medina, and R. Herrera. 1977. Studies on the ecology of Amazon caatinga forest in southern Venezuela. Part 1: General features. *Acta Científica Venezolana* 28: 270–276.

Knuth, R. 1928. Sammlungen und Literatur. *Repertorium Specierum Novarum Regni Vegetabilis, Beiheft* 43: 735–758.

Koch-Grünberg, T. 1917. *Vom Roroima zum Orinoco.* Volume 1. Berlin: Dietrich Reimer.

Kramer, K. U., and P. S. Green. 1990. *Pteridophytes and Gymnosperms.* Volume 1 in *The Families and Genera of Vascular Plants,* edited by K. Kubitzki. Berlin: Springer-Verlag.

Kubitzki, K. 1989a. Amazon lowland and Guayana highland: historical and ecological aspects of their floristic development. *Revista de la Academia Colombiana de Ciencias Exactas, Físicas y Naturales* 17(65): 271–276.

———. 1989b. The ecogeographical differentiation of Amazonian inundation forests. *Plant Systematics and Evolution* 162: 285–304.
———. 1990. The psammophilous flora of northern South America. *Memoirs of the New York Botanical Garden* 64: 248–253.
Lairet, R., and E. Rodríguez. 1989. *Región de Guayana.* In the series *Venezuela y su Geografía,* edited by S. Ovelar. Caracas: Editorial Minerva, C.A.
Lasser, T., and B. Maguire. 1950. A report on the plants of the Phelps Cerro Yaví expedition of 1947. *Brittonia* 7: 75–90.
Lichy, R. 1978. *Ya Kú: Expedición Franco-Venezolana del Alto Orinoco.* Caracas: Monte Avila.
Lizarralde, M. 1993. Indice y mapa de grupos etnolinguísticos autóctonos de América del Sur. *Antropologica,* supplemento 5: 6–200, 1 map.
Lizarralde, R. 1982. *Mapa Etnográfico de Venezuela y Zonas Adyacentes.* 2nd edition. Caracas: distributed by the author.
Lizot, J. 1988. Los Yanomami. Pages 479–583 in *Etnología Contemporánea II,* Monografía no. 35, edited by J. Lizot. Volume 3 in *Los Aborígenes de Venezuela,* general editor W. Coppens. Caracas: Fundación La Salle de Ciencias Naturales, Instituto Caribe de Antropología y Sociología.
Loefling, P. 1758. *Iter Hispanicum.* Stockholm.
Maguire, B. 1955. Cerro de la Neblina, Amazonas, Venezuela: a newly discovered sandstone mountain. *Geographical Review (New York)* 45: 27–51.
———. 1956. Distribution, endemicity, and evolution patterns among Compositae of the Guayana highland of Venezuela. *Proceedings of the American Philosophical Society* 100: 467–475.
———. 1957. Exploration. *A.I.B.S. Bulletin* (November 1957): 14–17.
———. 1959. Exploración botánica en Guayana. *El Farol* 21(185): 6–11.
———. 1964. Two decades of exploration in the American Tropics. *Garden Journal of the New York Botanical Garden* 14: 124–132.
———. 1966. Contributions to the botany of Guiana—I. *Memoirs of the New York Botanical Garden* 15: 50–128.
———. 1970. On the flora of the Guayana Highland. *Biotropica* 2: 85–100.
———. 1972a. Bonnetiaceae. In *The Botany of the Guayana Highland—Part IX,* edited by B. Maguire and J. J. Wurdack [and collaborators]. *Memoirs of the New York Botanical Garden* 23: 131–165.
———. 1972b. Guayana as a floristic province, its relationship within the Neotropics and to the Paleotropics. Pages 55–56 in *Resúmenes de los Trabajos.* México, D.F.: I Congreso Latinoamericano V Mexicano de Botánica.
———. 1979. Guayana, Region of the Roraima Sandstone Formation. Pages 223–238 in *Tropical Botany,* edited by K. Larsen and L. B. Holm-Nielsen. London: Academic Press.
Maguire, B., and K. Deery de Phelps. 1951. Botánica de las expediciones Phelps en la Guayana-Territorio Amazonas. *Boletín de la Sociedad Venezolana de Ciencias Naturales* 14: 5–19.
Maguire, B., and J. J. Wurdack [and collaborators], editors. 1957. *The Botany of the Guayana Highland—Part II. Memoirs of the New York Botanical Garden* 9: 235–392.
MARNR. 1979a. *Atlas de la Región Sur.* 2nd edition. Caracas: MARNR-CODESUR.
———. 1979b. *Atlas de Venezuela.* 2nd edition. Caracas: MARNR.

———. 1982. *Mapa de la Vegetación Actual de Venezuela.* 75 sheets, 1:250.000. Caracas: MARNR.
———. 1983–1984. *Sistemas Ambientales Venezolanos (Proyecto VEN/79/001). Región Guayana, Territorio Federal Amazonas.* 3 volumes. Caracas: MARNR.
———. 1991. Areas bajo régimen de administración especial. Map without scale. Caracas: MARNR-SEFORVEN-DGSPOA (Dirección General Sectorial de Planificación y Ordenamiento del Ambiente).
MARNR-ORSTOM. [1987]. *Atlas del Inventario de Tierras del Territorio Federal Amazonas.* Caracas: MARNR, DGSIIA.
Martius, C. F. P. von. 1840–1869. Tabulae physiognomicae. Brasiliae regiones iconibus expressas descripsit deque vegetatione illius terrae uberibus. Pages i–cx in *Flora Brasiliensis,* volume 1, part 1. Munich: R. Oldenbourg.
Mattick, F. 1964. Übersicht über die Florenreiche und Florengebiete der Erde. Pages 626–629 in *A. Engler's Syllabus der Pflanzenfamilien,* volume 2, edited by H. Melchior. Berlin: Gebrüder Borntraeger.
Maury, P. 1889. Enumeration des plantes du Haut-Orénoque recoltés par M. J. Chaffanjon et A. Gaillard. *Journal de Botanique* 3: 129–136, 157–164, 196–200, 209–212, 260, 266–273.
Mayr, E., and W. H. Phelps, Jr. 1955. Origin of the bird fauna of Pantepui. Pages 399–400 in *Acta XI Congressus Internationalis Ornithologici,* edited by A. Portmann and E. Sutter. Basel.
———. 1967. The origin of the bird fauna of the south Venezuelan highlands. *Bulletin of the American Museum of Natural History* 136: 269–328.
———. 1971. Origen de la avifauna de las altiplanicies del Sur de Venezuela. *Boletín de la Sociedad Venezolana de Ciencias Naturales* 29: 309–401.
Meade, R. H., F. H. Weibezahn, W. M. Lewis, Jr., and D. Pérez Hernández. 1990. Suspended-sediment budget for the Orinoco river. Pages 55–79 in *El Río Orinoco Como Ecosistema,* edited by F. H. Weibezahn, H. Alvarez, and W. M. Lewis, Jr. Caracas: EDELCA, Fondo Editorial Acta Científica Venezolana, CAVN, Universidad Simón Bolívar.
Medina, E. 1969. Expedición AsoVAC al Alto Orinoco. *Acta Científica Venezolana* 20: 9–13.
———. 1971. Expedición hover-craft al Río Negro-Casiquiare-Orinoco. *Defensa de la Naturaleza* 1(4): 24–35.
———. 1983. Adaptations of tropical trees to moisture stress. Pages 225–237 in *Tropical Rain Forest Ecosystems: Structure and Function,* edited by F. Bourlière. Volume 14A in *Ecosystems of the World,* editor-in-chief D. W. Goodall. Amsterdam: Elsevier Scientific.
Medina, E., V. García, and E. Cuevas. 1990. Sclerophylly and oligotrophic environments: relationships between leaf structure, mineral nutrient content, and drought resistance in tropical rain forests of the upper Río Negro region. *Biotropica* 22: 51–64.
Mendoza, V. 1977. Evolución tectónica del Escudo de Guayana. *Boletín de Geología. Publicación Especial* 7(3): 2237–2270.
Metzger, D. J., and R. V. Morey. 1983. Los Hiwi (Guahibo). Pages 125–216 in *Etnología Contemporánea I,* Monografía no. 29, edited by R. Lizarralde and H. Seijas. Volume 2 in *Los Aborígenes de Venezuela,* general editor W. Coppens. Caracas: Fundación La Salle de Ciencias Naturales, Instituto Caribe de Antropología

y Sociología.
Michelangeli, A. 1989. Biósfera del Marahuaka y zonas adyacentes (Territorio Federal Amazonas-Venezuela). Introducción. *Acta Terramaris* 1: 1–3.
Montoya Lirola, C. 1958. *Expedición al Río Paragua.* Caracas: Ministerio de Minas e Hidrocarburos.
Mori, S. A. 1991. The Guayana Lowland Floristic Province. *Compte Rendu des Séances de la Société de Biogéographie* 67(2): 67–75.
Mori, S. A., and G. T. Prance. 1987. Phytogeography. In *The Lecythidaceae of a Lowland Neotropical Forest: La Fumée Mountain, French Guiana,* edited by S. A. Mori [and collaborators]. *Memoirs of the New York Botanical Garden* 44: 55–71.
Moyersoen, B. 1993. Ectomicorrizas y micorrizas vesículo-arbusculares en Caatinga Amazónica del Sur de Venezuela. *Scientia Guaianae* 3: vii–xxvii, 1–82.
Müller, P. 1973. The dispersal centres of the terrestrial vertebrates in the Neotropical Realm. *Biogeographica (The Hague)* 2: 1–244.
OCEI. 1985. *División Político-territorial de Venezuela.* Caracas: OCEI.
———. 1992. *El Censo 90 en Bolívar: Resultados Básicos.* Caracas: OCEI.
———. 1993a. *El Censo 90 en Venezuela.* Caracas: OCEI.
———. 1993b. *Censo Indígena de Venezuela 1992.* Caracas: OCEI.
Orejas Miranda, B., and A. Quesada. 1976. Ecosistemas frágiles. *Ciencia Interamericana* 17(1): 9–15.
OTEHA (Oficina Técnica para Estudios Hidraúlicos y de Agua). 1971. *Estudio a Nivel Exploratorio de Recursos Forestales en el Distrito Cedeño del Estado Bolívar.* Bloque DC-H3, Estudio Técnico. Caracas. Mimeograph.
Overing, J., and M. R. Kaplan. 1988. Los Wóthuha (Piaroa). Pages 307–411 in *Etnología Contemporánea II,* Monografía no. 35, edited by J. Lizot. Volume 3 in *Los Aborígenes de Venezuela,* general editor W. Coppens. Caracas: Fundación La Salle de Ciencias Naturales, Instituto Caribe de Antropología y Sociología.
Palmer, J. R. 1991. Jarí: lessons for land managers in the tropics. Pages 419–429 in *Rain Forest Regeneration and Management,* edited by A. Gómez-Pompa, T. C. Whitmore, and M. Hadley. Volume 6 in *Man and the Biosphere.* Paris: UNESCO.
Pannier, F., and R. Fraino de Pannier. 1989. *Manglares de Venezuela.* Caracas: Cuadernos LAGOVEN.
Paolini, J. 1978. Charakterisierung und Dynamik von Huminstoffsystemen in der Amazonischen caatinga bei San Carlos de Río Negro, Venezuela. Ph.D. thesis, Georg-August-Universität Göttingen, Germany.
Passarge, S. 1903. Bericht über eine Reise im venezolanischen Guyana. *Zeitschrift der Gesellschaft für Erdkunde zu Berlin* 1903(1): 5–43.
———. 1933. Wissenschaftliche Ergebnisse einer Reise im Gebiet des Orinoco, Caura und Cuchivero im Jahre 1901–1902. *Abhandlungen aus dem Gebiet der Auslandskunde, Reihe C: Naturwissenschaften* 12: i–xii, 1–281.
Patouillard, N., and A. Gaillard. 1888. Champignons du Venezuela et principalement de la région du haut-Orénoque, recoltés en 1887 par M. A. Gaillard. *Bulletin de la Société Mycologique de France* 4: 7–46, 92–129.
Pelayo López, F., editor. 1990. *Pehr Löfling y la Expedición al Orinoco.* Madrid: Sociedad Estatal Quinto Centenario.
Pelayo López, F., and M. A. Puig-Samper. 1992. *La Obra Científica de Loefling en Venezuela.* Caracas: Cuadernos LAGOVEN.
Perera, M. A., editor. 1986. *El Orinoco y el Caura.* Caracas: Fundación Cultural Ori-

noco. Originally published as *L'Orénoque et le Caura,* by J. Chaffanjon (Librairie Hachette et Cie., Paris, 1889).
———. 1990. Actividad cauchera e impacto ambiental en el Territorio Federal Amazonas, Venezuela. *Revista Española de Antropología Americana* 20: 221–250.
Perkins, H. I. 1885. Notes on a journey to Mount Roraima, British Guiana. *Proceedings of the Royal Geographical Society,* n.s., 7(8): 522–534.
Pernía, J. E. 1985. *Mapa de Fisiografía y Vegetación del Área de Inundación de la Tercera Etapa del Embalse de Guri, Estado Bolívar. Memoria Descriptiva.* Mérida, Venezuela: ULA, Facultad de Ciencias Forestales. Mimeograph.
Pires, J. M., and G. T. Prance. 1985. The vegetation types of the Brazilian Amazon. Pages 109–145 in *Key Environments: Amazonia,* edited by G. T. Prance and T. E. Lovejoy. Oxford: Pergamon.
Pires, J. M., and J. S. Rodrigues. 1964. Sobre a flora das caatingas do Rio Negro. *Anais do 13° Congresso da Sociedade de Botânica do Brasil* (Recife 1962), 242–262.
Pittier, H. 1920. *Esbozo de las Formaciones Vegetales de Venezuela con una Breve Reseña de los Productos Naturales y Agrícolas.* (Complemento explicativo del Mapa ecológico del mismo autor). Caracas: Litografía del Comercio.
Prance, G. T. 1971. An index of plant collectors in Brazilian Amazonia. *Acta Amazonica* 1: 25–65.
———. 1977. The phytogeographic subdivisions of Amazonia and their influence on the selection of biological reserves. Pages 195–213 in *Extinction Is Forever,* edited by G. T. Prance and T. S. Elias. New York: New York Botanical Garden.
———. 1979. Notes on the vegetation of Amazonia. 3. The terminology of Amazonian forest types subject to inundation. *Brittonia* 31: 26–38.
———. 1982. Forest refuges: Evidence from woody angiosperms. Pages 137–158 in *Biological Diversification in the Tropics,* edited by G. T. Prance. New York: Columbia University Press.
Prance, G. T., and H. O. R. Schubart. 1978. Notes on the vegetation of Amazonia I. A preliminary note on the origin of the open white sand campinas of the lower Rio Negro. *Brittonia* 30: 60–63.
Quelch, J. J. 1895. A journey to the summit of Roraima. *Timehri* 9: 107–188.
Raleigh [Ralegh], W. 1596. *The Discoverie of the Large, Rich and Bewtiful Empyre of Gviana.* London: Robert Robinson.
Ramia, M. 1961. Sabanas. Volume 3 in *Reconocimiento Agropecuario Forestal del Oriente de la Guayana Venezolana.* Caracas: CVG-CBR-MAC.
Ramírez, N., C. Gil, M. López, O. Hokche, and Y. Brito. 1988. Caracterización florística y estructural de una comunidad arbustiva en la Guayana Venezolana (Gran Sabana, Estado Bolívar). *Acta Científica Venezolana* 39: 457–469.
Ramos Pérez, D. 1946. *El Tratado de Límites de 1750 y la Expedición de Iturriaga al Orinoco.* Madrid: Consejo Superior de Investigaciones Científicas, Instituto Juan Sebastian Elcano.
Reichenbach, H. G. 1873. Zum geographischen Verständniss der amerikanischen Reisepflanzen des Herrn Dr. Spruce. *Botanische Zeitung* 31(2): 28–29.
Reid, A. R. 1974. Stratigraphy of the type area of the Roraima Group, Venezuela. *Memoria de la Novena Conferencia Geológica Inter-Guayanas, Publicación Especial* 6: 343–353.
Richards, P. W. 1952. *The Tropical Rain Forest: An Ecological Study.* Cambridge: Cambridge University Press.

Rísquez-Iribarren, F. 1962. *Donde Nace el Orinoco.* Caracas: Ediciones Grecco.

Rodrigues, W. A. 1961a. Aspectos fitossociológicos das catingas do Rio Negro. *Boletim do Museu Paraense Emílio Goeldi, Botânica,* n.s., 15: 1–41.

———. 1961b. Aspects phytosociologiques des pseudo-catingas et forêts de várzea du Rio Négro. Pages 209–265 in *Etude Écologique des Principales Formations Végétales du Brésil,* edited by A. Aubréville. Nogent-sur-Marne, France: Centre Technique Forestier Tropical.

Romero, G. A. 1993. Unique orchid habitats in southern Venezuela. II. Arbustales or white-sand shrublands. *American Orchid Society Bulletin* 62: 811–818.

Rosales, J. 1988. Análisis florístico-estructural y algunas relaciones ecológicas en un bosque inundable en la boca del Río Mapire, Estado Anzoátegui. Master's thesis, Instituto Venezolano de Investigaciones Científicas, Caracas. Mimeograph.

Rosales, J., and E. Briceño. 1990. *Estudio Integrado del Área de Influencia Inmediata del Embalse Guri.* Volume 5, *Vegetación.* Puerto Ordaz, Venezuela: CVG-EDELCA, División de Cuencas e Hidrología. Mimeograph.

Rosales, J., E. Briceño, B. Ramos, and G. Picón. 1993. Los bosques ribereños en el área de influencia del Embalse de Guri. *Pantepui* 5: 4–23.

Rull, V. 1991. Contribución a la paleoecología de Pantepui y la Gran Sabana (Guayana Venezolana): clima, biogeografía, ecología. *Scientia Guaianae* 2: i–xxii, 1–133.

Rusby, H. H. 1896. Concerning exploration upon the lower Orinoco. *The Alumni Journal (New York)* 3(8): 185–191.

Sánchez P., H., J. I. Hernández C., J. V. Rodríguez M., and C. Castaño U. 1990. *Nuevos Parques Nacionales, Colombia.* Bogotá: Instituto Nacional de Recursos Naturales (INDERENA).

Sandwith, N. Y. 1925. Humboldt and Bonpland's itinerary in Venezuela. *Bulletin of Miscellaneous Information* 7: 295–310.

Sarmiento, G. 1983. The savannas of tropical America. Pages 245–288 in *Tropical Savannas,* edited by F. Bourlière. Volume 13 in *Ecosystems of the World,* editor-in-chief D. W. Goodall. Amsterdam: Elsevier Scientific.

Schomburgk, M. R. 1847–1848. *Reisen in Britisch-Guiana in den Jahren 1840–1844.* 3 volumes. Leipzig: Verlagsbuchhandlung von J. J. Weber.

Schomburgk, R. H. 1840a. *A Description of British Guiana.* London: Simpkin, Marshall.

———. 1840b. Journey from Esmeralda, on the Orinoco, to San Carlos and Moura on the Rio Negro, and thence by Fort San Joaquim to Demerara, in the spring of 1839. *Journal of the Royal Geographical Society* 10: 248–267.

———. 1840c. Journey from Fort San Joaquim, on the Río Branco, to Roraima, and thence by the rivers Parima and Merewari to Esmeralda, on the Orinoco, in 1838–1839. *Journal of the Royal Geographical Society* 10: 191–247.

Schubert, C., and H. Briceño. 1987. Origen de la topografía tepuyana: una hipótesis. *Pantepui* 2: 11–14.

Schubert, C., and P. Fritz. 1985. Radiocarbon ages of peat, Guayana Highlands (Venezuela). *Naturwissenschaften* 72: 427–429.

Schubert, C., and O. Huber. 1990. *The Gran Sabana: Panorama of a Region.* Caracas: LAGOVEN Booklets.

Schubert, C., H. O. Briceño, and P. Fritz. 1986. Paleoenvironmental aspects of the Caroní-Paragua river basin (southeastern Venezuela). *Interciencia* 11: 278–289.

Schubert, C., P. Fritz, and R. Aravena. 1989. Investigaciones paleoambientales en el

Macizo del Chimantá (Escudo de Guayana, Venezuela). *Memorias, VII Congreso Geológico Venezolano* 3: 1320–1342.
Schulz, J. P. 1960. Ecological studies on rain forest in northern Suriname. Volume 2 in *The Vegetation of Suriname.* Amsterdam: van Eedenfonds.
Sioli, H. 1965. Bemerkungen zur Typologie amazonischer Flüsse. *Amazoniana* 1: 74–83.
Smole, W. 1976. *The Yanoama Indians: A Cultural Geography.* Austin: University of Texas Press.
Spruce, R. 1908. *Notes of a Botanist on the Amazon and Andes Being Records of Travel on the Amazon and Its Tributaries, the Trombetas, Rio Negro, Vaupes, Casiquiare, Pacimoni, . . .,* edited by A. R. Wallace. 2 volumes. London: Macmillan. Reprint, New York/London: Johnson, 1970.
Stern, K. M. 1970. Der Casiquiare-Kanal, einst und jetzt. *Amazoniana* 2: 401–416.
Steyermark, J. A. [and collaborators]. 1951. Contributions to the flora of Venezuela. *Fieldiana, Botany* 28: 1–242.
——— [and collaborators]. 1952. Contributions to the flora of Venezuela. *Fieldiana, Botany* 28: 243–447.
——— [and collaborators]. 1953. Contributions to the flora of Venezuela. *Fieldiana, Botany* 28: 449–678.
——— [and collaborators]. 1957. Contributions to the flora of Venezuela. *Fieldiana, Botany* 28: 679–1190.
———. 1966. Contribuciones a la flora de Venezuela, parte 5. *Acta Botanica Venezuelica* 1(3/4): 9–256.
———. 1967. Flora del Auyán-tepui. *Acta Botanica Venezuelica* 2(5–8): 5–370.
———. 1968. Contribuciones a la flora de la Sierra Imataca, Altiplanicie de Nuria y región adyacente del Territorio Federal Delta Amacuro al sur del Río Orinoco. *Acta Botanica Venezuelica* 3(1–4): 49–175.
———. 1974. The summit vegetation of Cerro Autana. *Biotropica* 6: 7–13.
———. 1975. Informe sobre la flora del Cerro Autana. *Acta Botanica Venezuelica* 10: 219–233.
———. 1977. Future outlook for threatened and endangered species in Venezuela. Pages 128–135 in *Extinction Is Forever,* edited by G. T. Prance and T. S. Elias. Bronx: New York Botanical Garden.
———. 1979a. Flora of the Guayana Highland: endemicity of the generic flora of the summits of the Venezuela tepuis. *Taxon* 28: 45–54.
———. 1979b. Plant refuge and dispersal centres in Venezuela: their relict and endemic element. Pages 185–221 in *Tropical Botany,* edited by K. Larsen and L. B. Holm-Nielsen. London: Academic Press.
———. 1982. Relationships of some Venezuelan forest refuges with lowland tropical floras. Pages 182–220 in *Biological Diversification in the Tropics,* edited by G. T. Prance. New York: Columbia University Press.
———. 1984. Flora of the Venezuelan Guayana—I. *Annals of the Missouri Botanical Garden* 71: 297–340.
———. 1986a. Expedition to the lost world. *Missouri Botanical Garden Bulletin* 74(7): 5–7.
———. 1986b. Speciation and endemism in the flora of the Venezuelan tepuis. Pages 317–373 in *High Altitude Tropical Biogeography,* edited by F. Vuilleumier and M. Monasterio. New York: Oxford University Press.

Steyermark, J. A., and C. Brewer-Carías. 1976. La vegetación de la cima del Macizo de Jaua. *Boletín de la Sociedad Venezolana de Ciencias Naturales* 22(132/133): 179–405.
Steyermark, J. A., and G. C. K. Dunsterville. 1980. The lowland floral element on the summit of Cerro Guaiquinima and other cerros of the Guayana highland of Venezuela. *Journal of Biogeography* 7: 285–303.
Steyermark, J. A., and B. K. Holst [and collaborators]. 1989. Flora of the Venezuelan Guayana—VII. Contributions to the flora of the Cerro Aracamuni, Venezuela. *Annals of the Missouri Botanical Garden* 76: 945–992.
Steyermark, J. A., and B. Maguire [and collaborators]. 1967. Botany of the Chimantá Massif—II. *Memoirs of the New York Botanical Garden* 17: 440–464.
——— [and collaborators]. 1984a. Informe preliminar sobre la flora de la cumbre del Cerro Marahuaca. *Acta Botanica Venezuelica* 14: 53–89.
———. 1984b. Informe preliminar sobre la flora de la cumbre del Cerro Marutaní. *Acta Botanica Venezuelica* 14: 91–117.
Steyermark, J. A., and H. Meyer. 1945–1946. Informe de la misión de Cinchona en Venezuela. *Boletín de la Sociedad Venezolana de Ciencias Naturales* 10(65/66): 163–189.
Steyermark, J. A., and S. Nilsson. 1962. Botanical novelties in the region of Sierra de Lema, Estado Bolívar. I. *Boletín de la Sociedad Venezolana de Ciencias Naturales* 23: 59–95.
Takeuchi, M. 1961. The structure of the Amazonian vegetation. III. Campina forest in the Rio Negro region. *Journal of the Faculty of Science, University of Tokyo, Section 3. Botany* 8: 27–35.
———. 1962. The structure of the Amazonian vegetation. IV. High campina forest in the upper Rio Negro. *Journal of the Faculty of Science, University of Tokyo, Section 3. Botany* 8: 279–288.
Takhtajan, A. 1986. *Floristic Regions of the World.* Translated by T. J. Crovello, edited by A. Cronquist. Berkeley: University of California Press.
Tamayo, F. 1958. Notas explicativas del ensayo del mapa fitogeográfico de Venezuela (1955). *Revista Forestal Venezolana* 1: 7–31.
———. 1961. Exploraciones botánicas en el Estado Bolívar. *Boletín de la Sociedad Venezolana de Ciencias Naturales* 22(98/99): 25–180.
Tate, G. H. H. 1930. Notes on the Mount Roraima region. *Geographical Review (New York)* 20: 53–68.
———. 1931. Aspects of vegetation and plant associations. In *Botanical Results of the Tyler-Duida Expedition,* edited by H. A. Gleason. *Bulletin of the Torrey Botanical Club* 58: 287–298.
———. 1938a. Auyantepui: Notas sobre la Expedición Phelps (Phelps Venezuelan Expedition). *Boletín de la Sociedad Venezolana de Ciencias Naturales* 5(36): 96–125.
———. 1938b. Auyantepui: notes on the Phelps Venezuelan Expedition. *Geographical Review (New York)* 28: 452–474.
Tate, G. H. H., and C. Hitchcock. 1930. The Cerro Duida region of Venezuela. *Geographical Review (New York)* 20: 31–52.
Tavera-Acosta, B. 1905. *Anales de Guayana.* Volume 1. Ciudad Bolívar: Tipografía La Empresa.
Teggin, D., M. Martínez, and G. Palacios. 1985. Un estudio preliminar de las diabasas

del Estado Bolívar, Venezuela. *Memorias, VI Congreso Geológico Venezolano* 4: 2159–2206.

Terán, F., and R. Duno de Stefano. 1988. *Caracterización Fisionómica y Florística de los Morichales de la Cuenca del Río Yuruaní.* Dissertation, Universidad Central de Venezuela, Facultad de Ciencias, Caracas.

Texera Arnal, Y. 1991. *La Exploración Botánica en Venezuela.* Caracas: Fondo Editorial Acta Científica Venezolana.

Thomas, D. J. 1982. *Order Without Government. The Society of the Pemón Indians of Venezuela.* Illinois Studies in Anthropology, No. 13. Urbana: University of Illinois Press.

Tillett, S. S., and J. A. Steyermark. 1982. Contribuciones a la flora del Cerro Marahuaca, Territorio Federal Amazonas, Venezuela. *Ernstia* 9: 1–16.

Uhl, C., and I. C. Guimarães Vieira. 1989. Ecological impacts of selective logging in the Brazilian Amazon: a case study from the Paragominas region of the State of Pará. *Biotropica* 21: 98–106.

Ule, E. 1915. Die Vegetation des Roraima. *Botanische Jahrbücher für Systematik, Pflanzengeschichte und Pflanzengeographie* 52(Beiblatt 115): 42–53.

Urban, I. 1906. Vitae itineraque collectorum botanicorum, notae collaboratorum biographicae, Florae Brasiliensis ratio edendi chronologica, systema, index familiarum. Pages 1–154 in *Flora Brasiliensis,* 1(1), by K. F. P. von Martius.

Useche, L. M. 1987. *El Proceso Colonial en el Alto Orinoco-Río Negro (Siglos XVI a XVIII).* Bogotá: Fundación de Investigaciones Arqueológicas Nacionales, Banco de la República.

Vareschi, V. 1963a. Die Gabelteilung des Orinoco. *Petermanns Geographische Mitteilungen* 107: 241–248.

———. 1963b. La bifurcación del Orinoco—Observaciones hidrográficas y ecológicas de la expedición conmemorativa de Humboldt del año 1958. *Acta Científica Venezolana* 14: 98–106.

———. 1992a. *Ecología de la Vegetación Tropical.* Caracas: Sociedad Venezolana de Ciencias Naturales.

———. 1992b. Observaciones sobre la dinámica vegetal en el Macizo del Chimantá. Pages 179–188 in *El Macizo del Chimantá, Escudo de Guayana, Venezuela: Un Ensayo Ecológico Tepuyano,* edited by O. Huber. Caracas: Oscar Todtmann.

Vila, P. 1960. *Geografía de Venezuela. I: El Territorio Nacional y su Ambiente Físico.* Caracas: Ministerio de Educación.

Wagner, W. L., D. R. Herbst, and S. H. Sohmer. 1990. *Manual of the Flowering Plants of Hawai'i.* Bishop Museum Special Publication 83, 2 volumes. Honolulu: University of Hawaii Press.

Wallace, A. R. 1853. *Palm Trees of the Amazon and Their Uses.* London: John van Voorst. Reprint, Lawrence, Kansas: Coronado Press, 1971.

Walter, H. 1979. *Vegetation of the Earth and Ecological Systems of the Geo-biosphere.* Translated by J. Wieser. New York: Springer-Verlag.

Weibezahn, F. H. 1990. Hidroquímica y sólidos suspendidos en el Alto y Medio Orinoco. Pages 151–210 in *El Río Orinoco Como Ecosistema,* edited by F. H. Weibezahn, H. Alvarez, and W. M. Lewis, Jr. Caracas: EDELCA, Fondo Editorial Acta Científica Venezolana, CAVN, Universidad Simón Bolívar.

Weibezahn, F. H., and B. E. Janssen-Weibezahn. 1990. El Territorial Federal Amazonas, Venezuela: una bibliografía. *Scientia Guaianae* 1: ix + 294 pages.

Weibezahn, F. H., H. Alvarez, and W. M. Lewis, Jr., editors. 1990. *El Río Orinoco Como Ecosistema.* Caracas: EDELCA, Fondo Editorial Acta Científica Venezolana, CAVN, and Universidad Simón Bolívar.
Weibezahn, F., J. Marcano, C. Molina, and P. Zadrozny. 1983. *Bibliografía Relacionada Con la Hoya del Río Orinoco y Cuencas Vecinas.* Carcas: MARNR, DGSIIA/PDP-BOR/O1.
Williams, L. 1940. Botanical exploration in the middle and lower Caura. *Tropical Woods* 62: 1–20.
———. 1941. The Caura valley and its forests. *Geographical Review (New York)* 31: 414–429.
———. 1942. *Exploraciones botánicas en la Guayana Venezolana. I: El medio y bajo Caura.* Caracas: Servicio Botánico, Ministerio de Agricultura y Cría.
Wurdack, J. J. 1960. A historic portage. *Garden Journal of the New York Botanical Garden* 10: 8–9, 13.
Zinck, A. 1986. El inventario de los recursos naturales de Guayana en marcha. *Pantepui* 1: 2–16.

Index

—H—

—T—